Advances in

ORGANOMETALLIC CHEMISTRY

VOLUME 21

CONTRIBUTORS TO THIS VOLUME

Donald J. Darensbourg

Thomas P. Fehlner

William P. Hart

Catherine E. Housecroft

David W. Macomber

Marvin D. Rausch

Jacques Satgé

Gerard van Koten

Kees Vrieze

Advances in
Organometallic
Chemistry

EDITED BY

F. G. A. STONE

DEPARTMENT OF INORGANIC CHEMISTRY
THE UNIVERSITY
BRISTOL, ENGLAND

ROBERT WEST

DEPARTMENT OF CHEMISTRY
UNIVERSITY OF WISCONSIN
MADISON, WISCONSIN

VOLUME 21

1982

ACADEMIC PRESS

A Subsidiary of Harcourt Brace Jovanovich, Publishers

New York   London
Paris   San Diego   San Francisco   São Paulo   Sydney   Tokyo   Toronto

ACADEMIC PRESS, INC.
111 Fifth Avenue, New York, New York 10003

United Kingdom Edition published by
ACADEMIC PRESS, INC. (LONDON) LTD.
24/28 Oval Road, London NW1 7DX

LIBRARY OF CONGRESS CATALOG CARD NUMBER: 64–16030

ISBN 0–12–031121–6

PRINTED IN THE UNITED STATES OF AMERICA

82 83 84 85   9 8 7 6 5 4 3 2 1

Contents

Functionally Substituted Cyclopentadienyl Metal Compounds

DAVID W. MACOMBER, WILLIAM P. HART, and MARVIN D. RAUSCH

Metalloboranes: Their Relationships to Metal–Hydrocarbon Complexes and Clusters

CATHERINE E. HOUSECROFT and THOMAS P. FEHLNER

Mechanistic Pathways for Ligand Substitution Processes in Metal Carbonyls

DONALD J. DARENSBOURG

Contents

1,4-Diaza-1,3-butadiene (α-Diimine) Ligands: Their Coordination Modes and the Reactivity of Their Metal Complexes

GERARD VAN KOTEN and KEES VRIEZE

Multiply Bonded Germanium Species

JACQUES SATGÉ

Contributors

Numbers in parentheses indicate the pages on which the authors' contributions begin.

DONALD J. DARENSBOURG* (113), *Department of Chemistry, Tulane University, New Orleans, Louisiana 70118*

THOMAS P. FEHLNER (57), *Department of Chemistry, University of Notre Dame, Notre Dame, Indiana 46556*

WILLIAM P. HART (1), *Department of Chemistry, University of Massachusetts, Amherst, Massachusetts 01003*

CATHERINE E. HOUSECROFT (57), *Department of Chemistry, University of Notre Dame, Notre Dame, Indiana 46556*

DAVID W. MACOMBER (1), *Department of Chemistry, University of Massachusetts, Amherst, Massachusetts 01003*

MARVIN D. RAUSCH (1), *Department of Chemistry, University of Massachusetts, Amherst, Massachusetts 01003*

JACQUES SATGÉ (241), *Université Paul Sabatier, 118, Route de Narbonne, 31077 Toulouse Cedex, France*

GERARD VAN KOTEN (151), *Anorganisch Chemisch Laboratorium, J. H. van 't Hoff Instituut, University of Amsterdam, 1018 WV Amsterdam, The Netherlands*

KEES VRIEZE (151), *Anorganisch Chemisch Laboratorium, J. H. van 't Hoff Instituut, University of Amsterdam, 1018 WV Amsterdam, The Netherlands*

*Present address: Department of Chemistry, Texas A & M University, College Station, Texas 77843.

ADVANCES IN ORGANOMETALLIC CHEMISTRY, VOL. 21

Functionally Substituted Cyclopentadienyl Metal Compounds

DAVID W. MACOMBER, WILLIAM P. HART, and MARVIN D. RAUSCH

Department of Chemistry
University of Massachusetts
Amherst, Massachusetts

I

INTRODUCTION

The birth of cyclopentadienyl transition metal chemistry occurred in 1951 when Pauson and Kealy (*1*) discovered a most remarkable compound, bis(η^5-cyclopentadienyl)iron (ferrocene) (**1**). Ferrocene was found to be

(1)

unusually stable for an organoiron compound. Indeed, ferrocene does not

1

undergo Diels–Alder reactions, resists catalytic hydrogenation under normal conditions, and resists pyrolysis at 470°C (2, 3).

In 1952, Woodward, Rosenblum, and Whiting reported that ferrocene would undergo Friedel–Crafts acylation (4, 5). The reaction of 1 with acetyl chloride in the presence of aluminum chloride gave acetylferrocene (2), 1,2-diacetylferrocene (3), and 1,1'-diacetylferrocene (4). Shortly after this discovery ferrocene was shown to undergo other "aromatic-type" substitution reactions including alkylation (6, 7), formylation (8, 9), mercuration (10), and sulfonation (11).

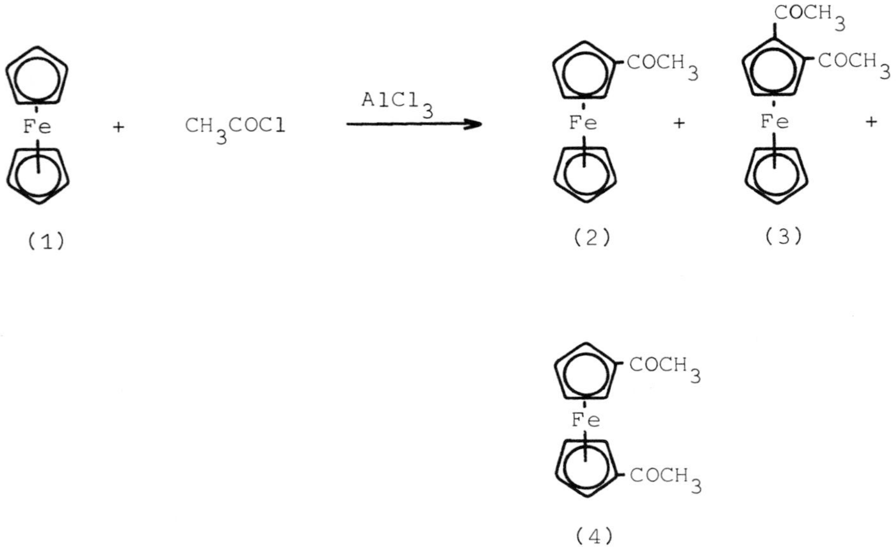

Since the discovery of ferrocene, many new η^5-cyclopentadienylmetal compounds have been synthesized, although only a few exhibit the same aromatic-type behavior as does 1. These compounds, which undergo electrophilic aromatic substitution, include ruthenocene (5) and osmocene (6) (12, 13), cymantrene (7) (14–16), its technetium (8) and rhenium (9) analogs (17–19), (η^5-cyclopentadienyl)tetracarbonylvanadium (10) (20–22), (η^5-cyclopentadienyl)dicarbonylnitrosylchromium (11) (23), (η^5-cyclopentadienyl)dicarbonylcobalt (12) (24, 25), and (η^5-cyclopentadienyl)(η^4-tetraphenylcyclobutadiene)cobalt (13) (26, 27).

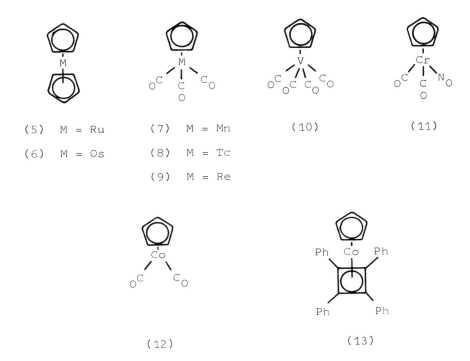

(5) M = Ru   (7) M = Mn   (10)   (11)

(6) M = Os   (8) M = Tc

                   (9) M = Re

        (12)   (13)

Compounds **1** and **5–13** are the only reported η^5-cyclopentadienylmetal compounds that undergo aromatic substitution reactions yielding functionally substituted derivatives. As a consequence, the synthesis and subsequent chemistry of functionally substituted η^5-cyclopentadienylmetal compounds has been severely limited in scope.

The present article deals with alternate methods for the formation of functionally substituted cyclopentadienylmetal compounds. No attempt has been made to review the functional group chemistry of ferrocene (**1**), or the other metalloaromatic systems (**5–13**) listed above. Of the known (η^5-C_5H_5)M(L_n) compounds, almost all have been synthesized from reactions of C_5H_5Tl, C_5H_5Na, C_5H_5Li, or C_5H_5MgX with an appropriate transition metal halide fragment, where M is a transition metal and L_n represents suitable ligands (*28*). It is logical, then, that the reaction of substituted cyclopentadienide reagents such as C_5H_4RTl, C_5H_4RNa, or C_5H_4RLi with metal halide fragments should lead to the corresponding substituted analogs (η^5-C_5H_4R)M(L_n), where R is a reactive functional group such as acyl, halogeno, etc.

For the purposes of this article, alkyl (*29*), aryl, and silyl substituents will not be considered as functional groups.

II

ALDEHYDES, KETONES, ESTERS, AND NITRILES

A. *Monosubstituted Compounds*

Sodium cyclopentadienides containing aldehyde, ketone, or ester substituents can be synthesized easily following a method developed originally by Thiele in 1900 (*30*). Condensation of cyclopentadiene with diethyl oxalate in the presence of sodium ethoxide gave sodium ethoxalylcyclopentadienide (**14**) (*30*). The reactions of sodium cyclopentadienide with either

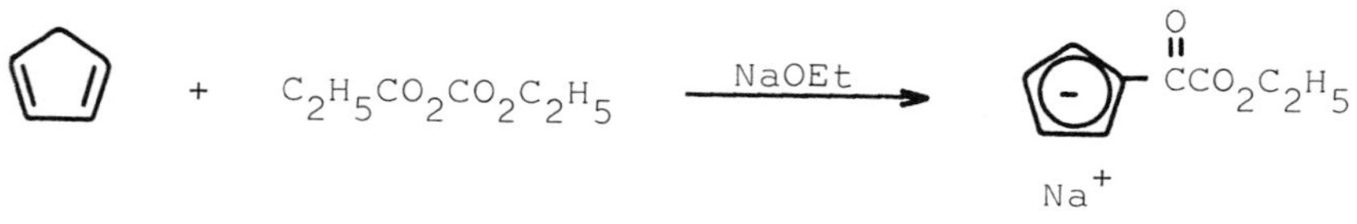

$$\text{[cyclopentadiene]} \quad + \quad C_2H_5CO_2CO_2C_2H_5 \quad \xrightarrow{\text{NaOEt}} \quad \text{[cyclopentadienide]} - \overset{\overset{O}{\|}}{C}CO_2C_2H_5$$

$$Na^+$$

(14)

ethyl formate, methyl acetate, or dimethyl carbonate have produced the respective compounds sodium formyl- (**15**), acetyl- (**16**), or carbomethoxycyclopentadienide (**17**) in yields of 60–90% (*31–33*). Potassium formyl-

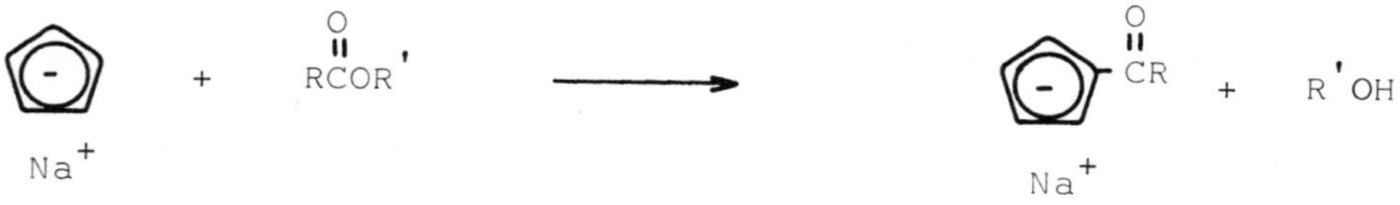

$$\text{[cyclopentadienide]} \quad + \quad \overset{\overset{O}{\|}}{R}COR' \quad \longrightarrow \quad \text{[cyclopentadienide]} - \overset{\overset{O}{\|}}{C}R \quad + \quad R'OH$$

$$Na^+ \qquad\qquad\qquad\qquad\qquad\qquad\qquad\qquad Na^+$$

(15)   R = H

(16)   R = CH_3

(17)   R = OCH_3

cyclopentadienide (**18**) has likewise been prepared from a reaction between potassium cyclopentadienide and ethyl formate (*34*).

$$\text{[cyclopentadienide]} \quad + \quad HCO_2C_2H_5 \quad \longrightarrow \quad \text{[cyclopentadienide]} - CHO \quad + \quad C_2H_5OH$$

$$K^+ \qquad\qquad\qquad\qquad\qquad\qquad\qquad\qquad K^+$$

(18)

El Murr has produced the acetyl- (**19**) and carbomethoxycyclopentadienide (**20**) anions from the electrolysis of $Fe(C_5H_4R)_2$ ($R = COCH_3$ or CO_2CH_3) in the presence of $(n\text{-}Bu_4N)(PF_6)$ (*35*).

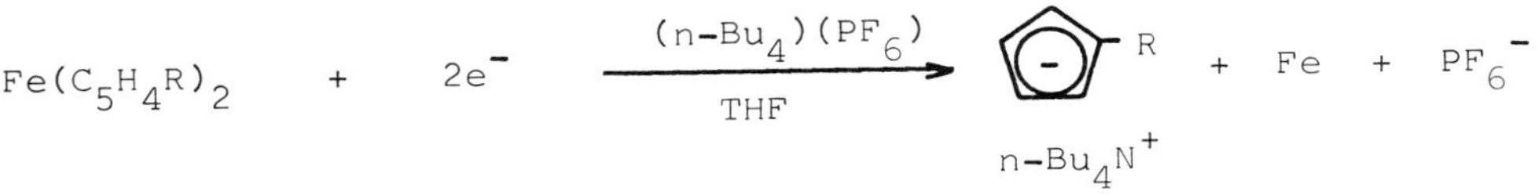

$$Fe(C_5H_4R)_2 \quad + \quad 2e^- \quad \xrightarrow[\text{THF}]{(n\text{-}Bu_4)(PF_6)} \quad \underset{n\text{-}Bu_4N^+}{\overset{}{\bigodot}} R \quad + \quad Fe \quad + \quad PF_6^-$$

(19) $R = COCH_3$

(20) $R = CO_2CH_3$

Various esters of thallium carboxycyclopentadienide have been prepared by Drew and Nelson (*34, 36*), starting from Thiele's dimeric acid (*37*). Thiele's acid was esterified by the method of Peters (*38, 39*) to give esters (**21–23**). These esters were then "cracked" at 220°C and 18–20 mm Hg into an aqueous solution of thallium acetate and potassium hydroxide to give compounds **24–26** in 60–65% yield.

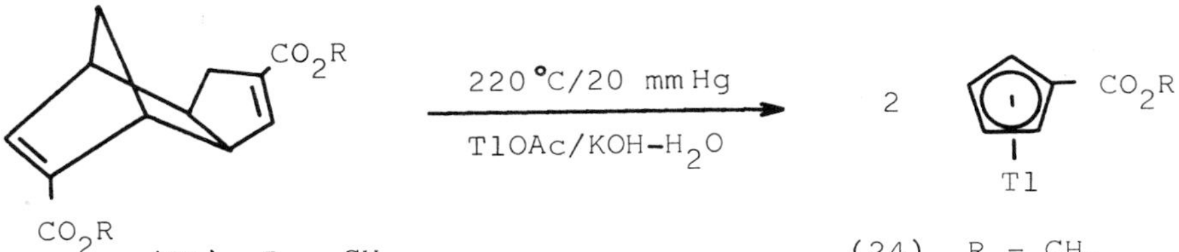

$$\xrightarrow[\text{TlOAc/KOH–H}_2\text{O}]{220\ ^\circ\text{C}/20\ \text{mm Hg}} \quad 2$$

(21) $R = CH_3$

(22) $R = C_2H_5$

(23) $R = i\text{-}C_3H_7$

(24) $R = CH_3$

(25) $R = C_2H_5$

(26) $R = i\text{-}C_3H_7$

Peters had also prepared other derivatives of Thiele's acid including the acid chloride (**27**), the nitrile (**28**), and the amides (**29** and **30**), which gave monomers **31–33** upon distillation. Organothallium derivatives of mono-

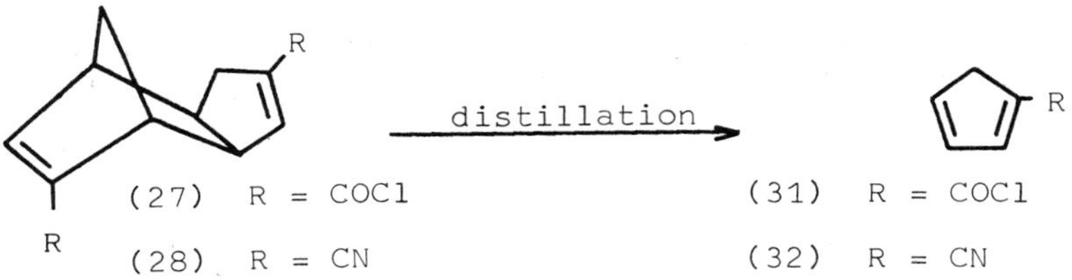

$$\xrightarrow{\text{distillation}}$$

(27) $R = COCl$

(28) $R = CN$

(29) $R = CON(CH_3)_2$

(30) $R = CONH_2$

(31) $R = COCl$

(32) $R = CN$

(33) $R = CON(CH_3)_2$

mers **31** and **33** were not prepared by this method, however. The dimer (**30**) does not depolymerize to the corresponding cyclopentadiene (**34**), because of the high melting point of **30** (202°C).

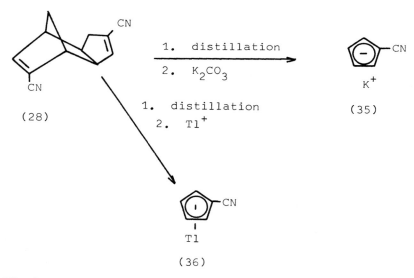

(34)

The reaction of equimolar amounts of cyclopentadiene and cyanogen chloride in the presence of sodium hydride gave dicyanocyclopentadiene dimer, presumably **28**, which upon distillation and treatment with potassium carbonate gave potassium cyanocyclopentadienide (**35**) (*40*). Cracking **28** into thallous ion gave the analogous thallium cyanocyclopentadienide (**36**) (*36, 41, 42*).

The fused cyclopentadienyl lactone anion (**38**) has been prepared by an intramolecular cyclization of the substituted cyclopentadiene (**37**) in the presence of sodium hydride (*43, 44*).

Many functionally substituted cyclopentadienide reagents have proved useful in the synthesis of functionally substituted η^5-cyclopentadienyl-transition metal compounds. Some examples will be cited below.

In 1957 Benson and Lindsey (*45*) generated *in situ* the alkoxy-substituted cyclopentadienide anion (**39**) and synthesized the heterodisubstituted ferrocene **40**.

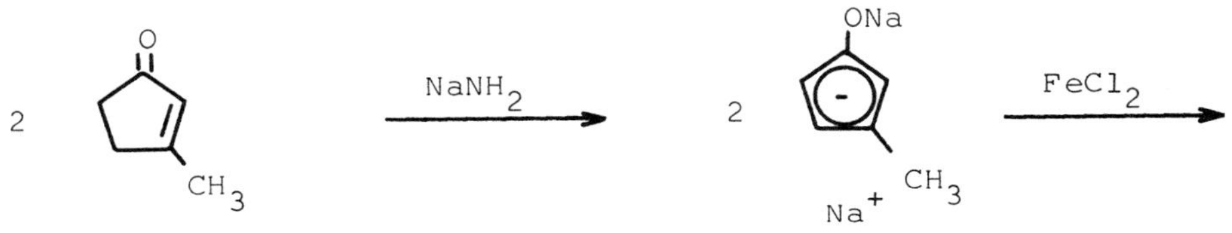

(39)

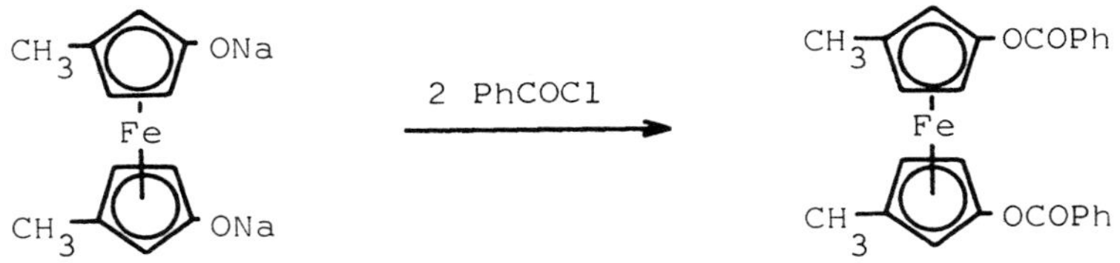

(40)

This same approach was subsequently used by Pauson and Osgerby (*46*) in the synthesis of 1,1'-dicarbomethoxyferrocene (**41**). Sodium carbomethoxycyclopentadienide (**17**), obtained from sodium cyclopentadienide, methyl chloroformate, and excess sodium, was allowed to react with ferrous chloride to give a 30% yield of **41**. The low yield of **41** was attributed to the formation of sodium 1,2-dicarbomethoxycyclopentadienide (**42**). Peters had earlier reported the isolation of **42** from the same reaction (*47*).

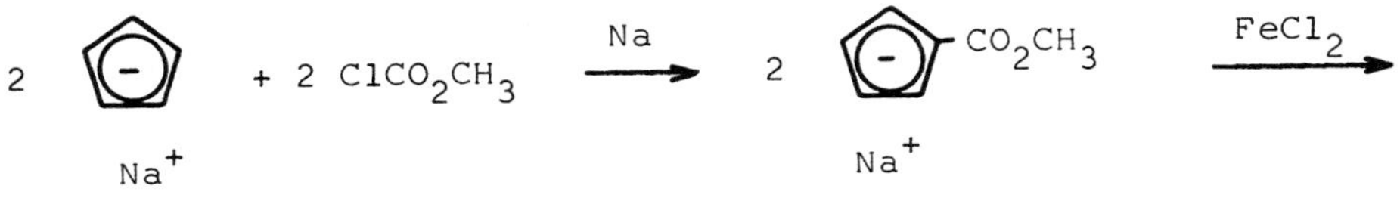

(17)

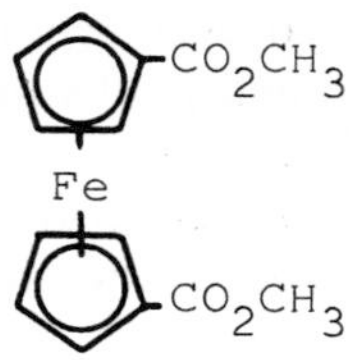

(41)

Pauson and Osgerby were also able to synthesize 1,1′-diacetylferrocene (**43**) in a similar manner, although only in 2% yield. Presumably, poly-

(16) (43)

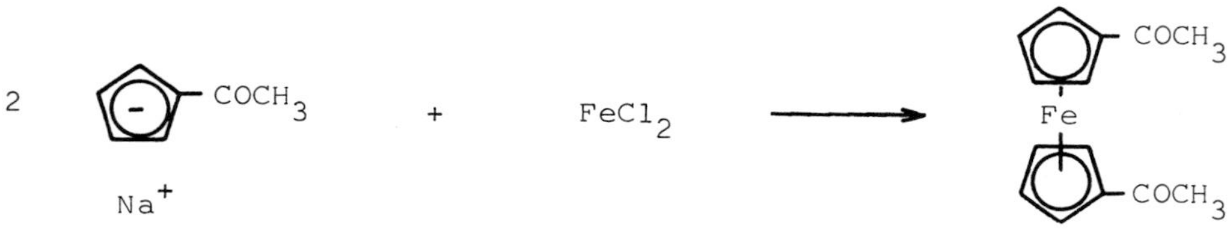

acylation occurs, giving cyclopentadienide anions that are incapable of reacting with ferrous chloride. Hart (*25*) has prepared 1,1′-diacetylferrocene in 30% yield from "isolated" sodium acetylcyclopentadienide and ferrous chloride.

Rausch and co-workers (*33*) have recently used organosodium reagents **15**–**17** to synthesize many new acyl-substituted organometallic compounds. Reactions between either **16** or **17** with $CoCl_2$ in THF solution produced 1,1′-diacetylcobaltocene (**44**) and 1,1′-dicarbomethoxycobaltocene (**45**), respectively, in 30–40% yields. Analogous reactions of **16** or **17** with $NiBr_2 \cdot$ 2DME in DME solution gave 1,1′-diacetylnickelocene (**46**) and 1,1′-dicarbomethoxynickelocene (**47**), respectively, in 40–45% yields.

(44) R = CH_3

(45) R = OCH_3

(46) R = CH_3

(47) R = OCH_3

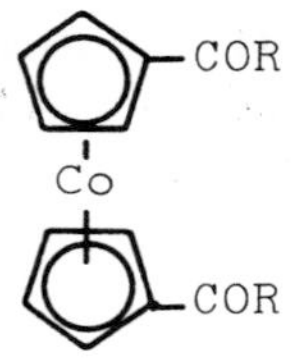

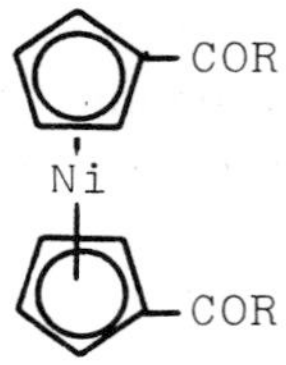

η^5-Cyclopentadienyldicarbonylmetal derivatives of cobalt and rhodium containing acyl substituents have also been obtained. Reactions of **15**, **16**, or **17** with an equimolar mixture of $Co_2(CO)_8$ and I_2 in THF produced the corresponding formyl (**48**), acetyl (**49**), and carbomethoxy (**50**) analogs.

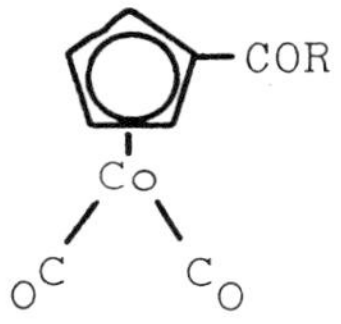

(48)	R = H
(49)	R = CH_3
(50)	R = OCH_3

(51)	R = H
(52)	R = CH_3
(53)	R = OCH_3

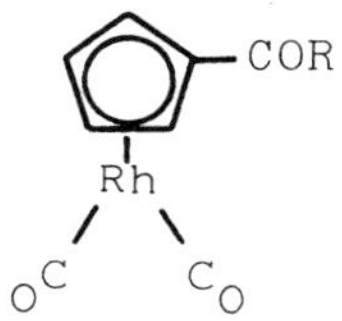

The rhodium counterparts have been synthesized via reactions of **15–17** with $[Rh(CO)_2Cl]_2$ in THF to give compounds **51–53**. These products were obtained in yields up to 50% (*25, 33*).

Reactions of **15–17** with group VIB metal hexacarbonyls have likewise led to a wide variety of functionally substituted η^5-cyclopentadienyl derivatives of these metals. Reactions of **17** with $Cr(CO)_6$ (DMF, reflux), $Mo(CO)_6$ (THF, reflux), or $W(CO)_6$ (DME, reflux), followed by acidification with acetic acid and subsequent nitrosylation with *N*-methyl-*N*-nitroso-*p*-toluenesulfonamide gave the respective carbomethoxy derivatives **54** (79%), **55** (12%), and **56** (41%) (*33, 48*).

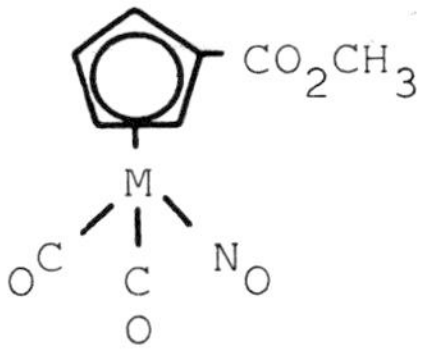

(54)	M = Cr
(55)	M = Mo
(56)	M = W

(57)	M = Cr
(58)	M = Mo
(59)	M = W

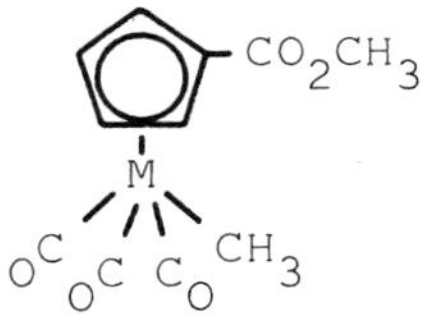

In a similar manner, treatment of the metal carbonyl anions, generated as above, with methyl iodide led to compounds **57–59** in 30–80% yields. Starting with the metal carbonyl anions derived from **15** or **16** and either

$Mo(CO)_6$ or $W(CO)_6$, the respective formyl (**60** and **61**) and acetyl (**62** and **63**) analogs have been prepared (*33, 48*).

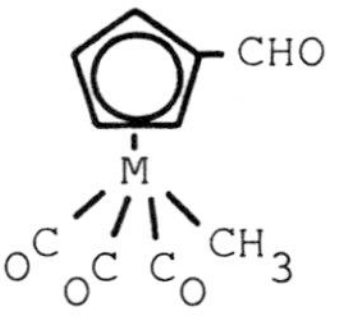

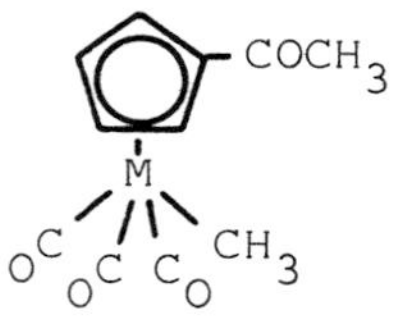

(60)   M = Mo

(61)   M = W

(62)   M = Mo

(63)   M = W

Numerous synthetic transformations occur at the functional groups of compounds **48–63**, yielding useful derivatives. For example, esters **50** and **59** could be saponified with aqueous potassium hydroxide in methanol to give, after acidification, the corresponding carboxylic acids **64** and **65** in yields of greater than 70%. Treatment of these acids with oxalyl chloride or PCl_5, respectively, has given the acid chlorides **66** and **67** in good yields (*25, 48*).

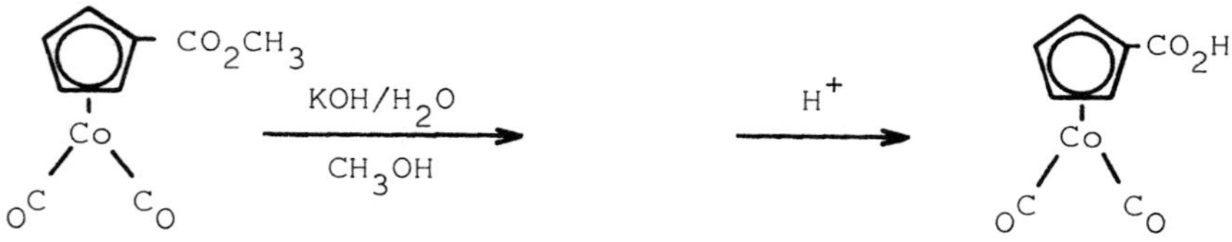

(50)

(64)

(66)

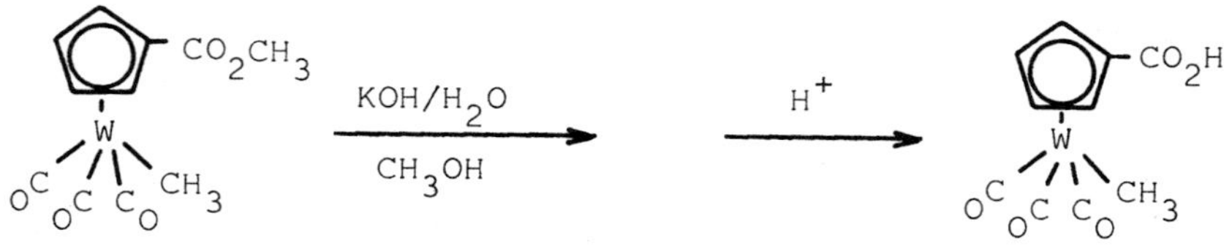

(59)

(65)

$$\xrightarrow{\text{PCl}_5}$$

(67)

Compounds **48–63** have also proved useful in the synthesis of organo-metallic vinyl monomers. Under phase-transfer conditions, aldehyde **61** was converted in excellent yield to the vinyl monomer **68** (*49*). Monomer

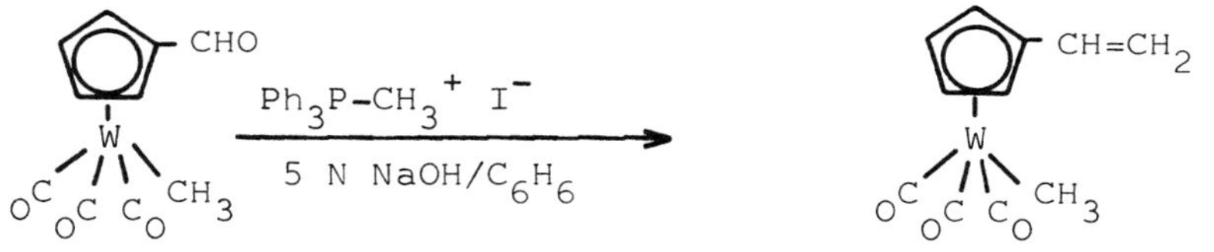

(61)

$$\xrightarrow[\text{5 N NaOH/C}_6\text{H}_6]{\text{Ph}_3\text{P-CH}_3{}^+ \text{ I}^-}$$

(68)

68 could be homopolymerized under free-radical initiation conditions, and copolymerized with organic monomers such as styrene, acrylonitrile, etc. A detailed polymerization study has shown **68** to be almost as electron-rich as vinylferrocene (*50*).

Using the substituted organothallium reagents **24** and **36**, Cramer and Mrowca prepared the cyano- and carbomethoxy-substituted cyclopenta-dienylbis(ethylene)rhodium compounds **69** and **70**, respectively (*41*). By

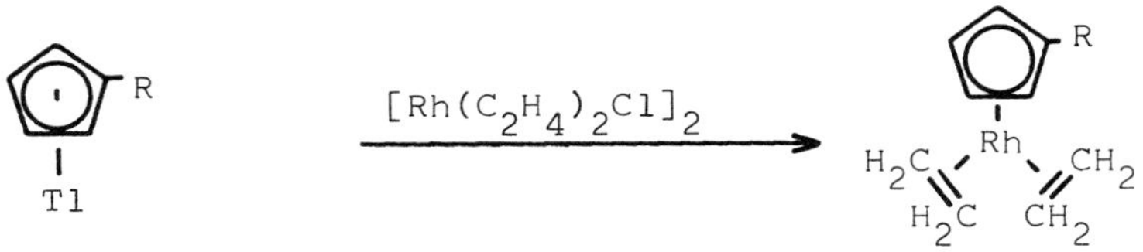

$$\xrightarrow{[\text{Rh}(\text{C}_2\text{H}_4)_2\text{Cl}]_2}$$

(24) R = CO_2CH_3

(36) R = CN

(69) R = CO_2CH_3

(70) R = CN

means of variable temperature proton NMR, it was determined that substitution of cyclopentadienyl protons with electronegative groups (viz., cyano or carbomethoxy) weakens the π bond between rhodium and ethylene moderately.

Drew and Nelson (*34, 36*) have prepared several compounds of the type $(\eta^5\text{-}C_5H_4X)RhL_2$, where X = CHO, CO_2R (R = CH_3, C_2H_5, $i\text{-}C_3H_7$), $COCO_2C_2H_5$, or CN, and L = CO, C_2H_4, or a symmetrical 1,5-, 1,4-, or 1,3-diene. These compounds were obtained from organothallium reagents **24–26** or **36**, organosodium reagent **14**, or organopotassium reagent **18** and a suitable organorhodium chloride dimer. For example, a reaction between thallium carbomethoxycyclopentadienide (**24**) and (cycloocta-1,5-diene)rhodium chloride dimer gave **71**. The molecular structure of **71** was confirmed by means of a single-crystal X-ray diffraction study.

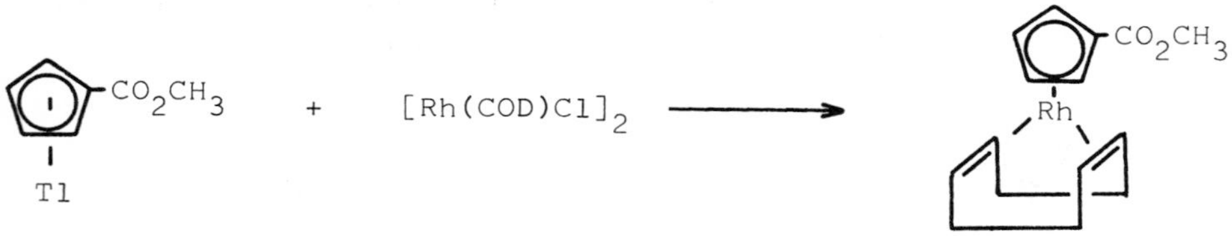

The anions **19** and **20** generated from the electroreduction of substituted ferrocenes, reacted with $Mo(CO)_6$ (THF, reflux) or with $W(CO)_3(DMF)_3$ (THF, reflux) to give the metal carbonyl anions **72–75** (*35*). Further reaction of these anions with methyl iodide gave compounds **58, 59, 62,** and **63**, respectively, previously described by Rausch *et al.* (*33*). Other functionally substituted compounds, including tricarbonyl iodide analogs **76** and dimers **77**, have been obtained from the reactions of anions **72–75** with I_2 and oxidizing agents, respectively.

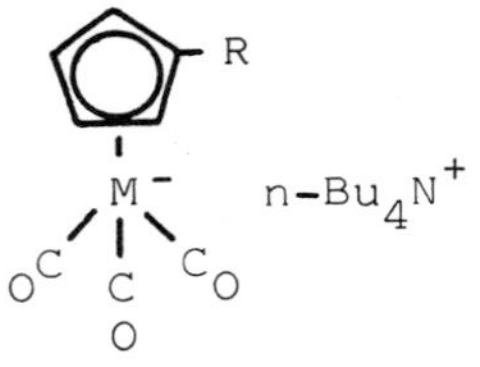

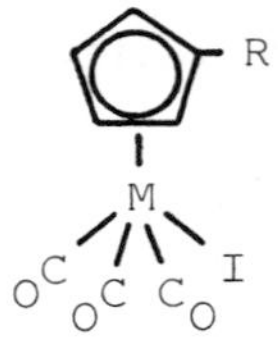

(72)   R = $COCH_3$; M = Mo

(73)   R = CO_2CH_3; M = Mo

(74)   R = $COCH_3$; M = W

(75)   R = CO_2CH_3; M = W

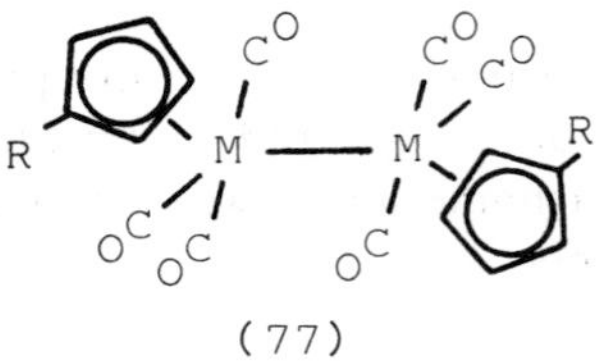

B. *Di- and Polysubstituted Compounds*

Peters has shown that a reaction between equimolar amounts of methyl chloroformate and sodium cyclopentadienide gave two main products (*47*). The expected product, carbomethoxycyclopenta-1,3-diene (**78**), was obtained in 20–30% yield upon distillation. A brick-red solid remained after distillation and was identified as sodium 1,2-dicarbomethoxycyclopentadienide (**42**), obtained in ∼20% yield. Peters' explanation for the formation

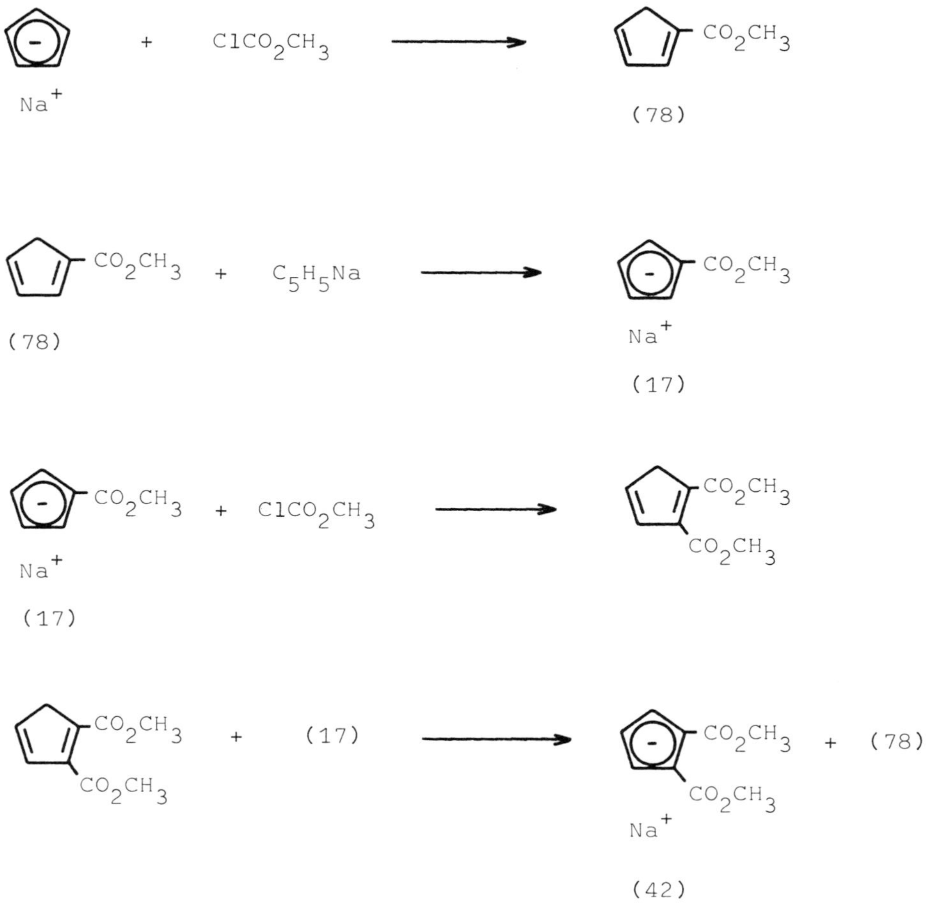

of **42** is that the cyclopentadiene **78** reacts with additional sodium cyclopentadienide giving the intermediate anion **17**, which reacts further with methyl chloroformate. A similar explanation was proposed by Linn and

Sharkey (*51*) for the reaction between sodium cyclopentadienide and benzoyl chloride.

Hafner *et al.* (*52*) have reported that 6-dimethylamino-2-formylfulvene (**79**) reacted with 2*N* sodium hydroxide to give sodium 1,2-diformylcyclopentadienide (**80**) in 63% yield. Treatment of **80** with acid gave 6-hydroxy-2-formylfulvene (**81**). Hafner also obtained sodium 1,2,4-triformylcyclopentadienide (**83**) from the dimethylaminofulvene (**82**) and 2*N* sodium hydroxide.

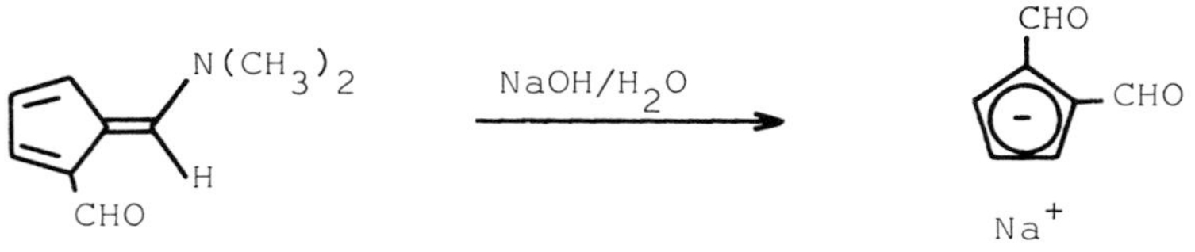

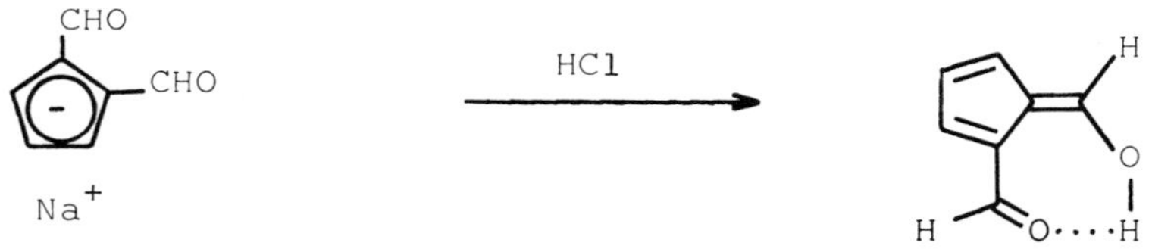

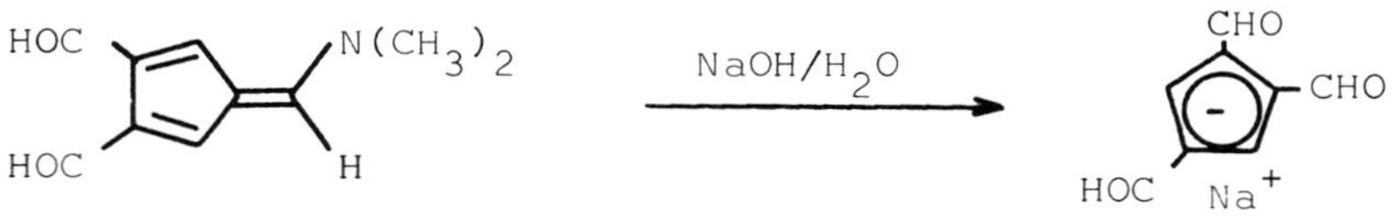

Webster (*40*) prepared both isomers of dicyanocyclopentadienide ion, and isolated them as salts of various cations. When sodium cyclopentadienide was treated with two equivalents of cyanogen chloride in the presence of excess sodium hydride, a mixture of sodium 1,2-dicyanocyclopentadienide (84) and sodium 1,3-dicyanocyclopentadienide (85) was obtained

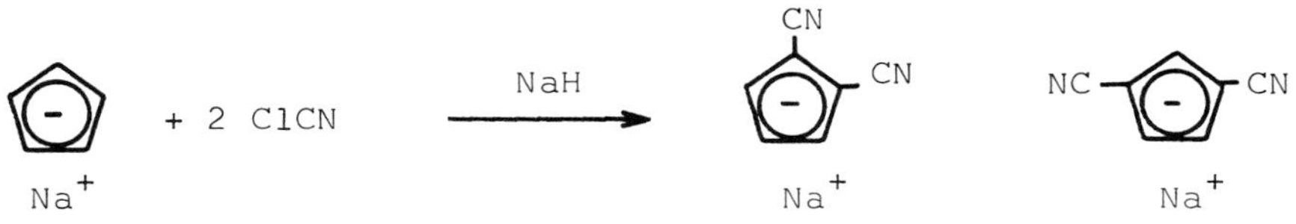

(84)   (85)

in about a six to one ratio. 1,3-Dicyanocyclopentadiene could also be separated from 84 and 85 by column chromatography of the reaction mixture on acidic alumina, and converted to its potassium salt with K_2CO_3. Various ammonium salts of 84 and 85 have also been produced.

Tricyanocyclopentadienide anions have been prepared by treating sodium cyclopentadienide with excess cyanogen chloride and sodium hydride. Potassium 1,2,4-tricyanocyclopentadienide (86) was isolated in 22% yield while potassium 1,2,3-tricyanocyclopentadienide (87) was isolated in 50% yield.

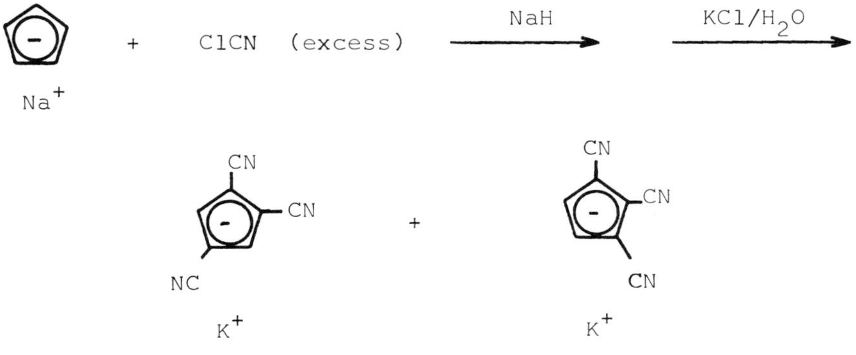

(86)   (87)

A Friedel–Crafts catalyst was required to further cyanate tricyanocyclopentadienide ion to tetra- and pentacyanocyclopentadienides. Thus a mixture of isomeric silver tricyanocyclopentadienides was treated with cyanogen chloride in the presence of aluminum chloride, then with tetraethylammonium chloride, to give tetraethylammonium tetracyanocyclopentadienide (88) and tetraethylammonium pentacyanocyclopentadienide (89) in about a 4 : 1 ratio.

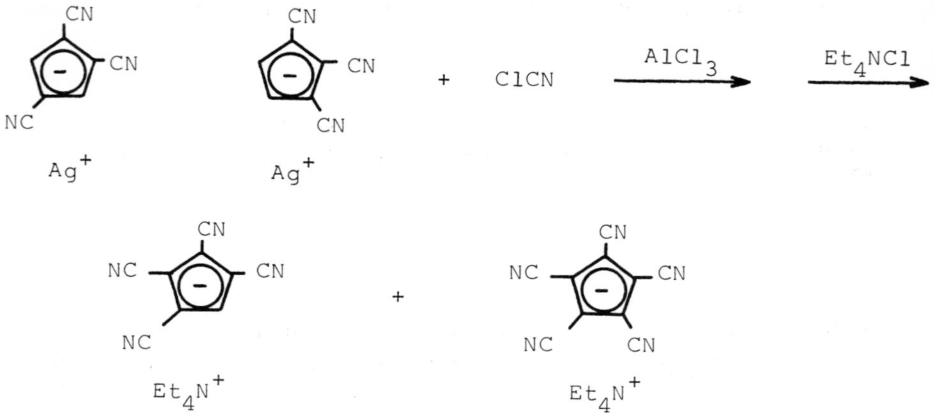

(88)   (89)

Dimethyl acetylenedicarboxylate condensed with dimethyl malonate in the presence of pyridine and acetic acid to give the cycloheptadiene (90) (53–55). When 90 was refluxed in aqueous potassium acetate, carbon dioxide was evolved and potassium 1,2,3,4,5-pentacarbomethoxycyclopentadienide (91) was formed in 62% yield. Treatment of anion 91 with acid gave 1,2,3,4,5-pentacarbomethoxycyclopentadiene (92) in 90% yield.

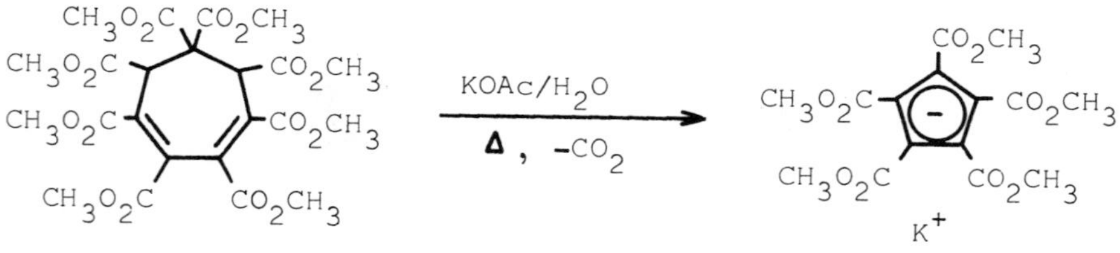

(90)   (91)

(91)   (92)

Bruce *et al.* (*56*) have shown that cyclopentadiene **92** reacted with either thallium carbonate or silver acetate to give thallium (**93**) or silver (**94**) salts, respectively.

These investigators have also used compounds **92–94** in the synthesis of several pentacarbomethoxy-substituted cyclopentadienyl transition metal complexes (**95–97**).

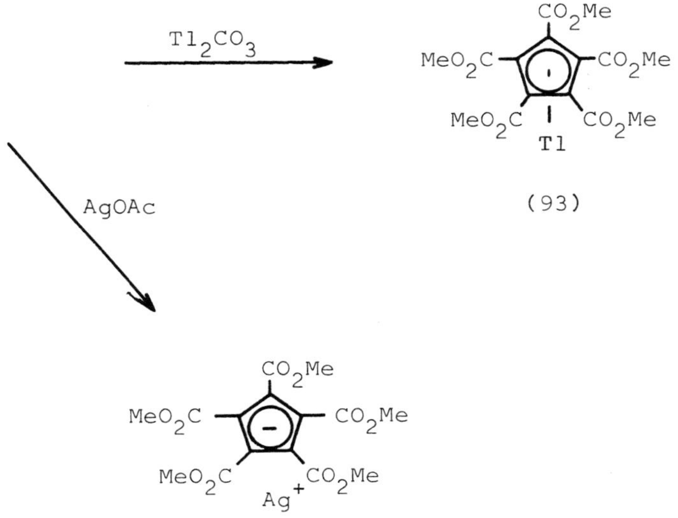

The molecular structure of **97** exhibited a metallocene type structure with the ruthenium atom almost equidistant from each ring. However, a reaction between **97** and PPh$_3$ in acetonitrile gave **95**, demonstrating that the Ru[C$_5$(CO$_2$CH$_3$)$_5$] bond is relatively weak. The molecular structure of Li[C$_5$(CO$_2$CH$_3$)$_5$](H$_2$O) has also been determined (*57*).

C. *Properties*

Cyclopentadienide anions possessing electron-withdrawing groups (viz., COCH$_3$, CO$_2$CH$_3$, CHO, or CN) generally have greater air stability than do the corresponding unsubstituted cyclopentadienide anions. Most likely

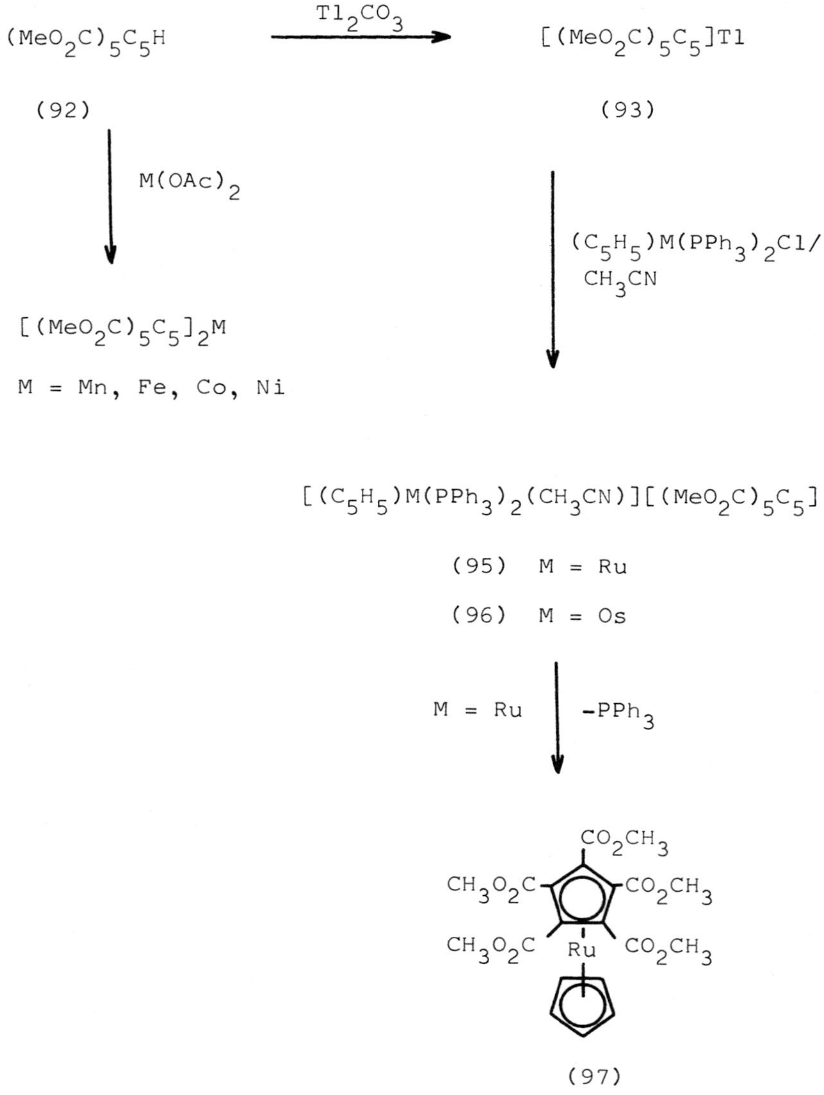
(MeO2C)5C5H
Tl2CO3
[(MeO2C)5C5]Tl
(92)
(93)
M(OAc)2
[(MeO2C)5C5]2M
M = Mn, Fe, Co, Ni
(C5H5)M(PPh3)2Cl/
CH3CN
[(C5H5)M(PPh3)2(CH3CN)][(MeO2C)5C5]
(95) M = Ru
(96) M = Os
M = Ru
-PPh3
CO2CH3
CH3O2C
CO2CH3
CH3O2C
Ru
CO2CH3
(97)

TABLE I

pK_a VALUES OF VARIOUS SUBSTITUTED CYCLOPENTADIENES IN AQUEOUS SOLUTIONS

Compound	Conjugate base	pK_a	Reference
C_5H_6		15	58
$C_5H_5CO_2CH_3$	Na^+ 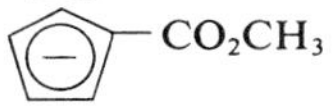 — CO_2CH_3	10.4	32
C_5H_5	K^+  — CN	9.8	40
$C_5H_5COCH_3$	Na^+ 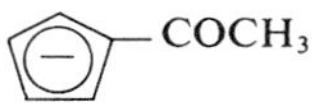 — $COCH_3$	8.8	32
C_5H_5CHO	Na^+ 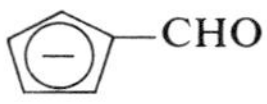 — CHO	7.4	32
$C_5H_4(CO_2CH_3)_2$	Na^+ 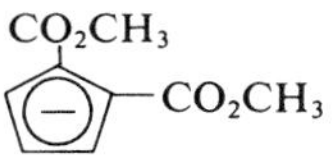 CO_2CH_3 — CO_2CH_3	5.0	32
$C_5H_4(CHO)_2$	Na^+ 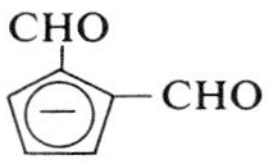 CHO — CHO	4.5	52
$C_5H_5NO_2$	Na^+  — NO_2	3.3	32
$C_5H_4(CN)_2$	K^+  NC — CN	2.5	40
$C_5H_3(CHO)_3$	Na^+ 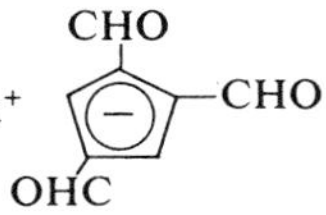 CHO — CHO, OHC	1.8	52
$C_5H_4(CN)_2$	K^+ 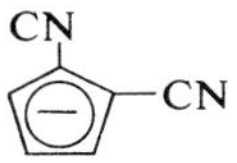 CN — CN	1.1	40

this is due to delocalization of electron density onto the substituent group, as shown below. These structures are merely resonance-stabilized enolate anions.

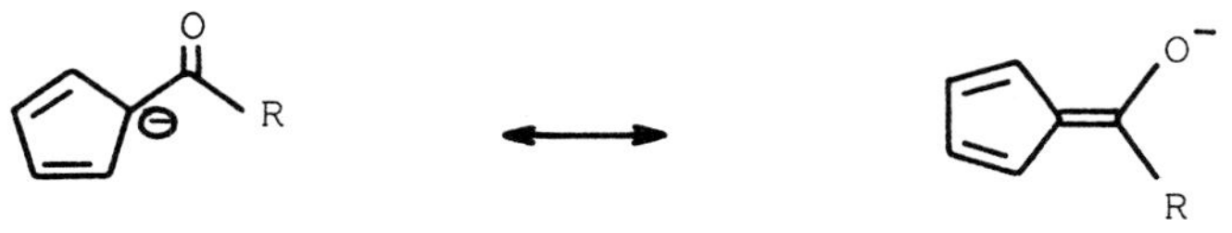

A   B

The same reasons that give these compounds increased stability also, in some cases, lower the reactivity, and in extreme instances prevent formation of η^5-cyclopentadienylmetal complexes. For example, the reaction between sodium carbomethoxycyclopentadienide and titanium tetrachloride did not give the expected titanocene dichloride (25). Presumably, sodium carbomethoxycyclopentadienide reacts through the oxygen atom (structure B) to give unidentifiable titanium alkoxides.

Table I gives pK_a values for some substituted cyclopentadiene compounds. Sodium nitrocyclopentadienide (151) has been included for comparison purposes and will be discussed in Section IV.

III

HALOGEN DERIVATIVES

A. *Compounds Derived from Diazocyclopentadienes*

Excluding ferrocene (1) and cymantrene (7), which can be halogenated indirectly via electrophilic substitution, there are two methods available for the synthesis of halocyclopentadienylmetal compounds. One method utilizes reactions of diazocyclopentadienes with metal carbonyl halides; the other involves reactions of halogen-substituted organothallium reagents. Shaver, Day, *et al.* (59, 60) have reported the syntheses and structures of some halogen-substituted η^5-cyclopentadienylmetal complexes of rhodium and manganese. Diazocyclopentadienes **98** and **99** were inserted into halogen-bridged dirhodium species to give compounds **100–104**. In similar re-

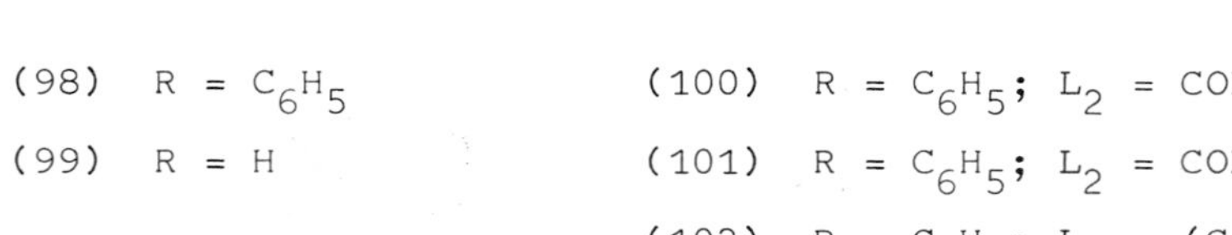

(98)   R = C_6H_5

(99)   R = H

(100)   R = C_6H_5; L_2 = COD; X = Cl

(101)   R = C_6H_5; L_2 = COD; X = Br

(102)   R = C_6H_5; L_2 = $(C_2H_4)_2$; X = Cl

(103)   R = C_6H_5; L_2 = $(CO)_2$; X = Cl

(104)   R = H; L_2 = COD; X = Cl

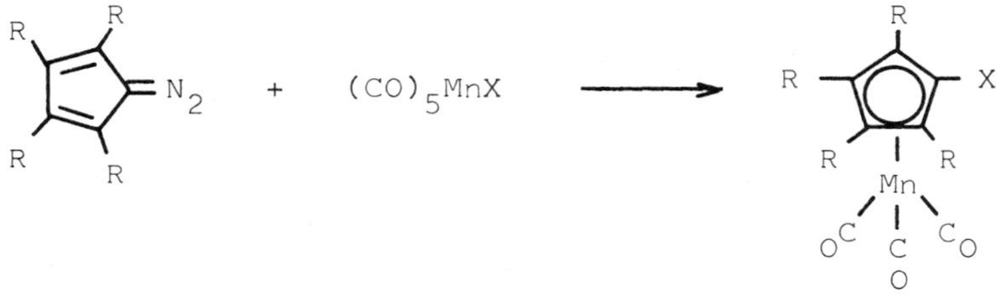

(98)   R = C_6H_5

(99)   R = H

(105)   R = H; X = Cl

(106)   R = H; X = Br

(107)   R = H; X = I

(108)   R = C_6H_5; X = Cl

actions, **98** and **99** reacted with manganese pentacarbonyl halides to give complexes (**105–108**). This new method of preparing halo-substituted cyclopentadienylmetal complexes **105–107** proved to be considerably easier than the multistep procedures previously employed (*61, 62*).

The crystal structure of **102** demonstrated that the cyclopentadienyl ring was bonded to the rhodium atom in a pentahapto fashion. The bond lengths around the five-membered ring, however, showed some similarities to those of σ-bonded cyclopentadienyl rings. Subsequently, Day and Shaver (*63, 64*) isolated an (η^1-C_5Cl_5)M(L$_n$) complex and determined its crystal structure. Manganese pentacarbonyl chloride reacted with 2,3,4,5-tetrachlorodiazocyclopentadiene (**109**) in pentane at room temperature to give (η^1-C_5Cl_5)Mn(CO)$_5$ (**110**) in 48% yield and (η^5-C_5Cl_5)Mn(CO)$_3$ (**111**) in

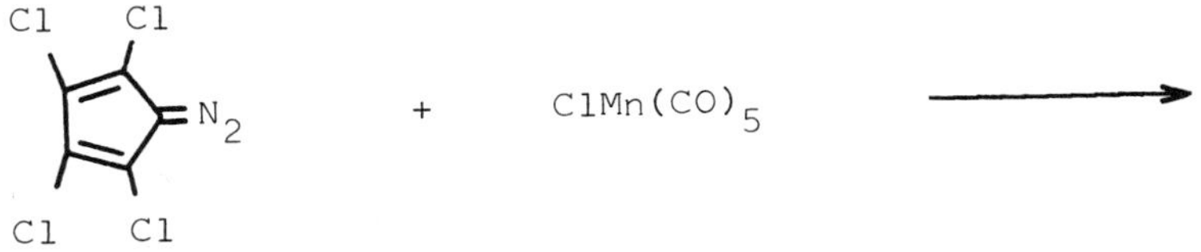

(109)

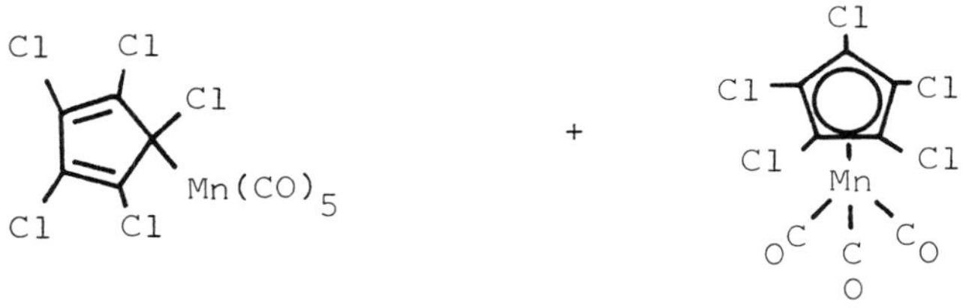

(110)   (111)

27% yield. The structure of **110** demonstrated that the manganese was octahedrally coordinated to five carbonyl groups and a σ-bonded cyclopentadienyl ring. The C_5Cl_5 group was bonded to the manganese through a single carbon atom at a distance of 2.204(6) Å. Complex **111** could be obtained free of **110** by running the reaction in octane at ~80°C.

Di-μ-chloro-bis(1,5-cyclooctadienerhodium) reacted with $C_5Cl_4N_2$ (**109**) in benzene to give the pentahapto complex **112** in 94% yield. The molecular structure of **112** has also been determined, and the data suggest appreciable contribution from a bonding model where the η^5-C_5Cl_5 ring is bonded to the metal by two π bonds and one σ-alkyl bond (*65*).

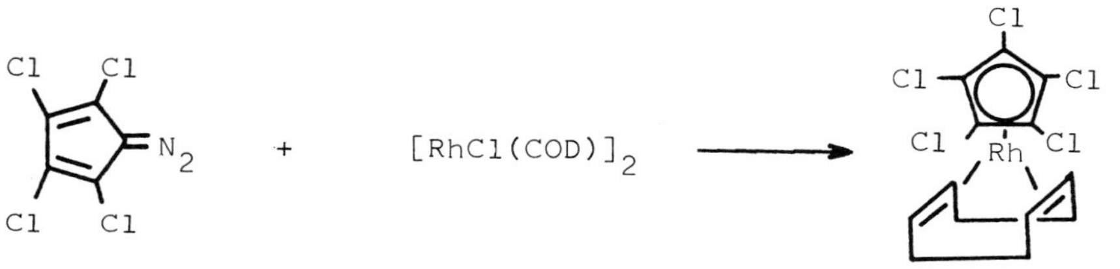

Herrmann and co-workers have independently used diazocyclopentadienes to synthesize halogen-substituted cyclopentadienylmetal complexes. Diazoindene reacted with manganese pentacarbonyl halides in a manner similar to diazocyclopentadiene to give compounds **113** (76%), **114** (48%), and **115** (16%) (*66*). Manganese pentacarbonyl halides reacted with 2,3,4,5-tetrabromodiazocyclopentadiene (**116**) or 2,5-diiododiazocyclopentadiene (**117**) to give the pentahapto complexes **118–120** in 70–76% yields

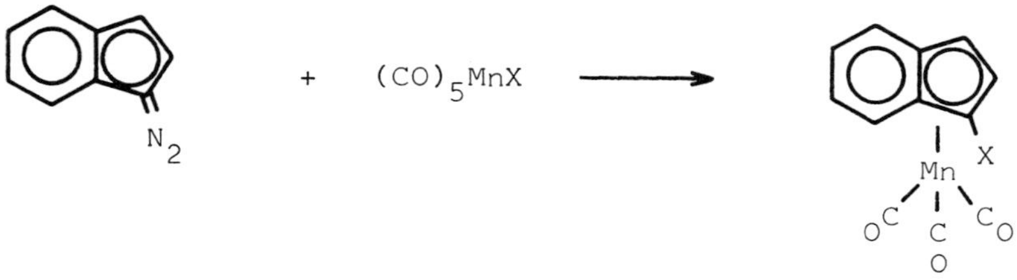

(113)   X = Cl

(114)   X = Br

(115)   X = I

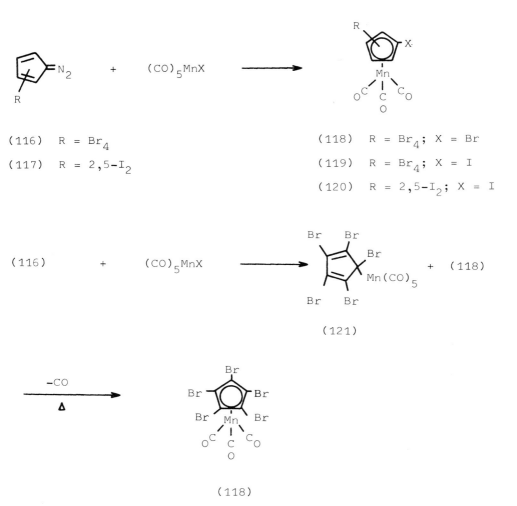

(116)   R = Br$_4$

(117)   R = 2,5-I$_2$

(118)   R = Br$_4$; X = Br

(119)   R = Br$_4$; X = I

(120)   R = 2,5-I$_2$; X = I

(121)

(118)

(67). In one example, the intermediate η^1-pentabromocyclopentadienyl-manganese pentacarbonyl (121) could be isolated. Upon melting 121 liberated CO and was converted to the pentahapto derivative 118. These results closely parallel the findings of Reimer and Shaver (64) for the pentachloro analogs.

Herrmann also found that diazocyclopentadiene (99) reacted with rhenium pentacarbonyl halides to give halo-substituted cyclopentadienylrhenium tricarbonyl compounds (122–124) in 80–86% yields (68). Compounds

122–124 had previously been prepared by Nesmeyanov *et al.* (*69, 70*) through a four- or five-step sequence, in overall yields of less than 50%.

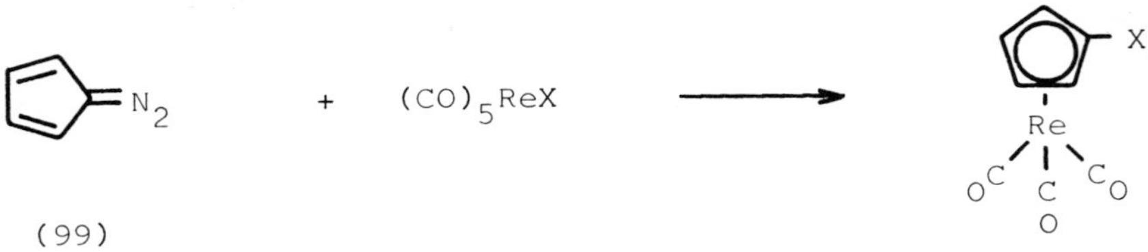

(99)

(122)   X = Cl

(123)   X = Br

(124)   X = I

Iron and ruthenium carbonyl halides have proved useful in the synthesis of various halo-substituted cyclopentadienyl compounds of these metals (*71*). For example, diazocyclopentadienes **99**, **109**, and **116** reacted with tetracarbonyliodo(σ-perfluoropropyl)iron to give complexes **125–127** in

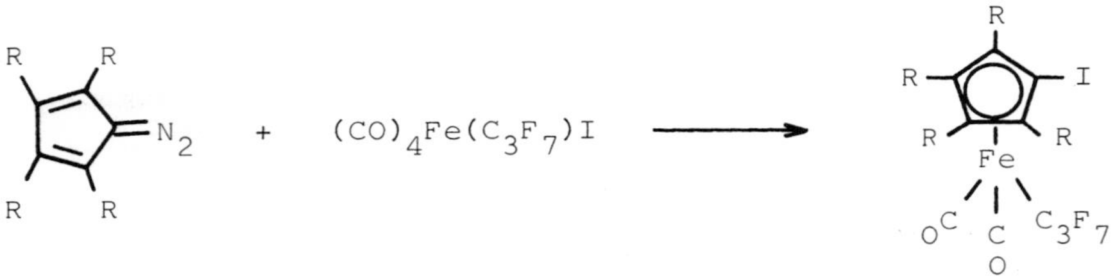

(99)   R = H

(109)   R = Cl

(116)   R = Br

(125)   R = H

(126)   R = Cl

(127)   R = Br

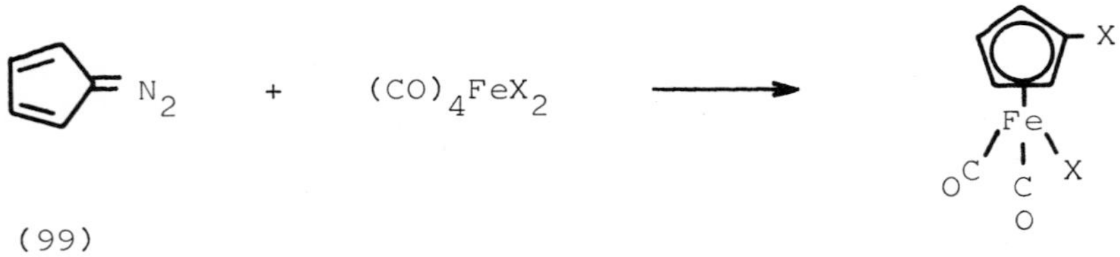

(99)

(128)   X = Br

(129)   X = I

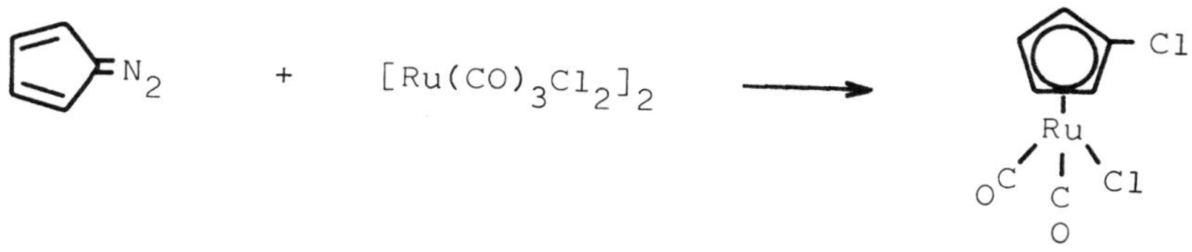

(99)

(130)

yields between 76 and 95%. Diazocyclopentadiene also reacted with $(CO)_4FeBr_2$ and $(CO)_4FeI_2$ to give **128** (1.6%) and **129** (58%), respectively. The analogous reaction for ruthenium occurred when $[Ru(CO)_3Cl_2]_2$ was treated with diazocyclopentadiene to give **130** in 18% yield. Compound **128** reacted further with diazocyclopentadiene, giving 1,1'-dibromoferrocene in 47% yield.

Compound **129** was characterized by reactions with C_5H_5Tl, CH_3MgI, and C_6H_5MgBr to give **131** (34%), **132** (17%), and **133** (52%), respectively. The chemistry of these halo-substituted cyclopentadienylmetal complexes has not as yet been fully explored (e.g., coupling reactions, nucleophilic substitutions, etc.).

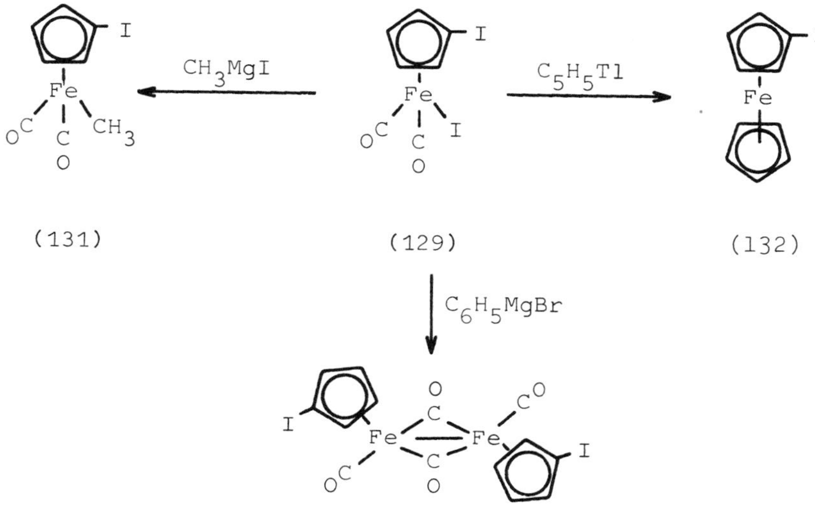

(131)   (129)   (132)

(133)

B. Halogen-Substituted Organothallium Reagents

Wulfsberg and West (72, 73) have synthesized thallium pentachloro-cyclopentadienide (134) as well as other $M^+C_5Cl_5^-$ salts, and have studied their properties. The thallium salt 134 was obtained by the addition of thallium ethoxide to C_5Cl_5H at $-78°C$ in pentane. The remaining salts (135–140) were prepared by abstraction of the acidic hydrogen on C_5Cl_5H. Compounds 134–140 are unstable above $-15°C$, especially 134, which ignites spontaneously in air or nitrogen with emission of orange light.

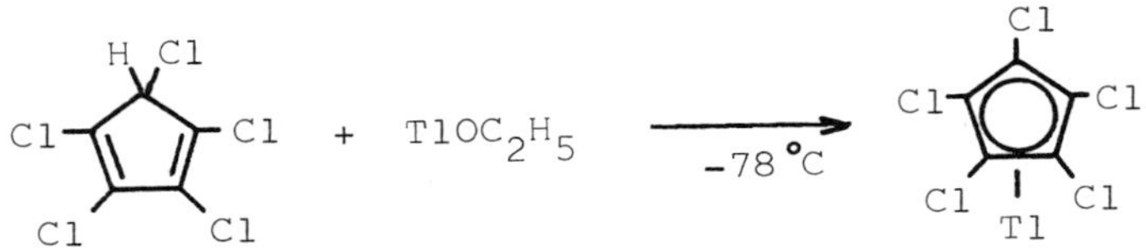

(134)

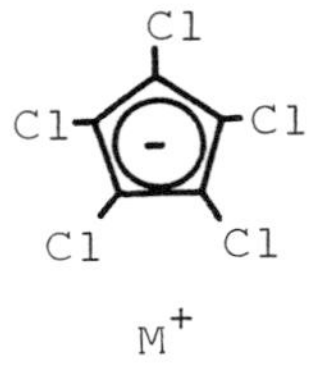

(135) $M = (n{-}C_3H_7)_4N$

(136) $M = (n{-}C_4H_9)_4N$

(137) $M = (n{-}C_7H_{15})_4N$

(138) $M = (C_2H_5)_3NH$

(139) $M = (n{-}C_4H_9)_4P$

(140) $M = C_5H_5N{-}CH_3$

Unfortunately, all attempts to convert these anions with transition metal halides to perchlorocyclopentadienyl complexes failed. However, 134 proved to be a useful reagent in the synthesis of some σ-bonded organomercury compounds. For example, $Hg(C_5Cl_5)_2$, C_5Cl_5HgCl, $C_5Cl_5HgCl \cdot HgCl_2$, and $C_6H_5HgC_5Cl_5$ were prepared by Wulfsberg and West.

Rausch and Conway (74, 75) have recently prepared organothallium reagents of the type C_5H_4XTl (where X = Cl, Br, or I), and used these in the synthesis of halo-substituted cyclopentadienylmetal complexes. Thallium cyclopentadienide reacted with N-chlorosuccinimide, N-bromosuccinimide, or iodine to give the intermediate halocyclopentadienes 141–143, which on treatment with thallous ethoxide produced thallium chlorocyclo-pentadienide (144) (72%), thallium bromocyclopentadienide (145) (96%), and thallium iodocyclopentadienide (146) (89%), respectively. The thermal and oxidative stabilities of the C_5H_4XTl compounds were found to fall in the order (for X = Cl, Br, I): Cl > Br > I.

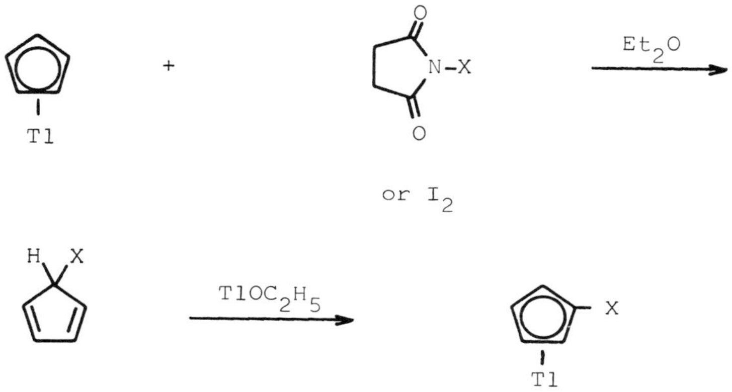

(141)   X = Cl   (144)   X = Cl

(142)   X = Br   (145)   X = Br

(143)   X = I   (146)   X = I

Organothallium reagents 144–146 reacted with an equimolar mixture of dicobalt octacarbonyl and iodine to produce the corresponding (η^5-halo-cyclopentadienyl)dicarbonylcobalt derivatives 147 (54%), 148 (57%), and 149 (12%), respectively.

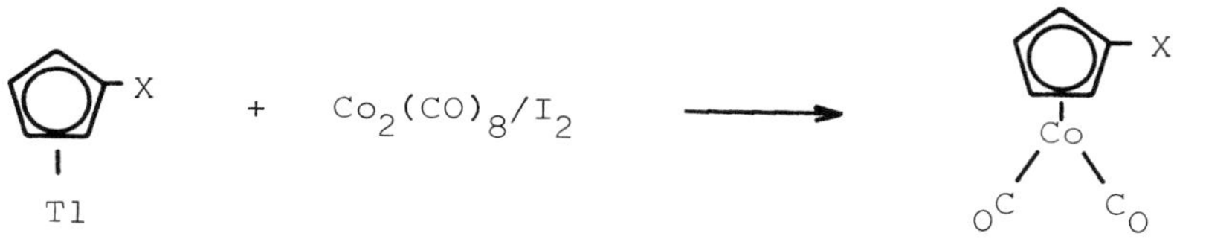

(144)   X = Cl   (147)   X = Cl

(145)   X = Br   (148)   X = Br

(146)   X = I   (149)   X = I

Treatment of **148** with *n*-butyllithium followed by cobaltous chloride gave a mixture of $(\eta^5\text{-}C_5H_5)Co(CO)_2$ (41%) and the fulvalene dicobalt compound **150** (28%). The molecular structure of **150**, as revealed by an X-ray structural analysis, showed the dicarbonylcobalt moieties on opposite sides of the planar fulvalene ring system (*76*).

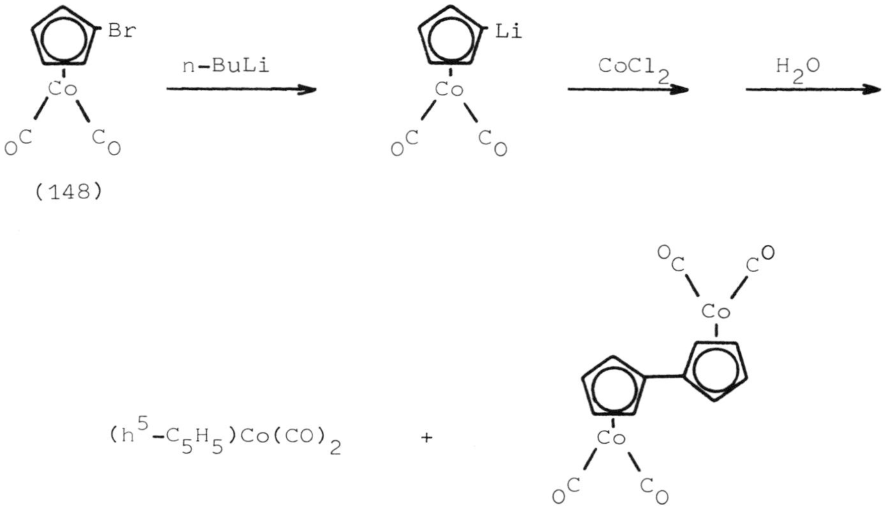

(148)

n–BuLi

CoCl$_2$

H$_2$O

$(h^5\text{-}C_5H_5)Co(CO)_2$ +

(12)   (150)

The utility of **144–146** in the synthesis of a wide variety of other (η^5-halocyclopentadienyl)-transition metal compounds has also been demonstrated, including derivatives of titanium, manganese, iron, rhodium, nickel, and copper (*74, 75*).

IV

NITROGEN- AND PHOSPHORUS-CONTAINING COMPOUNDS

A. *Nitrogen Compounds*

In 1900 Thiele (*30*) obtained the first cyclopentadienylmetal compound, sodium nitrocyclopentadienide **151**, from a reaction between cyclopentadiene and ethyl nitrate in the presence of sodium ethoxide. Kerber and Chick (77) used this same procedure to prepare 151 in yields of 16–21%. More recently, the lithium salt (152) was prepared from lithium cyclopentadienide and ethyl nitrate in THF (78).

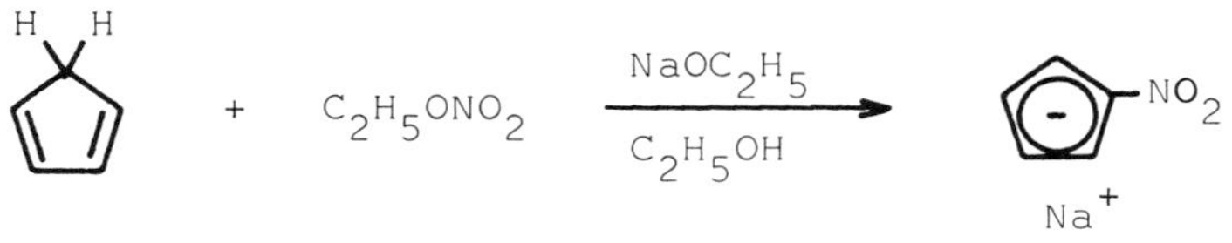

(151)

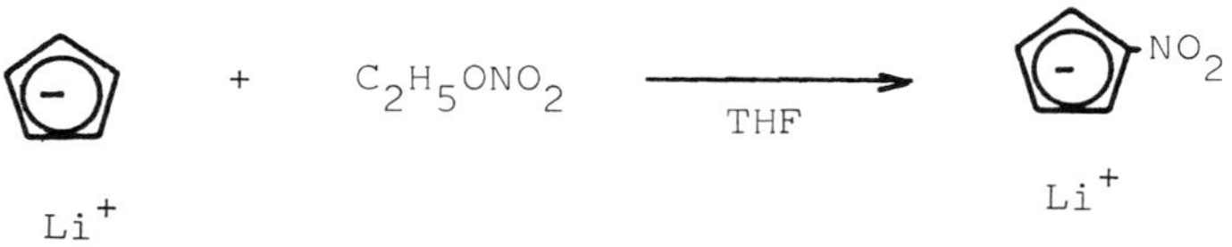

(152)

Sodium nitrocyclopentadienide (151) has not proved useful in the synthesis of η^5-nitrocyclopentadienyl-transition metal complexes. Attempts to synthesize 1,1'-dinitroferrocene by the reaction of ferrous chloride with 151 were unsuccessful (79). Compound 151 is much less reactive toward transition metal halides than is sodium cyclopentadienide itself. This is most likely due to the increased stability of a resonance stabilized anion, as shown below.

Compounds 151 and 152 did react with $[Rh(CO)_2Cl]_2$ to give (η^5-nitrocyclopentadienyl)dicarbonylrhodium (153) in yields of 22–30%, however. An X-ray diffraction study has shown that the nitrocyclopentadienyl ligand is significantly nonplanar (78).

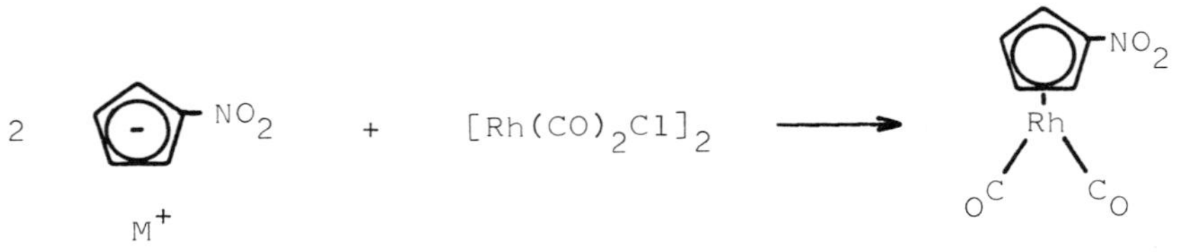

(151)   M = Na

(152)   M = Li

(153)

Bernheim and Boche (*80*) have prepared lithium *N,N*-dimethylamino-cyclopentadienide (**154**) in almost quantitative yield, and were able to convert this product to bis(dimethylamino)ferrocene (**155**). Compound **154** is the only amino-substituted cyclopentadienide salt known.

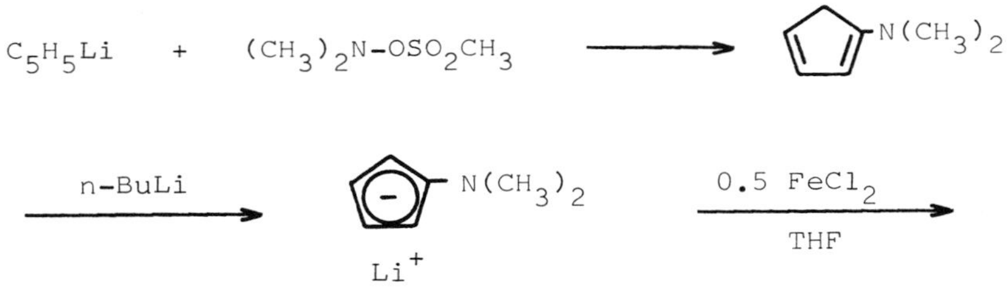

(154)

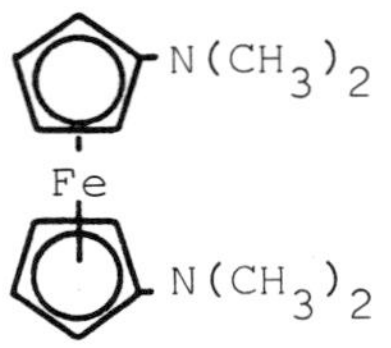

(155)

Pauson and Osgerby (*46*) found that dimethylaminomethylferrocene reacted with excess lithium metal to produce lithium cyclopentadienide and the amino-substituted anion **156**, which were subsequently treated with ferrous chloride to produce ferrocene (15%), dimethylaminomethyl-ferrocene (36%), and 1,1'-bis(dimethylaminomethyl)ferrocene (**157**) (15%).

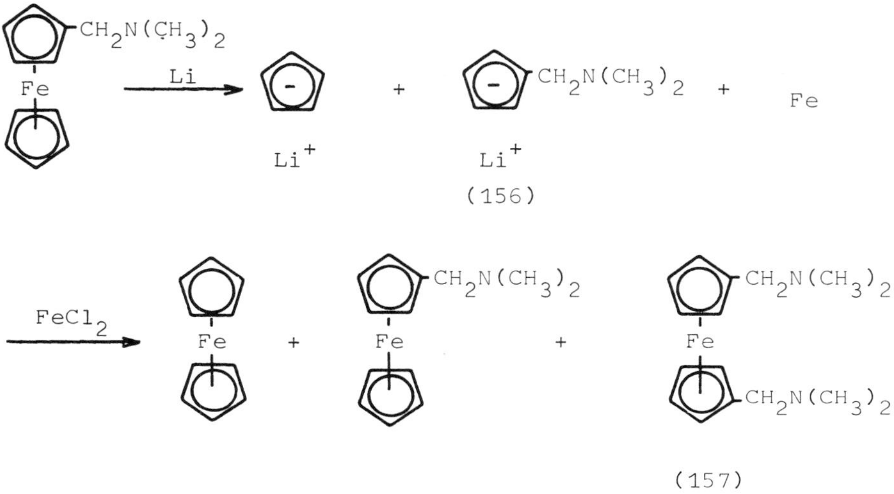

(156)

(157)

Pauson *et al.* (*81*) also generated **156** in ether solution from the reaction of 6-dimethylaminofulvene with lithium aluminum hydride. Treatment of this solution with either ferrous chloride or cobaltous chloride gave **157** and the cobalticinium analog **158** in yields of 58 and 78%, respectively.

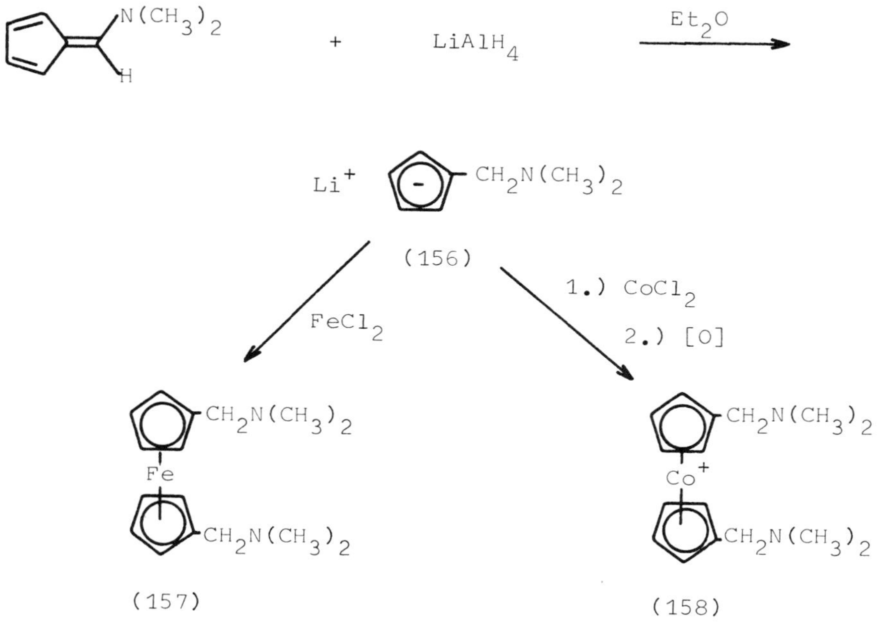

(157)   (158)

6-Dimethylaminofulvene also reacted with an ethereal solution of methyllithium to give **159**, which when treated with ferrous chloride gave the ferrocene derivative **160** in 71% yield.

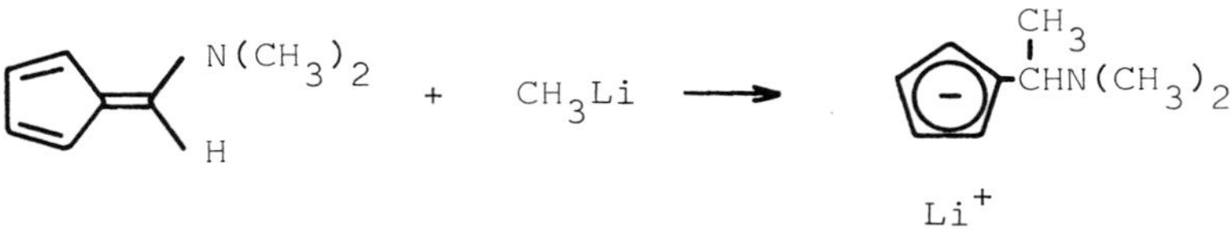

(159)

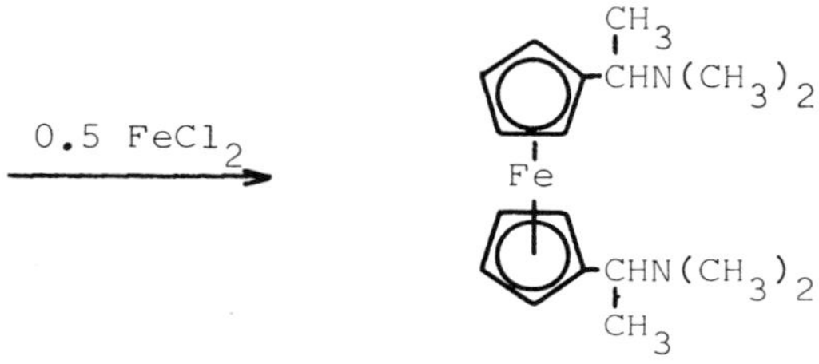

(160)

Knox and Pauson (82) obtained azoferrocenes (**161, 162**) via the azo-substituted cyclopentadienide anions (**163, 164**).

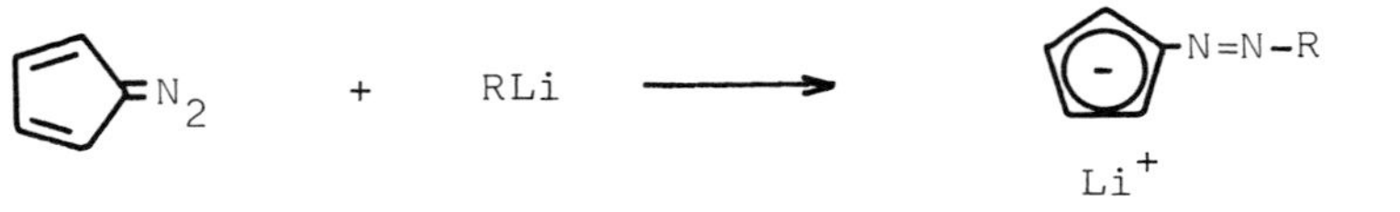

(163)   R = CH$_3$

(164)   R = C$_6$H$_5$

(161)   R = CH$_3$

(162)   R = C$_6$H$_5$

B. *Phosphorus Compounds*

Spiro[2.4]hepta-4,6-diene (**165**) has been used to prepare both nitrogen-
and phosphorus-containing cyclopentadiene compounds (*83*). Various nu-
cleophiles reacted with **165** to give, after hydrolysis, substituted cyclopentadienes as shown below. When the anionic intermediates were allowed
to react with ferrous chloride without prior hydrolysis, a series of substituted ferrocenes were obtained.

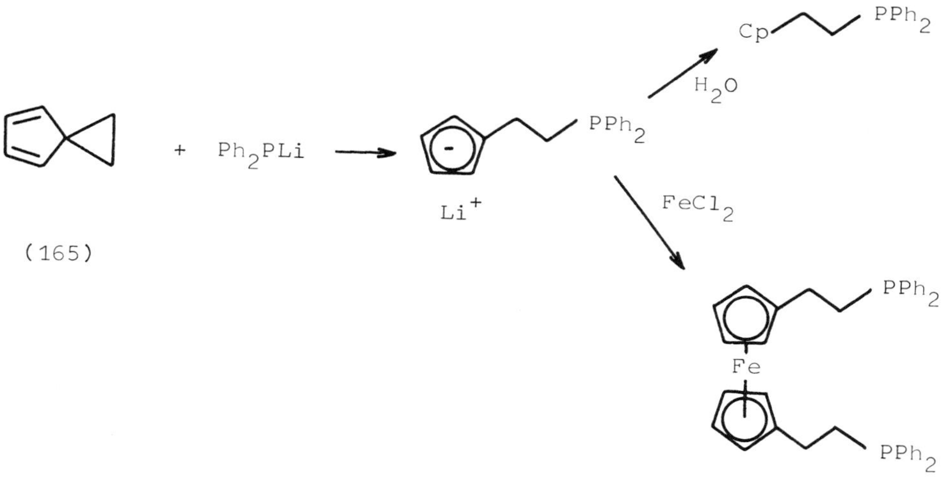

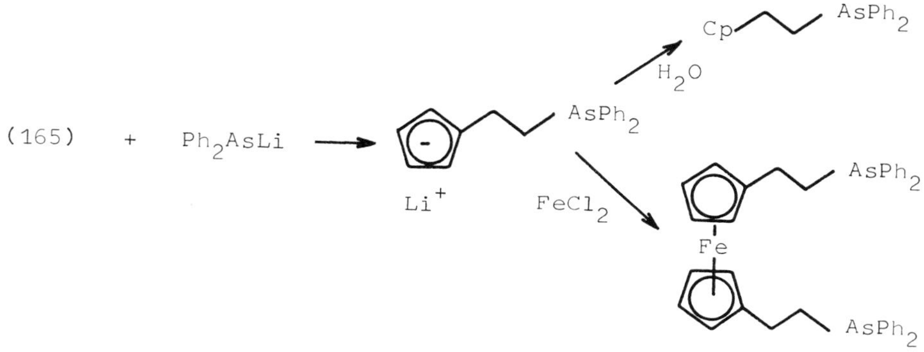

(165)

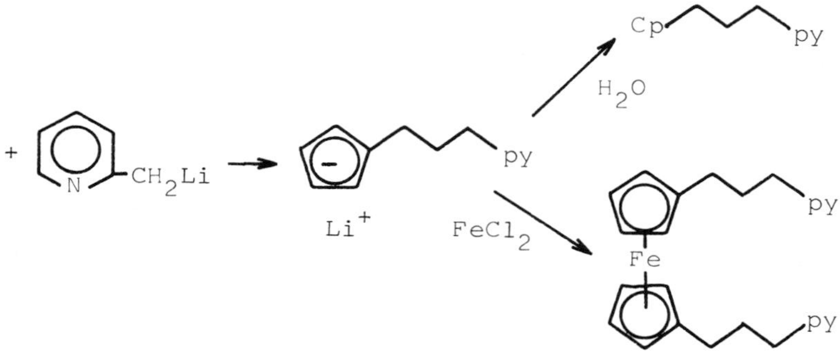

A list of various cyclopentadienylphosphines is given in Table II.

Mathey and Lampin (*86*) prepared *in situ* the phosphine-substituted cyclopentadienide anion **166**, which upon treatment with acetone at low temperature and subsequent hydrolysis gave **167** in ~30% yield.

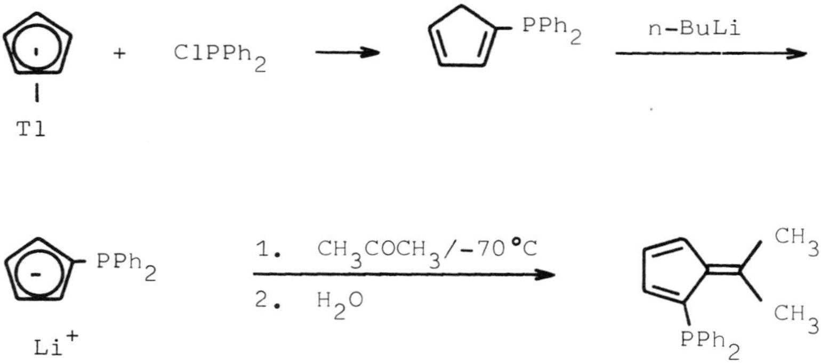

(166)   (167)

Davison *et al.* (*90*) also generated **166** *in situ*, but starting with (dimethoxyethane)sodium cyclopentadienide instead of thallium cyclopentadienide. Compound **166** was then treated with cobaltous chloride, and the resulting cobaltocene oxidized and isolated as the hexafluorophosphate adduct (**168**) in 20% yield.

TABLE II
Phosphine Substituted Cyclopentadienes

Preparation	Structure	Yield (%)	Properties	Reference
$C_5H_5MgBr + ClP(OC_4H_9)_2$	 ⬠—$P(OC_4H_9)_2$	33	Bp 80–81°C (1 mm)	84
$C_5H_5Tl + ClP(C_6H_5)_2$	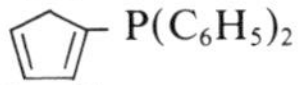 —$P(C_6H_5)_2$	100	Bp 65°C (5×10^{-5} mm)	85, 86
$C_5H_5Tl + Cl_2PC_6H_5$	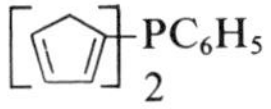 [⬠—PC_6H_5]$_2$	76	Unstable	85
$C_5H_5Tl + ClP(CH_3)_2$	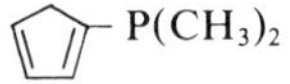 ⬠—$P(CH_3)_2$	75	Very unstable	85
$C_5H_5Tl + Cl_2PCH_3$	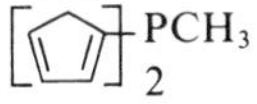 [⬠—PCH_3]$_2$	~40	Extremely unstable	85
$C_5H_5Na + ClCH_2P(C_6H_5)_2$	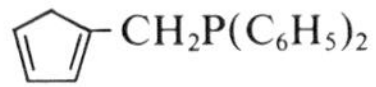 ⬠—$CH_2P(C_6H_5)_2$	50	Polymerizes at 25°C	87
$C_5H_5Tl + ClPF_2$	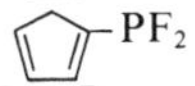 —PF_2	95–98	Bp −12°C (8 mm)	88, 89
$C_5H_5Tl + Cl_2PF$	 [⬠—PF]$_2$	85–90	Air and temperature sensitive	88

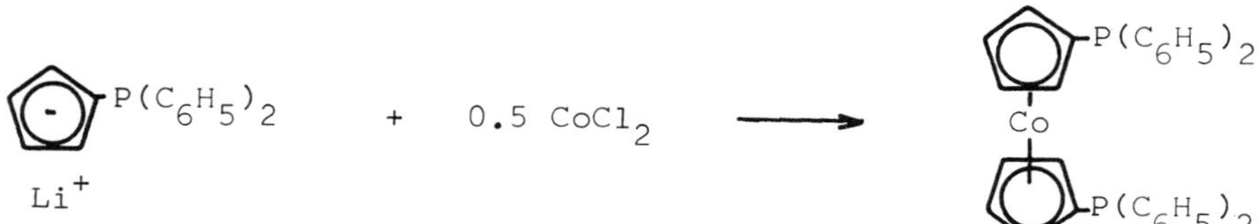

(166)

(168)

Compound **166** reacted with $Cr(CO)_6$ (diglyme, reflux), $Mo(CO)_6$ (dioxane, reflux), or $W(CO)_6$ (diglyme, reflux) to give a series of group VIB tetracarbonyl derivatives (**169–171**).

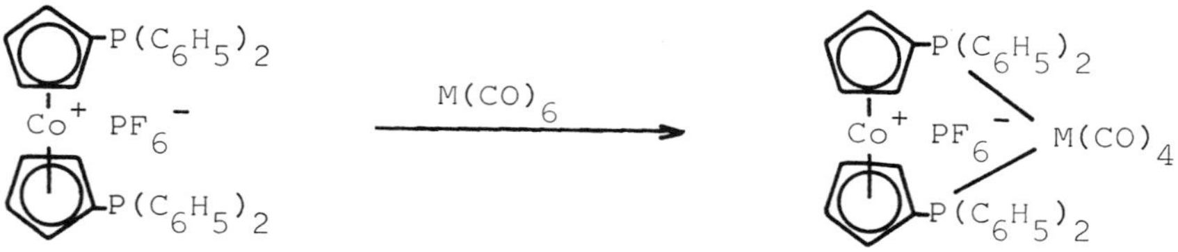

(168)

(169)   M = Cr

(170)   M = Mo

(171)   M = W

Schore (*91*) has recently isolated and studied the chemistry of some phosphine-substituted cyclopentadienide anions. Lithium [dimethyl-(diphenylphosphinomethyl)silyl]cyclopentadienide (**172**) was obtained in 63% overall yield.

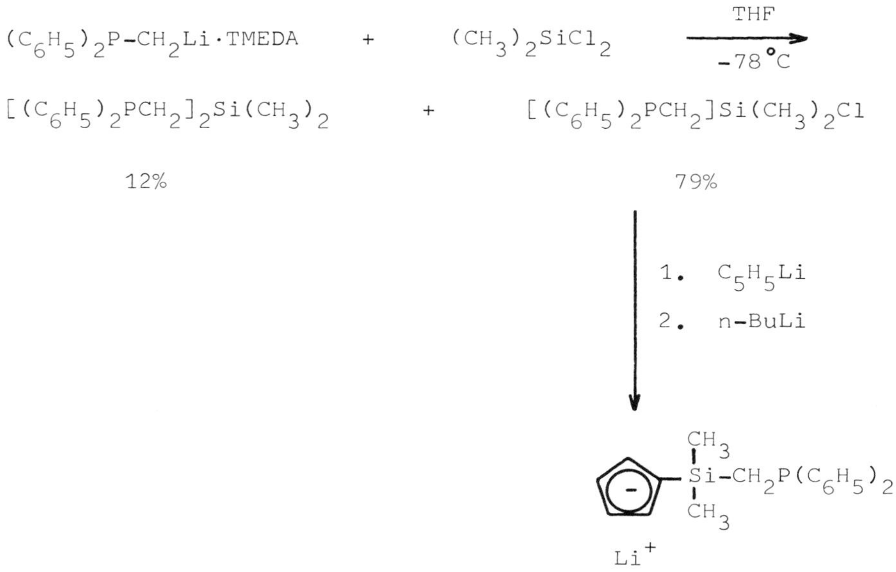

$(C_6H_5)_2P-CH_2Li \cdot TMEDA$ + $(CH_3)_2SiCl_2$ $\xrightarrow[-78\,°C]{THF}$

$[(C_6H_5)_2PCH_2]_2Si(CH_3)_2$ + $[(C_6H_5)_2PCH_2]Si(CH_3)_2Cl$

12%   79%

1. C_5H_5Li

2. $n-BuLi$

(172)

Compound **172** reacted with ferrous chloride in ethyl ether or with $(\eta^5\text{-}C_5H_5)ZrCl_3$ in THF to give complexes **173** and **174**, respectively.

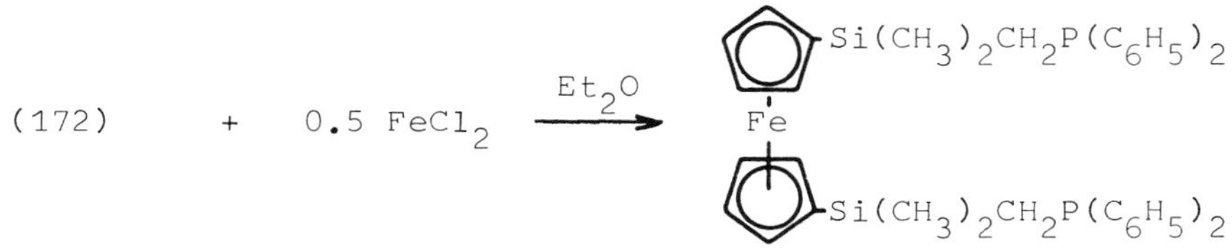

(172)   +   0.5 $FeCl_2$ $\xrightarrow{Et_2O}$

(173)

(172)   + $(h^5-C_5H_5)ZrCl_3$ $\xrightarrow{THF}$

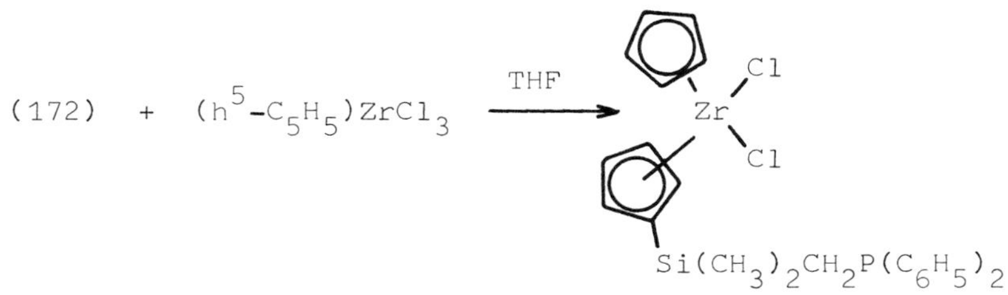

(174)

Bimetallic complexes were obtained by coordination of the free phosphine in **173** or **174** to a second metal center.

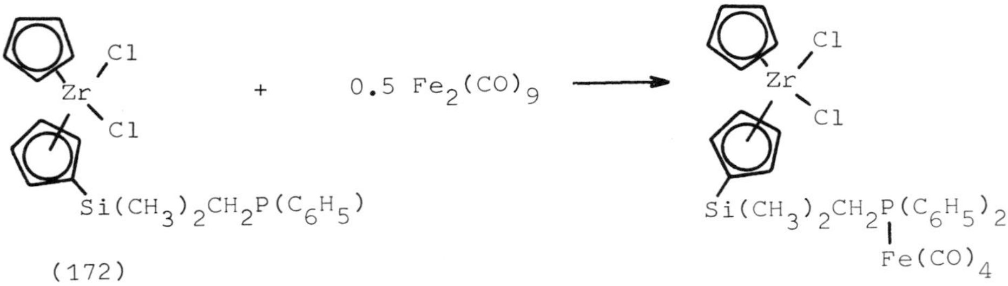

(172)

Recently Rausch and Edwards (75, 92) have synthesized a useful phosphine-substituted cyclopentadienylthallium reagent. Diphenylphosphinocyclopentadiene, which was generated *in situ* according to the method of Mathey and Lampin (86), was treated with thallium ethoxide to give diphenylphosphinocyclopentadienylthallium (**175**). This organothallium reagent (**175**) when treated with (η^5-C_5H_5)TiCl$_3$ gave the phosphine-substituted titanocene dichloride (**176**). The bimetallic complex (**177**) was subsequently prepared from **176** and (η^5-C_5H_5)Mn(CO)$_2$THF.

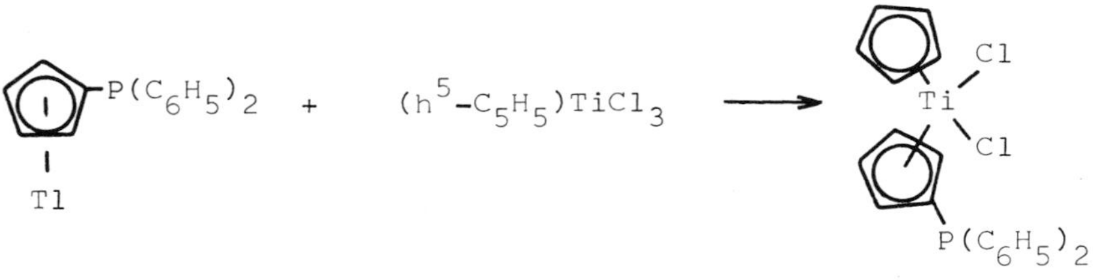

(175)   (176)

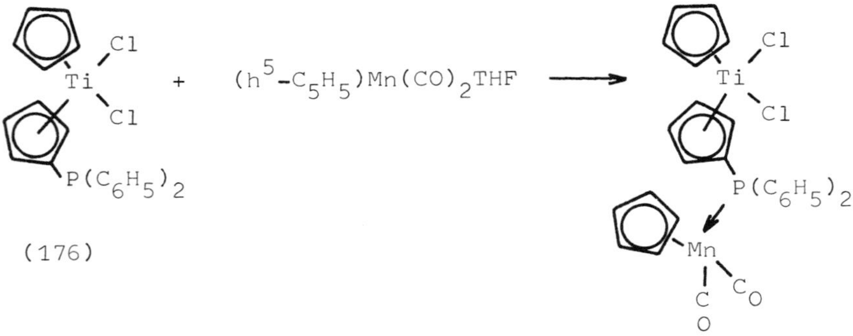

(176)   (177)

V

OLEFINS (ORGANOMETALLIC MONOMERS)

A. Olefins Derived from Fulvenes

Schlenk and Bergmann (93) first observed sodium isopropenylcyclopentadienide (**178**) as the product of a reaction between 6,6-dimethylfulvene and triphenylmethylsodium. Although **178** was not isolated or fully characterized in this study, triphenylmethane (**179**) was identified as one of the reaction products.

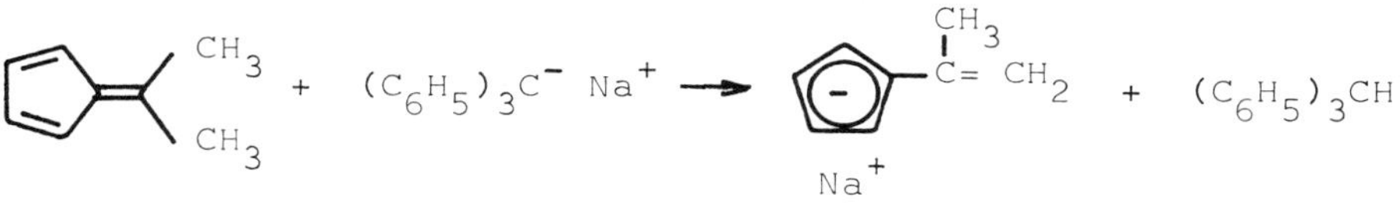

$$(178) \qquad\qquad (179)$$

6-Methyl-1,2,3,4-tetraphenylfulvene (**180**) condensed with quinoline-4-carbonal in the presence of sodium methoxide to give compound **181** (94). Presumably, sodium methoxide removed a proton from the 6-methyl group of **180** to give the intermediate **182** which rapidly condensed with the carbonyl compound.

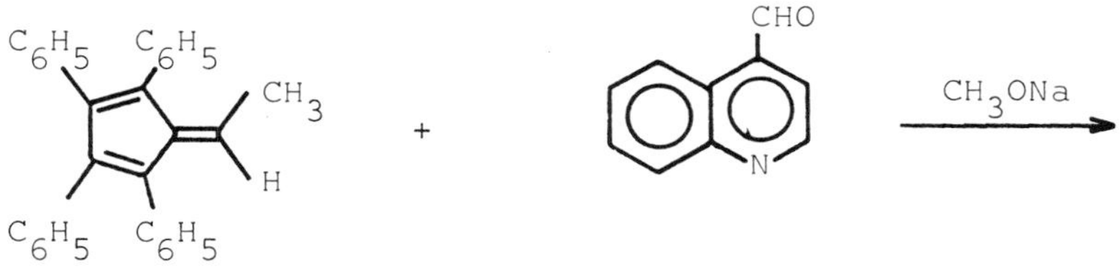

(180)

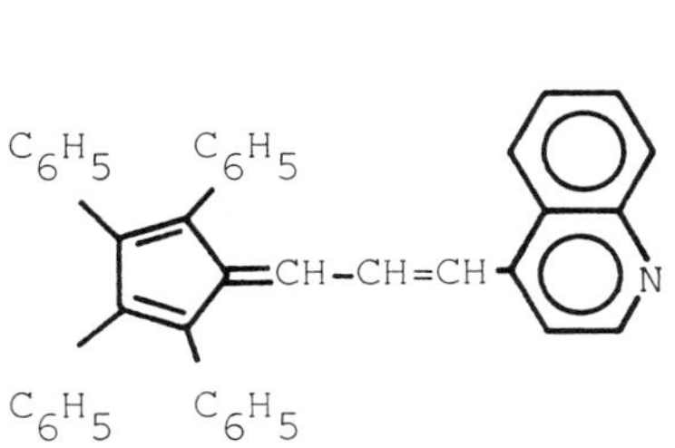

(181)

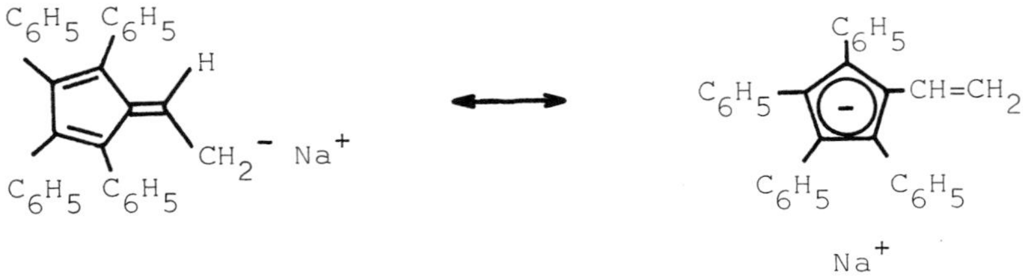

(182)

Knox and Pauson (95) have prepared 1,1'-diisopropenylferrocene in ~60% yield from 6,6-dimethylfulvene, sodium amide, and ferrous chloride. The intermediate, sodium isopropenylcyclopentadienide (178), was not isolated. A similar reaction to prepare 1,1'-divinylferrocene from 6-methylfulvene was reported to give only polymeric ferrocenes.

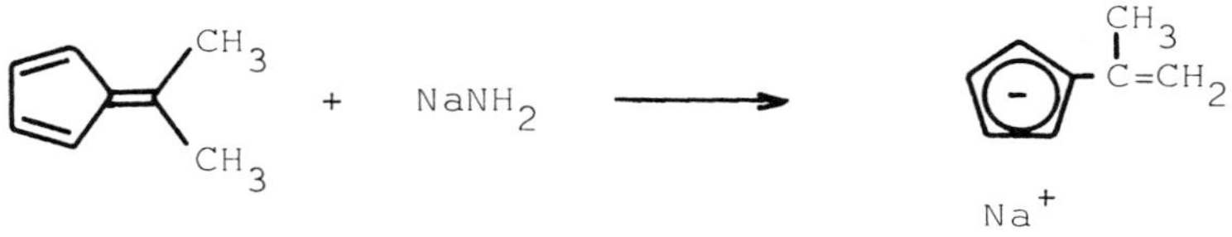

(178)

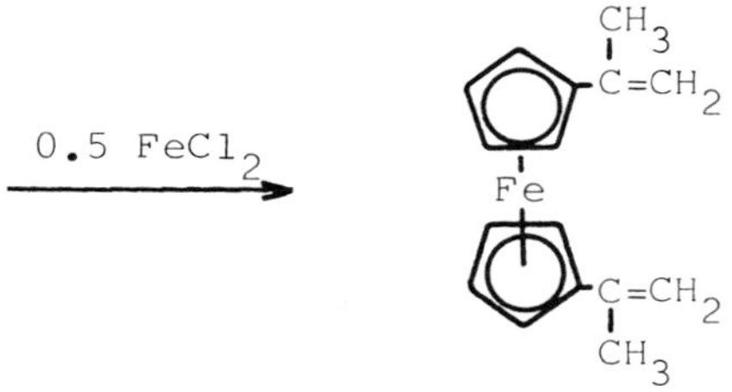

Hine and Knight (96) generated in situ potassium isopropenylcyclopentadienide (183) from 6,6-dimethylfulvene and potassium t-butoxide in diglyme. As with all the previous preparations, 183 was not isolated.

The isolation of lithium isopropenylcyclopentadienide (184) has been achieved by Schore and LaBelle (97) from a reaction between 6,6-dimethylfulvene and diphenylphosphinomethyllithium in ethyl ether or THF solution. Compound 184 was also formed in an NMR experiment from 6,6-dimethylfulvene and lithium diisopropylamide in THF-d_8.

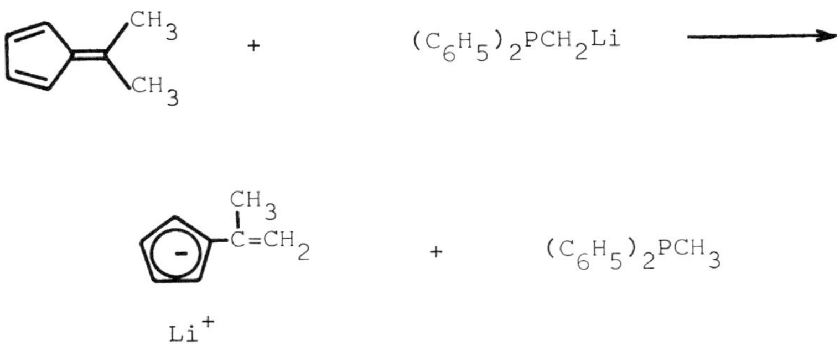

(184)

Recently, Rausch and Macomber (*48, 75, 98*) synthesized both **184** and lithium vinylcyclopentadienide (**185**), and have used these reagents in the synthesis of a wide variety of organometallic monomers. 6,6-Dimethylfulvene and 6-methylfulvene reacted with lithium diisopropylamide in THF at 25°C to produce lithium isopropenylcyclopentadienide (**184**) and lithium vinylcyclopentadienide (**185**), respectively, in yields of 80–90%. Both products could be isolated as air-sensitive white solids.

(184)

(185)

(η^5-Vinylcyclopentadienyl)metal compounds have been formed in 15–93% yields from reactions of **185** and various transition metal substrates, including titanium, molybdenum, tungsten, cobalt, rhodium, iridium, and copper. For example, the series of (η^5-$C_5H_4CH=CH_2$)$M(CO)_2$ compounds (**186–188**) has been prepared from **185** and the appropriate metal carbonyl halide (*48, 75, 98*).

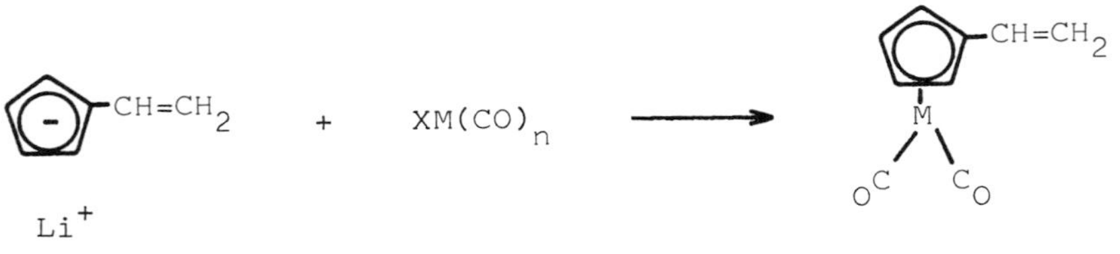

(185)

(186) M = Co

(187) M = Rh

(188) M = Ir

Some of these organometallic monomers have been homo- and copoly-merized under free-radical conditions. The resulting polymers have potential as catalysts, UV absorbers, or semiconductors.

Fulvenes have been found to react directly with various metal carbonyls to give olefins. Altman and Wilkinson (*99*) treated various fulvenes with dicobalt octacarbonyl to give mixtures of alkenyl- and alkylcyclopenta-dienylcobalt dicarbonyls. For example, 6,6-dimethylfulvene reacted with dicobalt octacarbonyl to give (η^5-isopropenylcyclopentadienyl)dicarbonylcobalt (**189**) and (η^5-isopropylcyclopentadienyl)dicarbonylcobalt (**190**) in a combined yield of 77%.

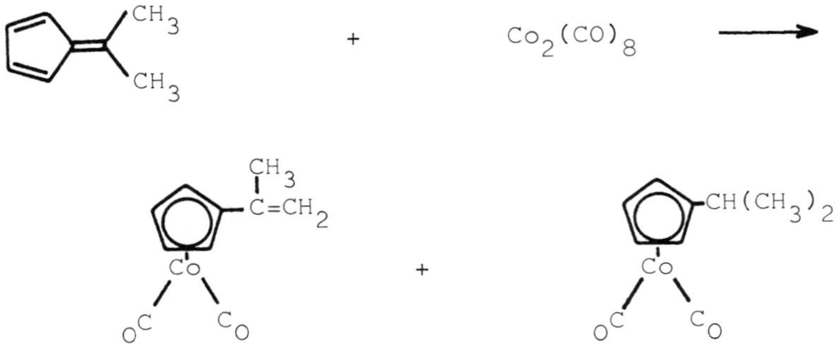

(189)

(190)

Hoffman and Weiss (*100*) found that a number of fulvenes reacted with vanadium hexacarbonyl to give good yields of (η^5-alkenylcyclopenta-dienyl)tetracarbonylvanadium compounds. For example, 6-methylfulvene reacted with vanadium hexacarbonyl to give the vinyl monomer (**191**) in 68% yield.

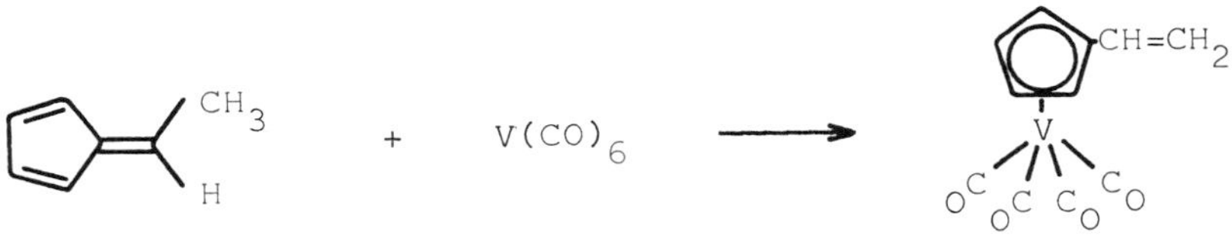

(191)

B. *Other Methods of Preparation*

Another less practical method for the synthesis of vinylcyclopentadienyl monomers has recently been described by two independent research groups. Eilbracht *et al.* (*101*) reported that spiro[2.4]hepta-4,6-diene (**165**) reacted with $Co_2(CO)_8$ to give **192** (35%), **186** (30%), and **193** (7%). These compounds were separated by column chromatography.

(165)   + $Co_2(CO)_8$ ⟶

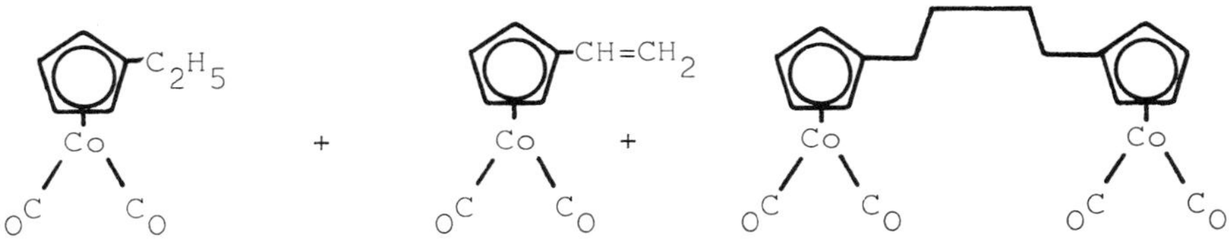

(192)   +   (186)   +   (193)

The spirohydrocarbon **165** was also used by Gladysz *et al.* (*102*). The cocondensation of iron atoms with **165** gave ferrocenes (**194–197**) in a combined yield of 44%. Likewise, when **165** was condensed with cobalt atoms followed by carbon monoxide, compounds **192** and **186** were obtained in a combined yield of no greater than 1%.

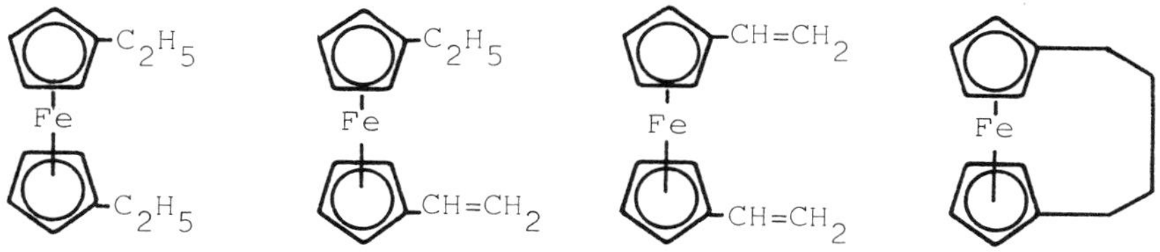

(194)   (195)   (196)   (197)

The above methods involving the use of **165** give mixtures of compounds which require tedious separation to obtain the pure monomers.

Sneddon *et al.* (*103, 104*) have reported that the reaction of tetracyanoethylene with thallium cyclopentadienide resulted in the formation of thallium tricyanovinylcyclopentadienide (**198**) in 90% yield.

$$C_5H_5Tl \quad + \quad (CN)_2C=C(CN)_2 \quad \longrightarrow \quad Tl^+[C_5H_4C(CN)C(CN)_2]^-$$

$$(198)$$

Reactions of **198** with $BrMn(CO)_5$, $ClCuP(C_6H_5)_3$, or $(\eta^5\text{-}C_5H_5)Fe(CO)_2I$ in THF gave, respectively, **199** (6.2%), **200** (8.1%), and **201** (2.7%).

$$(198) \quad + \quad BrMn(CO)_5 \quad \longrightarrow \quad [h^5\text{-}C_5H_4C(CN)C(CN)_2]Mn(CO)_3$$

$$(199)$$

$$(198) \quad + \quad ClCuP(C_6H_5)_3 \quad \longrightarrow \quad [h^5\text{-}C_5H_4C(CN)C(CN)_2]CuP(C_6H_5)_3$$

$$(200)$$

$$(198) \quad + \quad (h^5\text{-}C_5H_5)Fe(CO)_2I \quad \longrightarrow$$

$$[h^5\text{-}C_5H_4C(CN)C(CN)_2]Fe(h^5\text{-}C_5H_5)$$

$$(201)$$

Interestingly, when **201** was treated with C_5H_5Tl, then with $(\eta^5\text{-}C_5H_5)Fe(CO)_2I$, 1,1-dicyano-2,2-diferrocenylethylene (**203**) was obtained in 15.7% yield. Presumably, the reaction proceeds via the intermediate organothallium compound (**202**).

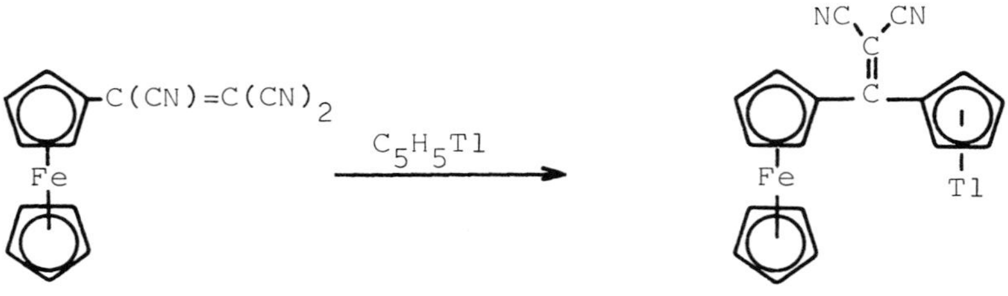

$$(201) \qquad\qquad\qquad\qquad (202)$$

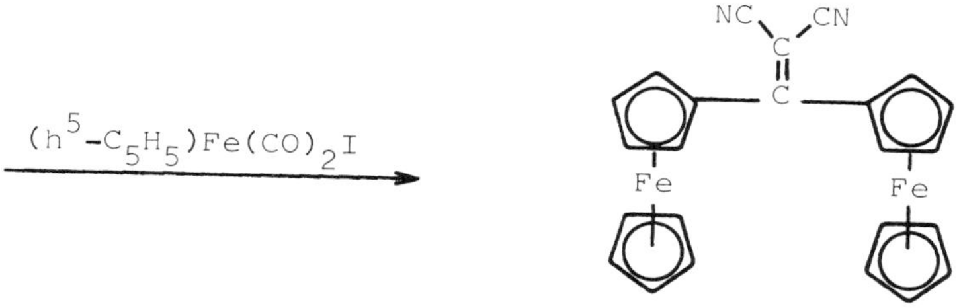

(203)

Katz and Mrowca (*105*) have found that dihydropentalene (**204**) reacted with thallous sulfate in aqueous potassium hydroxide to give thallium hydropentalenide (**205**) in quantitative yield.

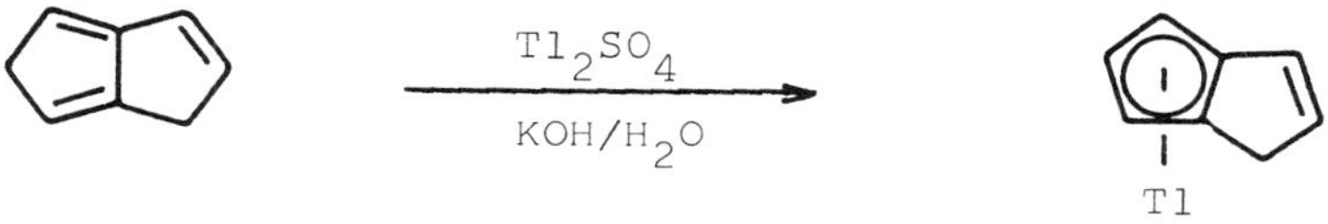

(205)

Similar to cyclopentadienylthallium, **205** reacted with various transition metal halides to give a number of derivatives. For example, the reaction of **205** with [RhCl(COD)]$_2$ gave complex **206** in 76% yield. When **206** was treated with *n*-butyllithium followed by either D$_2$O or benzophenone, compounds **208** and **209** were obtained. The pentalenylcycloocta-1,5-dienerhodium anion (**207**) was postulated as the reactive intermediate.

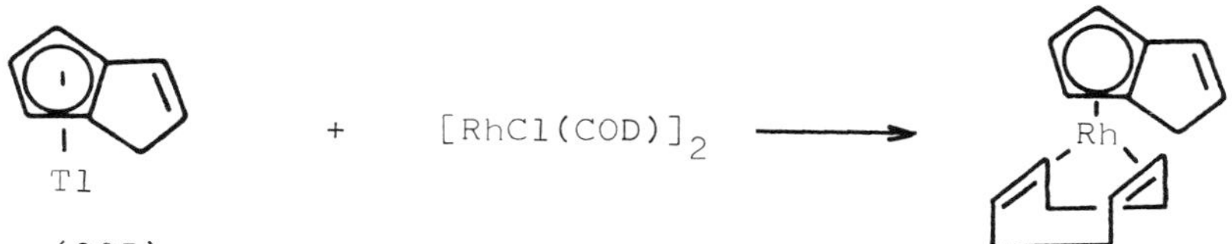

(205)   +   [RhCl(COD)]$_2$ ⟶

(206)

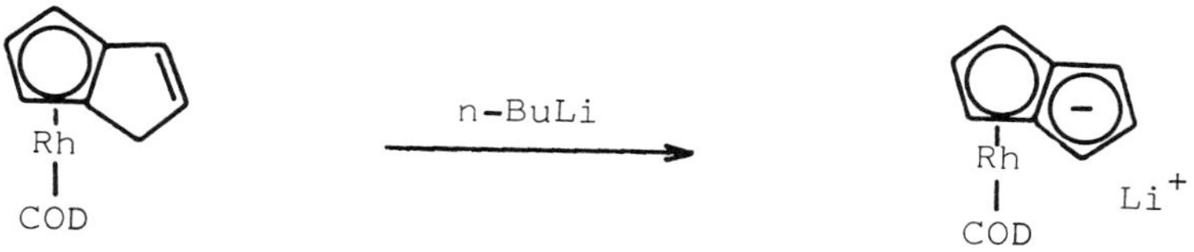

(206) $\xrightarrow{\text{n-BuLi}}$ (207)

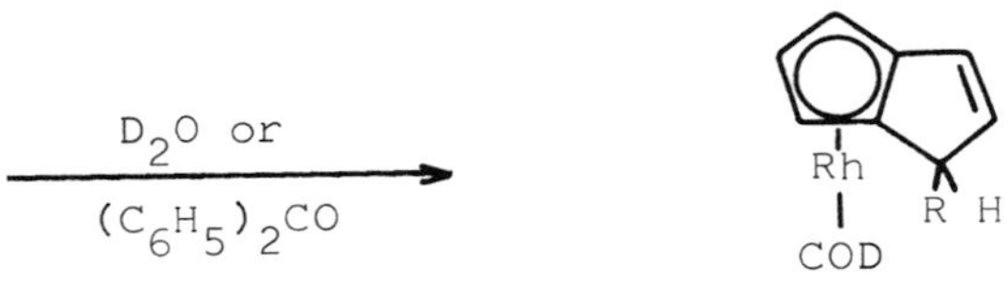

$$\xrightarrow{\begin{array}{c} D_2O \text{ or} \\ \hline (C_6H_5)_2CO \end{array}}$$

(208)   R = D

(209)   R = $(C_6H_5)_2$COH

VI

POLYMER-BOUND CYCLOPENTADIENYL COMPOUNDS

Polymer-supported cyclopentadienyl compounds have been synthesized mainly for possible catalytic applications. In general, polystyrene–divinyl-benzene copolymers have been used as the polymer supports. Metals incorporated into these polymers have included titanium, zirconium, hafnium, iron, cobalt, and rhodium.

The reaction of sodium cyclopentadienide with chloromethylated 20% cross-linked polystyrene–divinylbenzene copolymer gave complex **210** (*106*). Treating **210** with methyllithium followed by $(\eta^5\text{-}C_5H_5)TiCl_3$ gave the polymer-supported titanocene dichloride (**211**).

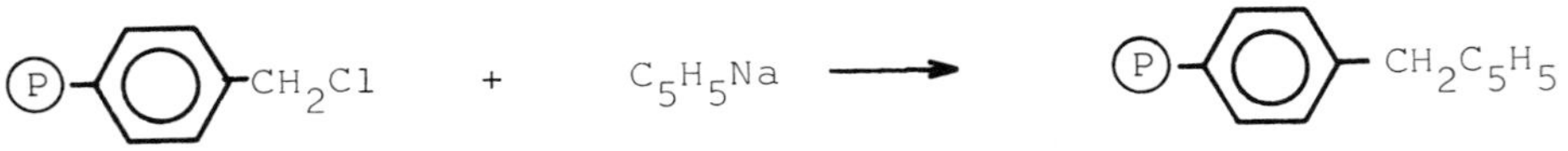

(210)

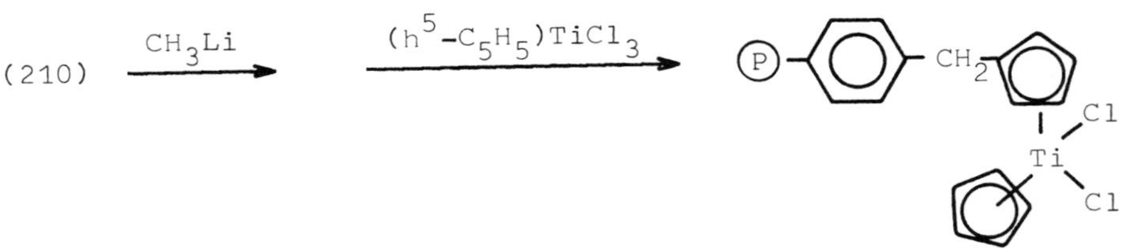

(211)

Another method for attaching cyclopentadienyl groups to polymer back-bones involved bromination of the copolymer, halogen–lithium exchange, and reaction of the lithiated intermediate with cyclopentenone. Treatment of the cyclopentadiene-substituted copolymer with methyllithium followed by $(\eta^5\text{-}C_5H_5)TiCl_3$ likewise gave a polymer-supported titanocene dichloride (212).

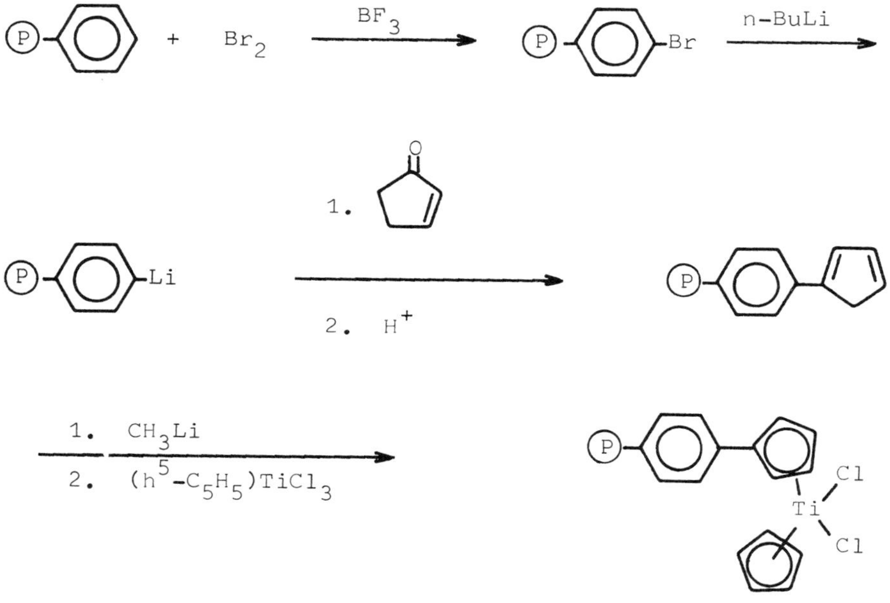

(212)

Polymers 211 and 212 following reduction with n-butyllithium were found to be 25–120 times more active than the corresponding reduced, nonattached titanocene dichloride for the hydrogenation of cyclohexene.

Brintzinger *et al.* (*107*) have obtained polymer-supported cyclopenta-dienylmetal carbonyl complexes of iron and cobalt. Treating a cyclopen-tadienyl-substituted polystyrene–divinylbenzene (18%) copolymer with $Fe_2(CO)_9$ or $Co_2(CO)_8$ gave complexes 213 and 214, respectively.

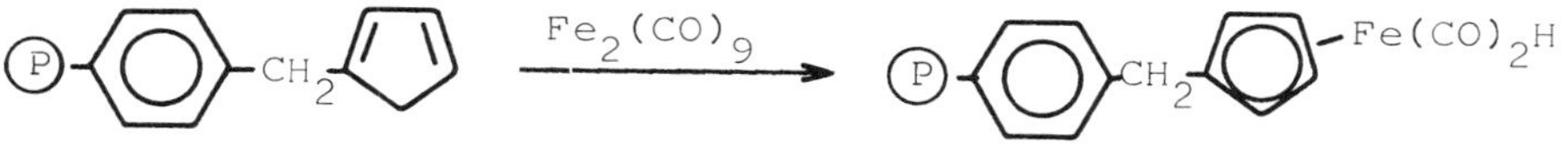

(213)

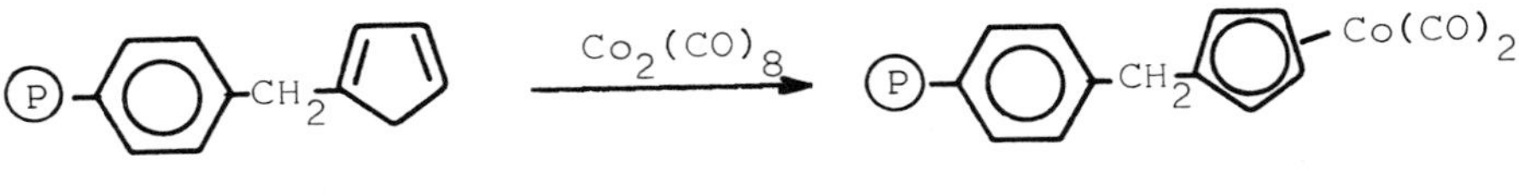

(214)

Brubaker *et al.* (*108*) have prepared 20% cross-linked polystyrene–divinylbenzene copolymer-attached (η^5-cyclopentadienyl)trichlorometal complexes of titanium (**215**), zirconium (**216**), and hafnium (**217**). Only the titanium polymer **215** showed catalytic activity in the hydrogenation of cyclohexene.

(215)

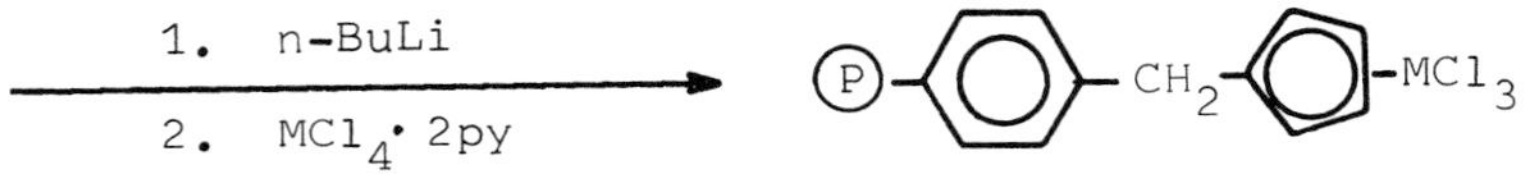

(216)   M = Zr

(217)   M = Hf

Polymers **211** and **215** have also been employed in the catalytic isomerization of allylbenzene and 1,5-cyclooctadiene. Likewise, **211** and **215** have effected the oligomerization of ethyl propiolate to a mixture of open and closed trimers (*109*).

Polymer-attached cyclopentadienyldicarbonylcobalt (**218**) and polymer-attached cyclopentadienyldicarbonylrhodium (**219**) have been prepared from **210** (*110*). Rhodium polymer **219** was found to be an active catalyst for hydrogenating olefins and ketones, isomerizing olefins, and for the hydroformylation of olefins. The cobalt polymer **218** had no catalytic activity in these processes.

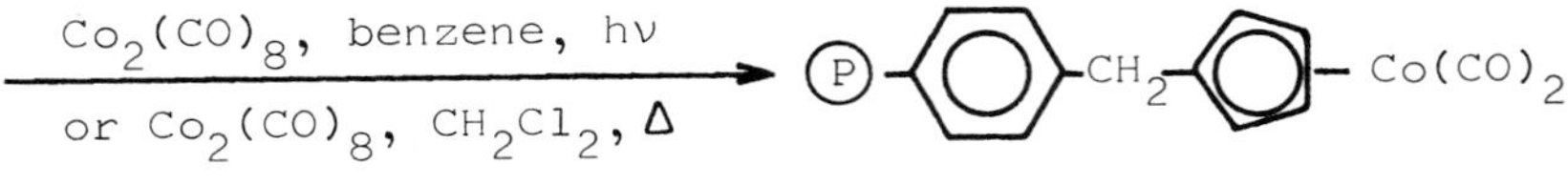

(218)

(210) 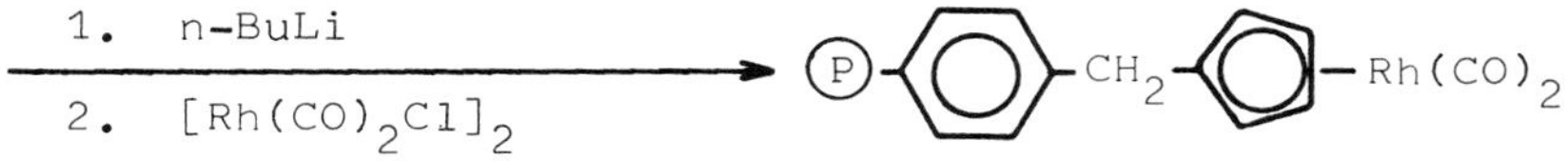 $\xrightarrow[\text{2. } [Rh(CO)_2Cl]_2]{\text{1. } n\text{–BuLi}}$ (P)—⟨◯⟩—CH_2—⟨◯⟩—$Rh(CO)_2$

(219)

Perkins and Vollhardt (111) formed a variant of polymer **218** by using 3% cross-linked polystyrene–divinylbenzene. This polymer also had limited activity as a hydroformylation catalyst. Interestingly, however, the polymer proved to be quite active as a methanation and Fischer–Tropsch catalyst. By means of this catalyst, methane as well as C_3–C_{20} hydrocarbons were produced from CO + H_2 (3 : 1, 75 psi, 25°C) at 190–200°C.

Sekiya and Stille (112) have recently developed a new method for the preparation of polystyrene-bearing cyclopentadienylmetal complexes. The reaction of styrylmagnesium bromide with 2-norbornen-7-one gave, after hydrolysis, a 68% yield of *syn-* and *anti-*7-(*p*-styryl)norborn-2-en-7-ol (**220**). Compound **220** was chlorinated with HCl to give *anti-*7-chloro-7-(*p*-styryl)norborn-2-ene (**221**) in 88% yield. The 7-methoxy analog (**222**) was prepared from the alkoxide of **220** and methyl iodide in 93% yield.

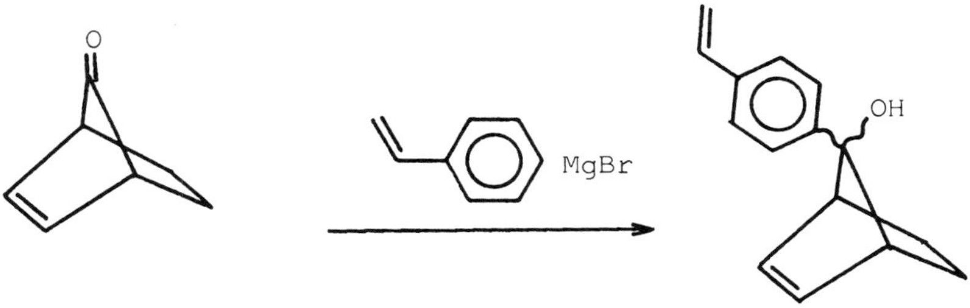

MgBr

(220)

HCl

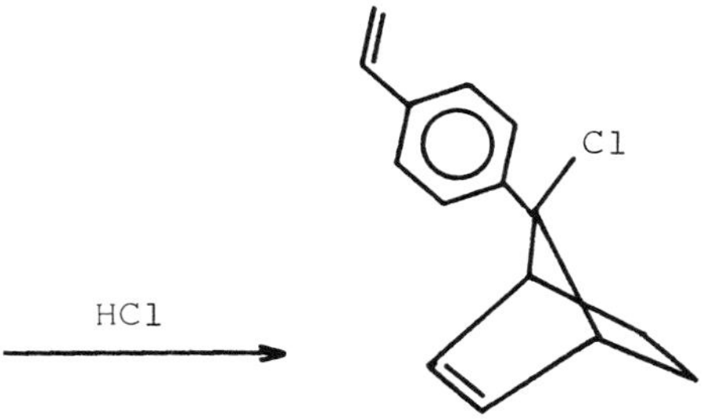

(221)

(220)

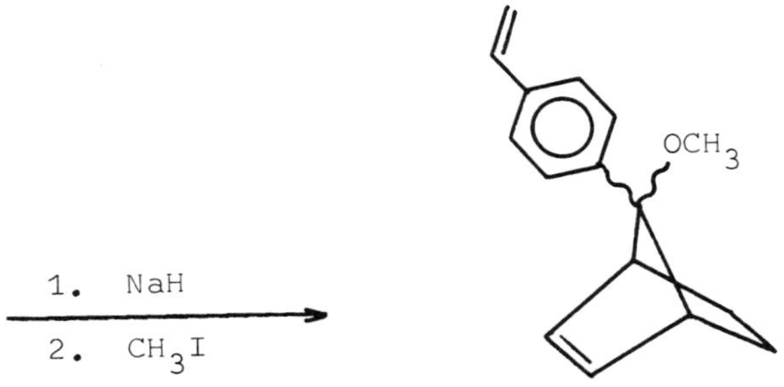

1. NaH

2. CH$_3$I

(222)

Both monomers **221** and **222** were copolymerized with styrene and di-vinylbenzene in the presence of a free-radical initiator to give copolymers **223** and **224**.

(221) or (222)   +   C$_6$H$_5$CH=CH$_2$ +   C$_6$H$_4$(CH=CH$_2$)$_2$ ⟶

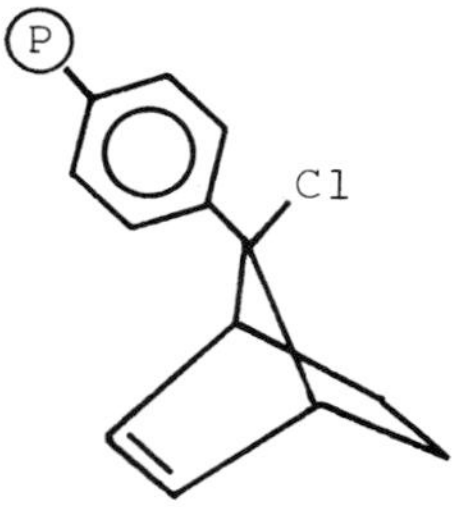

or

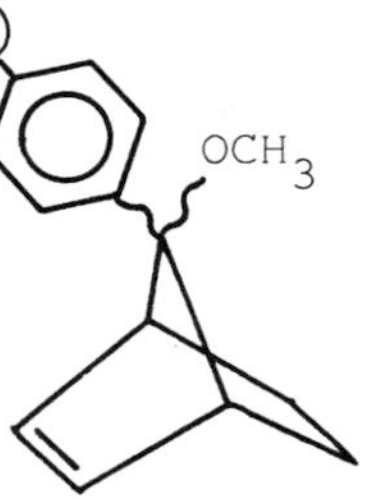

(223)   (224)

Treating **223** with *n*-butyllithium or **224** with sodium–potassium alloy gave the polymer-supported cyclopentadienide anion (**225**). Upon treating **225** with [Rh(CO)$_2$Cl]$_2$, polymer **226** was obtained. Hydrolysis of **225** with methanol followed by treatment of the resulting cyclopentadiene with Co$_2$(CO)$_8$ gave polymer **227**.

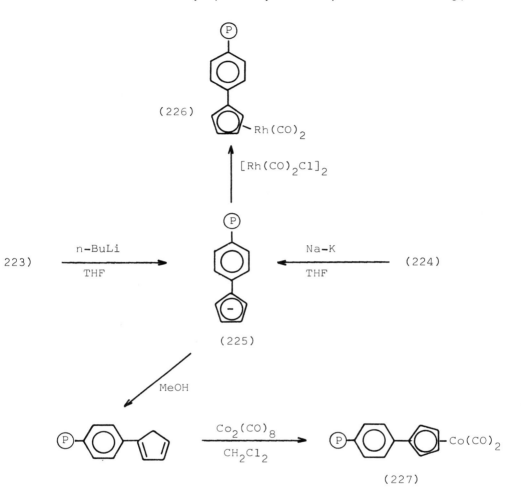

(226)

[Rh(CO)₂Cl]₂

(225)

223) —n–BuLi / THF→   ←Na–K / THF— (224)

MeOH

$\text{Co}_2(\text{CO})_8$ / CH_2Cl_2

(227)

Polymer **226** was found to be both a hydrogenation catalyst and a hydroformylation catalyst. The cobalt polymer **227** had only limited activity, and did not promote the Fischer–Tropsch synthesis of hydrocarbons from a $3:1$ $H_2 + CO$ mixture.

ACKNOWLEDGMENTS

The authors are very grateful to the National Science Foundation, to the Donors of the Petroleum Research Fund, administered by the American Chemical Society, and to the Materials Research Laboratory, University of Massachusetts, for grants that have made possible their contributions to the research described herein. They are likewise grateful to all their colleagues mentioned in the references, whose many original contributions have made this such a fascinating and rewarding area of research. Finally, the senior author (M.D.R.) is also deeply indebted to Dr. Harold Rosenberg, formerly of the Air Force Materials Laboratory, Wright-Patterson Air Force Base, Ohio, and to Prof. Dr. Ernst Otto Fischer, Technische Universität, Munich, Germany, who stimulated his interest in this type of chemistry so many years ago.

References

1. T. J. Kealy and P. L. Pauson, *Nature (London)* **168**, 1039 (1951).
2. M. D. Rausch, *Can. J. Chem.* **41**, 1289 (1963).
3. G. Wilkinson, M. Rosenblum, M. C. Whiting, and R. B. Woodward, *J. Am. Chem. Soc.* **74**, 2125 (1952).
4. R. B. Woodward, M. Rosenblum, and M. C. Whiting, *J. Am. Chem. Soc.* **74**, 3458 (1952).
5. M. Rosenblum and R. B. Woodward, *J. Am. Chem. Soc.* **80**, 5443 (1958).
6. A. N. Nesmeyanov and K. S. Kochetokova, *Dokl. Akad. Nauk SSSR* **109**, 543 (1956).
7. M. Vogel, M. D. Rausch, and H. Rosenberg, *J. Org. Chem.* **22**, 1016 (1957).
8. M. Rosenblum, *Chem. Ind. (London)* p. 72 (1957).
9. K. Schlögl, *Monatsh. Chem.* **88**, 601 (1957).
10. A. N. Nesmeyanov, E. G. Perevalova, R. V. Golovnya, and O. A. Nesmeyanova, *Dokl. Akad. Nauk SSSR* **97**, 459 (1954).
11. V. Weinmayer, *J. Am. Chem. Soc.* **77**, 3009 (1955).
12. M. D. Rausch, E. O. Fischer, and H. Grubert, *J. Am. Chem. Soc.* **82**, 76 (1960).
13. M. D. Rausch, E. O. Fischer, and H. Grubert, *Chem. Ind. (London)* p. 756 (1958).
14. F. A. Cotton and J. R. Leto, *Chem. Ind. (London)* p. 1368 (1958).
15. E. O. Fischer and K. Plesske, *Chem. Ber.* **91**, 2719 (1958).
16. J. Kozikowski, R. E. Maginn, and M. S. Klove, *J. Am. Chem. Soc.* **81**, 2995 (1959).
17. E. O. Fischer and W. Fellman, *J. Organomet. Chem.* **1**, 191 (1963).
18. A. N. Nesmeyanov, K. N. Anisimov, N. E. Kolobova, and L. I. Baryshnikova, *Dokl. Akad. Nauk SSSR* **154**, 646 (1964).
19. A. N. Nesmeyanov, N. E. Kolobova, K. N. Anisimov, and L. I. Baryshnikova, *Izv. Akad. Nauk SSSR, Ser. Khim.* p. 1135 (1964).
20. E. O. Fischer and K. Plesske, *Chem. Ber.* **93**, 1006 (1960).
21. R. Riemschneider, O. Goehring, and K. Kruger, *Monatsh. Chem.* **91**, 305 (1960).
22. R. Ercoli and E. Calderazzo, *Chim. Ind. (Milan)* **42**, 52 (1960).
23. E. O. Fischer and K. Plesske, *Chem. Ber.* **94**, 93 (1961).
24. J. Kozikowski, U. S. Patent 2,916,503 (1959).
25. W. P. Hart, Ph.D. Dissertation, University of Massachusetts, Amherst (1981).
26. M. D. Rausch and R. A. Genetti, *J. Org. Chem.* **35**, 3888 (1970).
27. M. D. Rausch and R. A. Genetti, *J. Am. Chem. Soc.* **89**, 5502 (1967).
28. For the synthesis of certain organometallics, substituted cyclopentadiene reagents need not be converted into their cyclopentadienide anions. N. E. Schore, *J. Organomet. Chem.* **173**, 301 (1979).
29. Alkyl groups attached to cyclopentadienyl rings of certain organometallic compounds can serve as reactive functional groups in some instances, however. For example, 1,1'-dimethylcobalticinium and 1,1'-dimethylrhodicinium ions can be oxidized with basic permanganate to the corresponding dicarboxylic acids. J. E. Sheats and M. D. Rausch,

J. Org. Chem. **35,** 3245 (1970); J. E. Sheats and T. Kirsch, *Synth. Inorg. Met.-Org. Chem.* **3,** 59 (1973).

30. J. Thiele, *Chem. Ber.* **33,** 666 (1900).
31. K. Hafner, G. Schultz, and K. Wagner, *Justus Liebigs Ann. Chem.* **678,** 39 (1964).
32. T. Okuyama, Y. Ikenovchi, and T. Fueno, *J. Am. Chem. Soc.* **100,** 6162 (1978).
33. W. P. Hart, D. W. Macomber, and M. D. Rausch, *J. Am. Chem. Soc.* **102,** 1196 (1980).
34. M. Arthurs, M. Sloan, M. G. B. Drew, and S. M. Nelson, *J. Chem. Soc., Dalton Trans.* p. 1794 (1975).
35. A. Chaloyard and N. El Murr, *Inorg. Chem.* **19,** 3217 (1980).
36. M. Arthurs, S. M. Nelson, and M. G. B. Drew, *J. Chem. Soc., Dalton Trans.* p. 779 (1977).
37. J. Thiele, *Chem. Ber.* **34,** 68 (1901).
38. D. Peters, *J. Chem. Soc., London* p. 1761 (1959).
39. D. Peters, *J. Chem. Soc., London* p. 1832 (1960).
40. O. W. Webster, *J. Am. Chem. Soc.* **88,** 3046 (1966).
41. R. Cramer and J. J. Mrowca, *Inorg. Chim. Acta* **5,** 528 (1971).
42. J. J. Mrowca, personal communication.
43. T. Fujisawa and K. Sakai, *Tetrahedron Lett.* p. 3331 (1976).
44. T. Fujisawa, T. Kobori, and H. Ohta, *J. Chem. Soc., Chem. Commun.* p. 186 (1976).
45. R. E. Benson and R. V. Lindsey, Jr., *J. Am. Chem. Soc.* **79,** 5471 (1957).
46. J. M. Osgerby and P. L. Pauson, *J. Chem. Soc., London* p. 4604 (1961).
47. D. Peters, *J. Chem. Soc., London* p. 1757 (1957).
48. D. W. Macomber, Ph.D. Dissertation, University of Massachusetts, Amherst (1982).
49. D. W. Macomber, M. D. Rausch, T. V. Jayaraman, R. D. Priester, and C. U. Pittman, Jr., *J. Organomet. Chem.* **205,** 353 (1981).
50. C. U. Pittman, Jr., T. V. Jayaraman, R. D. Priester, S. Spencer, M. D. Rausch, and D. W. Macomber, *Macromolecules* **14,** 237 (1981).
51. W. J. Linn and W. H. Sharkey, *J. Am. Chem. Soc.* **79,** 4970 (1957).
52. K. Hafner, K. H. Vopel, G. Ploss, and C. Konig, *Justus Liebigs Ann. Chem.* **661,** 52 (1963).
53. O. Diels, *Chem. Ber.* **75,** 1452 (1942).
54. E. Le Goff and R. B. LaCount, *J. Org. Chem.* **29,** 423 (1964).
55. R. C. Cookson, J. B. Henstock, J. Hudec, and B. R. D. Whitear, *J. Chem. Soc.* C p. 1986 (1967).
56. M. I. Bruce, B. W. Skelton, R. C. Wallis, J. K. Walton, A. H. White, and M. L. Williams, *J. Chem. Soc., Chem. Commun.* p. 428 (1981).
57. M. I. Bruce, J. K. Walton, M. L. Williams, B. W. Skelton, and A. H. White, *J. Organomet. Chem.* **212,** C35 (1981).
58. D. J. Cram, "Fundamentals of Carbanion Chemistry," p. 19. Academic Press, New York, 1965.
59. V. W. Day, B. R. Stults, K. J. Reimer, and A. Shaver, *J. Am. Chem. Soc.* **96,** 1227 (1974).
60. K. J. Reimer and A. Shaver, *J. Organomet. Chem.* **93,** 239 (1975).
61. M. Cais and N. Narkis, *J. Organomet. Chem.* **3,** 269 (1965).
62. A. N. Nesmeyanov, K. N. Anismov, and Z. P. Valuera, *Izv. Akad. Nauk SSSR, Otd. Khim. Nauk* p. 1683 (1962).
63. V. W. Day, B. R. Stults, K. J. Reimer, and A. Shaver, *J. Am. Chem. Soc.* **96,** 4008 (1974).
64. K. J. Reimer and A. Shaver, *Inorg. Chem.* **14,** 2702 (1975).
65. V. W. Day, K. J. Reimer, and A. Shaver, *J. Chem. Soc., Chem. Commun.* p. 403 (1975).

66. W. A. Herrmann, B. Reiter, and M. Huber, *J. Organomet. Chem.* **139,** C4 (1977).
67. W. A. Herrmann and M. Huber, *J. Organomet. Chem.* **140,** 55 (1977).
68. W. A. Herrmann, *Chem. Ber.* **111,** 2458 (1978).
69. A. N. Nesmeyanov, K. N. Anisimov, and Yu. V. Makavora, *Izv. Akad. Nauk SSSR, Ser. Khim.* p. 357 (1969).
70. A. N. Nesmeyanov, N. E. Kolobova, Yu. V. Makavora, and K. N. Anisimov, *Izv. Akad. Nauk SSSR, Ser. Khim.* p. 1992 (1969).
71. W. A. Herrmann and M. Huber, *Chem. Ber.* **111,** 3124 (1978).
72. G. Wulfsberg and R. West, *J. Am. Chem. Soc.* **93,** 4055 (1971).
73. G. Wulfsberg and R. West, *J. Am. Chem. Soc.* **94,** 6069 (1972).
74. B. G. Conway, M.S. Thesis, University of Massachusetts, Amherst (1981).
75. M. D. Rausch, W. P. Hart, B. G. Conway, and D. W. Macomber, *Abstr. Pap. Int. Conf. Organomet. Chem., 10th, 1981* p. 134 (1981).
76. R. D. Rogers, J. L. Atwood, B. G. Conway, and M. D. Rausch, unpublished studies.
77. R. C. Kerber and M. J. Chick, *J. Org. Chem.* **32,** 1329 (1967).
78. M. D. Rausch, W. P. Hart, J. L. Atwood, and M. J. Zaworotko, *J. Organomet. Chem.* **197,** 225 (1980).
79. A. I. Titov. E. S. Lisitsgna, and M. R. Shemtova, *Dokl. Akad. Nauk SSSR* **130,** 341 (1960).
80. M. Bernheim and G. Boche, *Angew. Chem., Int. Ed., Engl.* **19,** 1010 (1980).
81. G. R. Knox, J. D. Munro, P. L. Pauson, G. H. Smith, and W. E. Watts, *J. Chem. Soc.* p. 4619 (1961).
82. G. R. Knox and P. L. Pauson, *J. Chem. Soc., London* p. 4615 (1961).
83. T. Kaufmann, J. Ennen, H. Lhotak, A. Rensing, I. Steinseifer, and A. Woltermann, *Angew. Chem., Int. Ed., Engl.* **19,** 328 (1980).
84. M. I. Kabachnik and E. N. Tsnetkov, *Zh. Obshch. Khim.* **30,** 3227 (1960).
85. F. Mathey and J. P. Lampin, *Tetrahedron* **31,** 2685 (1975).
86. F. Mathey and J. P. Lampin, *J. Organomet. Chem.* **128,** 297 (1977).
87. F. Mathey and C. Charrier, *Tetrahedron Lett.* p. 2407 (1978).
88. J. E. Benthem, E. A. V. Ebsworth, H. Moretto, and D. W. A. Rankin, *Angew. Chem., Int. Ed. Engl.* **11,** 640 (1972).
89. R. T. Paine, R. W. Light, and D. E. Mair, *Inorg. Chem.* **18,** 368 (1979).
90. A. W. Rudie, D. W. Lichtenberg, M. L. Katcher, and A. Davison, *Inorg. Chem.* **17,** 2859 (1978).
91. N. E. Schore, *J. Am. Chem. Soc.* **101,** 7410 (1979).
92. B. H. Edwards and M. D. Rausch, unpublished studies.
93. W. Schlenk and E. Bergmann, *Justus Liebigs Ann. Chem.* **479,** 58 (1930).
94. S. M. Linden, E. I. Becker, and P. E. Spoerri, *J. Am. Chem. Soc.* **75,** 5972 (1953).
95. G. R. Knox and P. L. Pauson, *J. Chem. Soc.* p. 4610 (1961).
96. J. Hine and D. B. Knight, *J. Org. Chem.* **35,** 3946 (1970).
97. N. E. Schore and B. E. LaBelle, *J. Org. Chem.* **46,** 2306 (1981).
98. D. W. Macomber, W. P. Hart, M. D. Rausch, R. D. Priester, and C. U. Pittman, Jr., *J. Am. Chem. Soc.* **104,** 884 (1982).
99. J. Altman and G. Wilkinson, *J. Chem. Soc., London,* p. 5654 (1964).
100. N. Hoffman and E. Weiss, *J. Organomet. Chem.* **131,** 273 (1977).
101. P. Eilbracht, P. Dahler, and G. Tiedtke, *J. Organomet. Chem.* **185,** C25 (1980).
102. A. J. L. Harley, R. C. Ugolick, J. G. Fulcher, S. Togashi, A. B. Bocarsly, and J. A. Gladysz, *Inorg. Chem.* **19,** 1543 (1980).
103. M. B. Freeman, L. G. Sneddon, and J. C. Huffman, *J. Am. Chem. Soc.* **99,** 5194 (1977).
104. M. B. Freeman and L. G. Sneddon, *Inorg. Chem.* **19,** 1125 (1980).

105. T. J. Katz and J. J. Mrowca, *J. Am. Chem. Soc.* **89,** 1105 (1967).
106. W. D. Bonds, Jr., C. H. Brubaker, Jr., E. S. Chandrasekaran, C. Gibbons, R. H. Grubbs, and L. C. Kroll, *J. Am. Chem. Soc.* **97,** 2128 (1975).
107. G. Gubitosa, M. Boldt, and H. H. Brintzinger, *J. Am. Chem. Soc.* **99,** 5174 (1977).
108. E. S. Chandrasekaran, R. H. Grubbs, and C. H. Brubaker, Jr., *J. Organomet. Chem.* **120,** 49 (1976).
109. C. P. Lau, B. H. Chang, R. H. Grubbs, and C. H. Brubaker, Jr., *J. Organomet. Chem.* **214,** 325 (1981).
110. B. H. Chang, R. H. Grubbs, C. H. Brubaker, Jr., *J. Organomet. Chem.* **172,** 81 (1979).
111. P. Perkins and K. P. C. Vollhardt, *J. Am. Chem. Soc.* **101,** 3985 (1979).
112. A. Sekiya and J. K. Stille, *J. Am. Chem. Soc.* **103,** 5096 (1981).

ADVANCES IN ORGANOMETALLIC CHEMISTRY, VOL. 21

Metalloboranes: Their Relationships to Metal–Hydrocarbon Complexes and Clusters

CATHERINE E. HOUSECROFT and THOMAS P. FEHLNER

Department of Chemistry
University of Notre Dame
Notre Dame, Indiana

I

INTRODUCTION

The hydridic character of boranes reflects the polarity of the $B-H$ bond whereas the Lewis acidity of boranes reflects the fact that boron possesses fewer valence electrons than valence orbitals. However, polyhedral boranes exhibit more chemistry than simply forming hydrogen with protonic acids and donor–acceptor adducts with Lewis bases. The intracluster bonding in the polyhedral boranes results in behavior that might be considered uncharacteristic of monoboranes. That is, unbridged $B-B$ bonds in boranes act as if they possess excess electron density and function as electron donors to Lewis acids. Hydrogen bridged borons in polyhedral boranes function as proton donors to bases, thereby producing reactive, unbridged $B-B$ bonds. In addition, catenation in the boranes is characterized by a rich variety of cage structures ranging from single cages varying in "openness" to cages coupled by single bonds and to fused cages (Fig. 1). Such structural

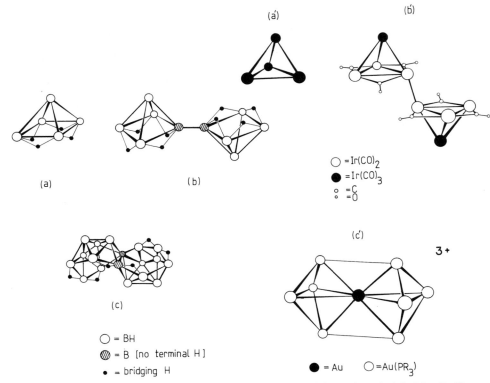

FIG. 1. Proposed or known structures of (a) B_5H_9 (1), (b) 2,2'-$(B_5H_8)_2$ (6), (c) n-$B_{18}H_{22}$ (1), and (a') $Ir_4(CO)_{12}$ (7), (b') $Ir_8(CO)_{22}^{2-}$ (8), (c') $[Au_9(PR)_8]$ $[PF_6]_3$ (9), demonstrating single clusters, coupled clusters, and fused clusters for borane and metal systems.

variation modulates and mixes the basic properties associated with boranes. This leads to a rich chemistry which is described in the standard works on this subject (1–5).

Despite possessing a varied chemistry and many unique properties, much of the usefulness of boranes has been found in the underlying relationships with other areas of chemistry. One area in which the established principles of borane chemistry have had an impact is that of transition metal clusters (10–13). The synthesis and characterization of metal clusters is an area that is throbbing with activity at present (14) partly because it is new and largely unexplored but also because of the hypothesis that the behavior of metals in a cluster can mimic that of metal atoms on a surface of the bulk metal (15, 16). Like boranes, metal clusters possess properties that go beyond the simple behavior that might be associated with the fragments that make up the clusters. Also like boranes, metal clusters are found in a variety of catenated arrays that (Fig. 1) are analogous to those exhibited by the boranes. The similarities between boranes and transition metal clusters were first defined by Wade (17) and Mingos (18). The result has been called the *borane analogy*, in which main group behavior is attributed to metal fragments; e.g., $B_6H_6^{2-}$ and $Ru_6(CO)_{18}^{2-}$ (Fig. 2) are considered analogous in that the cluster bonding properties of $Ru(CO)_3$ are similar to those of BH. As might have been expected, the cluster behavior of transition metals is much more varied than that of boranes, and various extensions and modifications of the borane analogy have since been proposed (19, 20). Still, the boranes serve to provide a well-defined context in which the vagaries of the metal systems can be understood. They also suggest initial approaches to problems in the metal cluster area. For example, techniques found useful in manipulating the boranes, e.g., deprotonation, have been found to be very effective in metal systems as well (21).

This article is concerned with compounds containing metal–boron bonds, i.e., metalloboranes. Considering the properties of boron, it is not surprising that the first metalloboranes to be prepared and characterized reflected the elementary properties of the boranes and the metals, i.e., the ability to form covalent and coordinate bonds. The earliest metal–boron compound with a coordinate bond is stated to be $Li[(C_6H_5)_3BGe(C_6H_5)_3]$ (22, 23), and simple covalent bond formation was first found in compounds like $(Me_2N)_2BMn(CO)_5$ (24). It is interesting to note that compounds of this type, e.g., $(TPE)_2Co(BR_2)_2$, may be viewed as related to carbenes and have been called *borenes* (22). Metal tetrahydroborates (BH_4^-) have a much longer history than metalloboranes, but compounds with both organic and hydroborate ligands bonded to a transition metal are relatively recent ad-

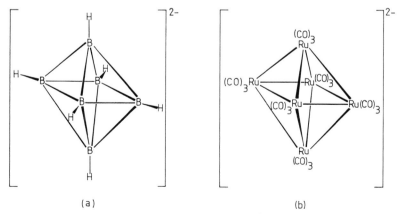

(a)   (b)

FIG. 2. Comparison of the skeletal structures of (a) $B_6H_6^{2-}$ and (b) $Ru_6(CO)_{18}^{2-}$, illustrating the "borane analogy."

ditions (25). Since the mid-1960s complex metalloboranes containing a great variety of transition metal fragments have been prepared and characterized (26). Examples of metalloboranes containing equal numbers of boron and metal atoms are now known (27). These metalloboranes are related to both borane and metal clusters and possess properties that reflect their parentage. The cluster view has been used to classify metalloboranes (28) and one objective of this article is to summarize the structures and behavior of these materials as clusters.

The above account may convey a false sense of the history of metalloboranes, though, since metalloborane chemistry grew both as a result of and in the shadow of metallocarborane chemistry. In the seminal work of Hawthorne (29) it was demonstrated that the five-atom open face created from an icosahedral carborane, e.g., $1,10\text{-}C_2B_{10}H_{12}$, by removal of one BH vertex bonds a transition metal fragment in a manner resembling the bonding of the $\eta^5\text{-}C_5H_5^-$ ligand (Fig. 3). Thus $C_2B_9H_{11}^{2-}$ is seen to function as a six-electron π donor toward an appropriate transition metal fragment, e.g., $(\eta^5\text{-}C_5H_5)Fe^+$ (30). When polyhedral metalloboranes appeared on the scene, it was clear that the borane fragments could be considered to function as electron donors toward metals in the same fashion (31). Thus they were classified in the same manner as hydrocarbon ligands (Fig. 4). Hence, another objective of this article is to review the ligand behavior of boranes and compare it with that of related hydrocarbons.

Because boron is an electropositive element, like a transition metal, a borane fragment when bound to a metal fragment constitutes a ligand that in fact simulates a metal fragment. On the other hand, since boron is a first-row main-group element adjacent to carbon, it also mimics in a real sense an organic ligand. For this reason the metalloboranes (and metallocarboranes) are seen to be related to both metal clusters and organometallic compounds with hydrocarbon ligands (10). This observation serves

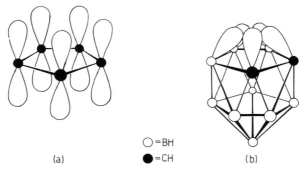

$\bigcirc$ = BH
$\bullet$ = CH

(a)   (b)

FIG. 3.   Schematic comparison of the bonding in $C_5H_5^-$ (a) to that in the open face of $1,2\text{-}C_2B_9H_{11}^{2-}$ (b) (29).

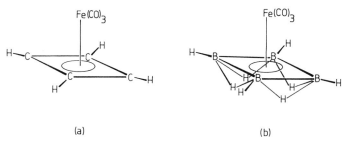

FIG. 4.   Comparison of η^4-$(C_4H_4)Fe(CO)_3$ (a) and η^4-$(B_4H_8)Fe(CO)_3$ (b).

rationally to connect two of the most vital areas of inorganic chemistry; organometallic π-complexes and metal clusters.

The paramount objective of this article is to dissect and analyze the relationship of metalloboranes to organometallic π-complexes. This has already been done to some extent in a review by Grimes, which also included metalloheteroboranes that are related to organometallic complexes (*32*).

A cursory survey of existing metalloboranes makes it obvious that a view derived simply from organometallic complexes is much too restrictive and that a comprehensive picture of metalloboranes must also consider their relationships to cluster systems (*14*) on the one hand and to metal hydroborates (*33*) on the other. Despite an intimate relationship, we shall ignore the large class of metallocarboranes (as well as other metalloheteroboranes). This permits space for an in-depth look at compounds possessing the simplicity of only two types of atoms (boron and a metal) involved in the bonding relationship of primary interest. Information on metallocarboranes may be obtained from several past and current reviews (*2, 27, 32, 34–36*). Though it is not the primary purpose of this article to provide an exhaustive compilation of metalloboranes and their properties, the review is reasonably comprehensive for the period from 1975 to mid-1981 and includes all pertinent examples from earlier work. Other reviews containing information on metalloboranes but not mentioned above are available (*37–49*).

II

CHARACTERIZED METALLOBORANES

A. *Classification*

As noted in the introduction, metalloboranes have been classified both according to ligand behavior attributed to the borane via the requirements of the metal fragment or cluster behavior attributed to the entire metal-loborane. Neither of these schemes comfortably includes all metallo-boranes, and thus a method of classification is used here that is based only on the number and type of metal–boron interactions. This phenomenolog-ical method of classification is independent of bonding model and so ac-commodates metal–ligand complexes as well as clusters.

This classification scheme is defined as follows with examples of each class being shown in Fig. 5. Class **1** contains compounds in which the metal is bound to the borane (either mono- or polyborane) by a single two-center bond. In the case of a polyhedral borane the metal fragment would be an exopolyhedral substituent. Class **2** contains compounds in which the metal is bound to the borane only by one or more $B-H-M$ bridge bonds and hence these compounds are formally related to metal hydroborates. Class **3** contains compounds in which the metal fragment formally replaces a bridging hydrogen on an edge of a polyhedral borane so that the primary

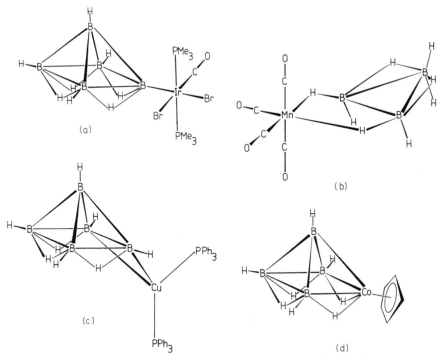

FIG. 5. Examples of the four classes of metalloboranes: (a) **1** $(R_3P)_2(CO)Br_2IrB_5H_8$ (*178*); (b) **2** $(CO)_4MnB_3H_8$ (*51*); (c) **3** $(R_3P)_2CuB_5H_8$ (*52–56*); and (d) **4** $\eta^5\text{-}(C_5H_5)CoB_4H_8$ (*57–59*).

metal–boron interaction involves two borons. Finally, class **4** contains compounds in which the metal fragment interacts with three or more borons per metal atom. In these compounds the metal can be naturally viewed as a part of the cluster defined by the boron atoms. Where appropriate the cluster designation according to the borane analogy (see Section IV) is indicated for metalloboranes of class **4**.

Table I lists characterized metalloboranes organized according to: (a) the number of borons in the principal borane unit in the compound; (b) the number of metal atoms in the compound; and (c) the type of metal

TABLE I

SELECTION OF KNOWN METALLOBORANES

	Compound[a]	Class[b]	Structure[c]	References
One boron				
M_1	$Be(BH_4)_2$	2	S, E, M	60–65
	$B_5H_{10}BeBH_4$	2(2), **4**	X, S, M	66–69
	$Al(BH_4)_3$	2(2)	S, E, P	70–74
	$HGa(BH_4)_2$	2(2)	S	75
	Cp_2TiBH_4	2(2)	E, S	76, 77
	$Zr(BH_4)_4$	2(3)	X, S	78–81
	$Cp_2Zr(BH_4)_2$	2	S	82, 83
	Cp_2VBH_4	2(2)	S	84
	Cp_2NbBH_4	2(2)	S	84
	$Cr(BH_4)_2 \cdot 2THF$	2(2)	S	85
	$(CO)_4MoBH_4^-$	2(2)	X, S	86
	$(CO)_5MnBH_4$	2	S	87
	$(CO)_5MnBH_3^-$		S	88
	$(CO)_5ReBH_3^-$		S	88
	$(R_3P)_2Rh(H)_2(BH_4)$	2(2)	S	89
	$(R_3P)_2Ir(H)_2(BH_4)$	2(2)	S	89, 90
	$(R_3P)_2Ni(H)(BH_4)$	2(2)	X, S	91
	$(R_3P)_3CuBH_4$	2(1)	X, S	92, 93
	$(R_3P)_2CuBH_4$	2(2)	X, S	94–96
	$Zn(BH_4)_2$	2	S	97, 98
	$Hf(BH_4)_4$	2(3)	S	99
	$Cp_2Hf(BH_4)_4$	2	S	100
	$U(BH_4)_4 \cdot nOR$	2(2, 3)	X, S	101, 104
	$Cp_2U(BH_4)_2$	2(3)	X, S	105
	$Cp_3U(H_3BR)$	2(3)	S	106
	$Np(BH_4)_4$	2(3)	X, S	107
M_3	$(CO)_9Co_3BNR_3$	4	X, S	108, 109
M_4	$(CO)_{12}Ru_4(H)BH_2$	4	S	110
	$(CO)_{12}Fe_4(H)BH_2$	4	X, S	111
M_6	$(CO)_{18}Co_6B$	4	S	108

Continued

TABLE I (*Continued*)

Compound[a]	Class[b]	Structure[c]	Reference
Two boron			
M_1 $(CO)_4FeB_2H_5^-$	**3**	S	*112*
$Cp(CO)_2FeB_2H_5$	**3**	X, S, M, P	*113, 114*
M_2 $Cp_2'Nb_2(B_2H_6)_2$	**2(4)**	X, S	*115*
$(CO)_6Fe_2B_2H_6$	**2(4)**	S, M, P	*116–118*
M_3 $(CO)_{10}Mn_3H(B_2H_6)$	**2(6)**	X, S	*119*
$(CO)_9Ru_3(B_2H_6)$		S	*110*
Three boron			
M_1 $Be(B_3H_8)_2$	**2(2)**	X, S	*120–122*
$(CH_3)_2AlB_3H_8$	**2(2)**	S	*123*
$(CH_3)_2GaB_3H_8$	**2(2)**	S	*123*
$Cp_2TiB_3H_8$	**2**	S	*124*
$(CO)_4CrB_3H_8^-$	**2(2)**	X, S	*125, 126*
$(CO)_4MoB_3H_8^-$	**2**	S	*125*
$(CO)_4WB_3H_8^-$	**2**	S	*125*
$(CO)_3MnB_3H_8$	**2(3)**	X, S	*51, 127, 128*
$(CO)_4MnB_3H_8$	**2(2)**	S, M	*51, 129*
$(CO)_4MnB_3H_7Br$	**2(2)**	X, S	*130*
$(CO)_4ReB_3H_8$	**2(2)**	S	*51*
$Cp(CO)_2MoB_3H_8$	**2(2)**	S	*51*
$Cp(CO)_2WB_3H_8$	**2(2)**	S	*51*
$Cp(CO)FeB_3H_8$	**2(2)**	S	*51*
$(CO)_4(H)FeB_3H_8$	**2(2)**	S	*51*
$(R_3P)_2(H)(CO)IrB_3H_7$	**4(a)**	S	*131*
$(R_3P)_2NiB_3H_7$	**4(n)**	S	*132, 133*
$(R_3P)_2PtB_3H_7$	**4(n)**	X, S, M, P	*129, 132, 133*
$(R_3P)_2PdB_3H_7$	**4(n)**	S	*132, 133*
$(R_3P)_2CuB_3H_8$	**2(2)**	X, S	*134–139*
M_2 $(CO)_6(Br)Mn_2B_3H_8$	**2(4)**	X, S	*140*
$(CO)_6Fe_2B_3H_7$	**4(n)**	X, S, M, P	*118, 141, 142*
M_3 $(\mu_3\text{-}CO)\text{-}1,2,3\text{-}Cp_3Co_3B_3H_3$	**4(c)**	S	*143*
$1,2,3\text{-}Cp_3Co_3B_3H_5$	**4(c)**	X, S	*144, 145*
$1,2,3\text{-}[Cp_2Co_2(CO)_4Fe]B_3H_3$	**4(c)**	S	*146*
Four boron			
M_1 AlB_4H_{11}	**4**	S	*147*
$1\text{-}(CO)_3FeB_4H_8$	**4(n)**	S, M, P	*41, 148–152*
$1\text{-}CpCoB_4H_8$	**4(n)**	S	*57, 58*
$2\text{-}CpCoB_4H_8$	**4(n)**	X, S	*57–59, 146*
$(R_3P)_2(CO)IrB_4H_9$	**4(a)**	X, S	*153*
$(R_3P)_2RhB_4H_9$	**4(n)**	S	*153*
$(R_3P)_2CuB_4H_9$	**4(a)**	S	*154*
M_2 $1,2\text{-}Cp_2Co_2B_4H_6$	**4(c)**	X, S	*57, 146, 155*
M_3 $Cp_3Co_3B_4H_4$	**4**	X, S	*156a, b*
$Cp_3'Co_3B_4H_4$	**4**	X, S	*157b*
M_4 $Cp_4Co_4B_4H_4$	**4(c)**	X, S, M	*158, 159*
$Cp_4Ni_4B_4H_4$	**4(c)**	X, S, M	*159–161*

Continued

TABLE I (*Continued*)

Compound[a]	Class[b]	Structure[c]	Reference
Five boron			
M_1 CpBeB$_5$H$_8$	3	X, S, M	*66, 162, 163*
B$_5$H$_{10}$BeB$_5$H$_{10}$	4	X, S, M	*66–69*
AlB$_5$H$_{12}$	4	S	*147*
2-(CO)$_3$MnB$_5$H$_{10}$	4(n)	S	*164*
2-(CO)$_5$MnB$_5$H$_8$	1	S	*165*
2-(CO)$_5$ReB$_5$H$_8$	1	S	*165*
2-Cp(CO)$_2$FeB$_5$H$_8$	1	S	*166*
2,2′-Fe(B$_5$H$_{10}$)$_2$	4	S	*167*
2-(CO)$_3$FeB$_5$H$_9$	4(n)	S, M, P	*149, 168–171*
2-(CO)$_3$FeB$_5$H$_8^-$	4(n)	X, S	*167*
2-CpFeB$_5$H$_{10}$	4(n)	S	*172–173*
1-CpFeB$_5$H$_{10}$	4(n)	S	*172–173*
(CO)$_3$Fe(CO)$_2$B$_5$H$_3$	4(c)	S, P, M	*149, 174, 175*
1-CpCoB$_5$H$_9$	4(n)	S	*176*
2-(CO)$_3$CoB$_5$H$_8$	4(n)	S	*167, 177*
(R$_3$P)$_2$(CO)IrB$_5$H$_8$	4(n)	X, S	*50*
(R$_3$P)$_2$(CO)Br$_2$IrB$_5$H$_8$	1	X, S	*178, 179*
(dppe)(Cl)NiB$_5$H$_8$	3	S	*180, 181*
(CO)$_3$Ni[B$_5$H$_8$P(CF$_3$)$_2$]	1	S	*182*
(R$_3$P)$_2$ClPtB$_5$H$_8$	3	S	*183*
(R$_3$P)$_2$ClPdB$_5$H$_8$	3	S	*183*
(R$_3$P)$_2$CuB$_5$H$_8$	3	X, S, M	*52–56, 170, 184*
(R$_3$P)$_2$AgB$_5$H$_8$	3	S	*184*
(R$_3$P)ClCdB$_5$H$_8$	3	S	*185*
Hg(B$_5$H$_8$)$_2$	3	S	*186*
M_2 2,4-[Cp(CO)$_2$Fe]$_2$B$_5$H$_8$	1	S	*166*
(R$_3$P)$_2$CuB$_5$H$_8$Fe(CO)$_3$	2, 3, 4	X, S	*187*
M_3 Cp$_3$Co$_3$B$_5$H$_5$	4	S	*188*
M_4 Cp$_4$Ni$_4$B$_5$H$_5$	4(n)	S	*161*
Six boron			
M_1 AlB$_6$H$_{13}$	4	S	*147*
(THF)$_2$Mg(B$_6$H$_9$)$_2$	3	X, S	*189, 190*
(R$_3$P)$_2$CuB$_6$H$_9$	2, 3	S	*191*
(CO)$_4$FeB$_6$H$_{10}$	3	S, M	*171, 192, 193*
Cl$_2$Pt(B$_6$H$_{10}$)$_2$	3	X, S	*193, 194*
M_2 (R$_3$P)$_2$Pt$_2$(B$_6$H$_9$)$_2$	4	X, S	*195*
Seven boron			
M_1 (CO)$_4$FeB$_7$H$_{12}^-$	3	X, S	*196*
Eight boron			
M_1 (CO)$_3$MnB$_8$H$_{13}$	2(3)	X, S	*197*
4-(R$_3$P)$_2$PtB$_8$H$_{12}$	4(a)	X, S	*198*
M_2 6,9-[(R$_3$P)$_2$Pt]$_2$B$_8$H$_{12}$	4(a)	X, S	*198*
6,9- and 5,7-Cp$_2'$Co$_2$B$_8$H$_{12}$	4(n)	X, S	*157a, c*
1,6-Cp$_2$Ni$_2$B$_8$H$_8$	4(c)	S	*161*

Continued

TABLE I (*Continued*)

Compound[a]	Class[b]	Structure[c]	Reference
Nine boron			
M_1 6-$(CO)_3MnB_9H_{12} \cdot THF$	**4(n)**	X, S	*199*
6-$CpCoB_9H_{13}$	**4(n)**	S	*200*
6-$Cp'CoB_9H_{13}$	**4(n)**	X, S	*157a, c*
5-$CpCoB_9H_{13}$	**4(n)**	X, S	*57, 200–202*
2-$CpCoB_9H_{13}$	**4(n)**	S	*176*
2-$CpNiB_9H_9^-$	**4(c)**	S	*203*
1-$CpNiB_9H_9^-$	**4(c)**	S	*203*
Ten boron			
M_1 $H_2AlB_{10}H_{12}^-$	**4**	S	*204*
$(CH_3)_2InB_{10}H_{12}^-$	**4**	S	*205*
$(CH_3)_2TlB_{10}H_{12}^-$	**4**	X, S	*206*
6-$Cp(CO)_2FeB_{10}H_{13}$	**1**	S	*207*
2-$CpFeB_{10}H_{15}$	**4**	S	*172*
$(R_3P)_2(CO)IrB_{10}H_{12}$	**4(n)**	S	*208*
$Ni(B_{10}H_{12})_2^{2-}$	**4**	X, S, M	*209, 210*
$(R_3P)_2PtB_{10}H_{11} \cdot OC_2H_5$	**4(n)**	S, P	*211*
$Zn(B_{10}H_{12})_2^{2-}$	**4**	X, S	*212, 213*
M_2 $Cp_2Ni_2B_{10}H_{10}$	**4(c)**	S	*161, 214*
$[(R_3P)_2Cu]_2B_{10}H_{10}$	**4**	X, S	*138a, b*
Eleven boron			
M_1 $CpNiB_{11}H_{11}^-$	**4(c)**	S	*214*
Eighteen boron			
M_1 $(CO)_3CoB_{18}H_{22}^-$	**4**	S	*215a*
Twenty boron			
M_1 $(R_3P)_2PtB_{20}H_{24}$		X, S	*215b*

[a] $Cp \equiv \eta^5\text{-}C_5H_5$; $Cp' \equiv \eta^5\text{-}C_5(CH_3)_5$; dppe $\equiv$ 1,2-bis(diphenylphosphine)-ethane; R, see specific reference.

[b] a $\equiv$ arachno; n $\equiv$ nido; c $\equiv$ closo: For class 2 the number in parentheses is the number of BHM interactions per ligand.

[c] Geometric and electronic structural data: X $\equiv$ single crystal X-ray diffraction: S $\equiv$ spectroscopic, usually ^{11}B NMR and IR; E $\equiv$ electron diffraction, gas phase; M $\equiv$ molecular orbital calculations; P $\equiv$ UV or X-ray photoelectron spectra.

atom. For each the basic type of metal–boron interaction is indicated by the classification scheme defined above. Note that even with this general scheme there is the occasional odd compound that does not quite fit. The table does not list every derivative prepared but does give an example of each type of metalloborane known. A cursory look at Table I demonstrates that although most metalloboranes contain a single metal atom, the number of compounds containing multiple metal atoms is significant and has increased substantially in recent years. There is at least one example of a metalloborane containing one through eleven borons although the most numerously exemplified categories are B_1, B_3, B_5, and B_{10}. In support of the introductory remarks, there is great variety in the nature of the metal–boron interaction, particularly in the B_5 category. The table demonstrates that boron interacts with a large number of different metals and there is no indication that boron refuses to bond in some manner to any given metal.

Inspection of the table also shows that in terms of numbers of compounds there exist preferences for bonding types that depend on the size of the borane. For example, class **2** interactions are illustrated for compounds containing one, two, three, and eight borons and class **3** interactions are shown for two, five, six, and seven borons. This apparent dependence of the metal–boron bonding on borane size has been pointed out in an earlier review (*41*). There is, however, sufficient variation in bonding type within each category to lead one to suspect that no type is intrinsically unfavored for any size of borane. The question of whether the apparent preferences observed reflect real bonding preferences as a function of size or whether they reflect presently available preparative routes remains to be answered.

B. *Preparative Methods*

The methods by which the metalloboranes in Table I have been synthesized are many and varied. Although it cannot be said that metalloboranes can be prepared at will, there are rational routes to some compounds, usually those containing a single metal atom. A significant number of the compounds resulted from serendipitous routes but it must be kept in mind that these "unusual" methods contain the seeds of future syntheses. A sampling of the various routes to metalloboranes is given below.

1. *Direct Ligand Substitution*

This method utilizes the Lewis basicity of the B—B bond in either a neutral borane or the salt of a borane anion prepared by deprotonation of a neutral borane as indicated in Eqs. (1)–(3) (*52–56, 192, 193*).

$$Fe_2(CO)_9 + B_6H_{10} \rightarrow \mu\text{-}[(CO)_4Fe]B_6H_{10} \tag{1}$$

$$B_5H_9 + KH \rightarrow KB_5H_8 + H_2 \tag{2}$$

$$[(C_6H_5)_3P]_3CuCl + KB_5H_8 \rightarrow \mu\text{-}[(C_6H_5)_3P]_2CuB_5H_8 + P(C_6H_5)_3 + KCl \tag{3}$$

The yield of reaction (1) is 77% while that of reaction (3) is 80%. Deprotonation also provides a route to the expansion of the borane portion of a previously existing metalloborane, Eqs. (4) and (5) (*196*)

$$(CO)_4FeB_6H_{10} + KH \rightarrow K(CO)_4FeB_6H_9 + H_2 \tag{4}$$

$$(CO)_4FeB_6H_9^- + \tfrac{1}{2}B_2H_6 \rightarrow (CO)_4FeB_7H_{12}^- \tag{5}$$

as well as to metalloboranes containing two metal atoms, Eqs. (6) and (7) (*187*)

$$(CO)_3FeB_5H_9 + KH \rightarrow K(CO)_3FeB_5H_8 + H_2 \tag{6}$$

$$K(CO)_3FeB_5H_8 + [(C_6H_5)_3P]_3CuCl \rightarrow (CO)_3FeB_5H_8Cu[P(C_6H_5)_3]_2 + P(C_6H_5)_3 + KCl \tag{7}$$

Note, however, that reaction of $B_5H_8^-$ with a source of a metal fragment does not always yield a μ-derivative. In some cases rearrangement to a more stable isomer readily occurs under the reaction conditions, e.g., to a class **1** derivative, Eq. (8) (*166*)

$$KB_5H_8 + (\eta^5\text{-}C_5H_5)(CO)_2FeI \rightarrow 2\text{-}(\eta^5\text{-}C_5H_5)(CO)_2FeB_5H_8 + KI \tag{8}$$

or to a class **4** derivative, Eq. (9) (*50*).

$$KB_5H_8 + Ir(CO)Cl[P(C_6H_5)_3]_2 \rightarrow [(C_6H_5)_3P]_2(CO)IrB_5H_8 + KCl \tag{9}$$

Deprotonated boranes have also been successfully treated with the precursors of the metal fragments as in Eq. (10) (*57, 161*).

$$B_5H_8^- + CoCl_2 + C_5H_5^- \rightarrow 2\text{-}(\eta^5\text{-}C_5H_5)CoB_4H_8 \tag{10}$$

Note that in this reaction $FeCl_2$, $CoCl_2$, and $NiBr_2$ give totally different isolated products. In this preparative approach cage degradation is more prevalent in the smaller boranes than the larger, i.e., B_5 boranes yield few B_5 products while B_{10} boranes yield mostly B_{10} products. Hydroborates, both with one and three borons, can be prepared by the direct reaction of the salt of the borane with an appropriate metal halide as in Eq. (11) (*51*).

$$(CO)_4MnCl + NaB_3H_8 \rightarrow (CO)_4MnB_3H_8 + NaCl \tag{11}$$

Such reactions proceed with reasonable yields. The same type of reaction carried out in the presence of a base yields the "borallyl" complexes [Eq. (12)] (*132, 133*)

$$(C_2H_5)_3N + (R_3P)_2PtCl_2 + B_3H_8^- \rightarrow (R_3P)_2PtB_3H_7 + Cl^- + (C_2H_5)_3NHCl \tag{12}$$

2. *Activation of Metal Carbonyls*

Thermal activation of $Fe(CO)_5$ yields a variety of ferraboranes. The direct preparation of $(CO)_3FeB_4H_8$ proceeds in about 20% yield [Eq. (13)] (*148*).

$$B_5H_9 + Fe(CO)_5 \xrightarrow{\Delta} (CO)_3FeB_4H_8 + \text{other products} \tag{13}$$

even though it was later demonstrated that the initial product of this reaction is $(CO)_3FeB_5H_9$ [Eq. (14)] (*168, 169*).

$$B_5H_9 + Fe(CO)_5 \xrightarrow{\Delta} (CO)_3FeB_5H_9 + \text{other products} \tag{14}$$

The apparent activation of $Fe(CO)_5$ with hydride provides a route to $(CO)_6Fe_2B_2H_6$ in about 20% yield [Eq. (15)] (*115*) as well as other

$$H^- + Fe(CO)_5 + B_5H_9 \rightarrow (CO)_6Fe_2B_2H_6 + \text{other products} \tag{15}$$

products (*216*). Metal carbonyl anions that are basic are sufficiently active to produce metalloboranes in high yield as, for example, in Eq. (16) (*112*).

$$Fe(CO)_4^{2-} + 3BH_3 \cdot THF \rightarrow (CO)_4FeB_2H_5^- \tag{16}$$

As a mirror image, as it were, of method (8), the reaction of metal carbonyl anions with haloboranes yields metalloboranes as well [Eq. (17)] (*165*). It is curious to note that the observed product of this reaction is independent of the position of the halogen substitution on B_5H_9.

$$1\text{- or }2\text{-}ClB_5H_8 + NaMn(CO)_5 \rightarrow 2\text{-}(CO)_5MnB_5H_8 \tag{17}$$

This type of compound can also be prepared by the direct oxidative addition of a B—H bond by a transition metal, as, for example, in Eq. (18) (*178*).

$$t\text{-}IrCl(CO)[P(CH_3)_3]_2 + B_5H_9 \rightarrow 2\text{-}IrClH(CO)[P(CH_3)_3]_2B_5H_8 \tag{18}$$

3. *Direct Reaction with Metal Hydrides or Alkyls*

This method uses the formation of hydrogen or alkane as the driving force for the formation of the metalloborane and an example is given in Eq. (19) (*31*).

$$2B_{10}H_{14} + 2(C_2H_5)_2Cd \xrightarrow{(C_2H_5)_2O} \{[(C_2H_5)_2O]_2CdB_{10}H_{12}\}_2 + 2C_2H_6 \tag{19}$$

4. *From the Bulk Metal*

Initial work suggests that the condensation of metal atoms with boranes and appropriate organic ligands, as illustrated in Eq. (20) (*188*), is a promising route to metalloboranes.

$$B_5H_9 + C_5H_5^- + Co(\text{atoms}) \rightarrow (\eta^5\text{-}C_5H_5)_3Co_3B_5H_5 \tag{20}$$

The anodic dissolution of metals has also been found to yield metalloboranes as in Eq. (21) (*139*).

$$Cu + B_3H_8^- + P(C_6H_5)_3 \rightarrow [(C_6H_5)_3P]_2CuB_3H_8 \tag{21}$$

Recently it has been demonstrated that the direct oxidation of elemental transition metals by $B_{10}H_{14}$ leads directly to the formation of the metalloboranes $M(B_{10}H_{12})_2^{2-}$ in modest yields (M = Zn, Ni, Co) (213).

C. Probes of the Metal–Boron Interaction

An important objective of this work is to compare and contrast the metal–boron interaction in metalloboranes with the metal–carbon interaction in organometallic compounds containing hydrocarbon ligands. The two principal tools that are used for this comparison and for examining the metal–boron interaction are listed here and briefly described.

1. Geometry

A fundamental tenet of chemistry is that the equilibrium positions of the nuclei of a molecule reflect the nature of the bonding between the nuclei. This being the case, the first tool for examining the metal–boron interaction is the difference in observed geometries of the metal and borane fragments. More than one point of view is possible. The change in the equilibrium cage structure of the metal-free borane fragment in going to the metalloborane reflects the changes in the electronic structure of the borane caused by association with the metal fragment. On the other hand, the difference in the structure of the metal fragment in a metalloborane versus another metal–ligand complex reflects the ligand properties of the borane. Finally, comparison of the borane fragment structure in the metalloborane to that in a known borane analogous to the metalloborane reflects the cluster behavior of the metalloborane.

2. Quantum Chemistry

Most modern bonding models are cast in the language of quantum chemistry. The valence bond treatment, utilizing the concept of a three-center two-electron bond, rationalized the structure and bonding found in polyhedral boranes (1). However, the molecular orbital (MO) treatment is almost a requirement for closed borane cages, and it also provides the simplest accurate description of hydrocarbon–transition metal π-complexes. Hence, as the MO treatment encompasses both limiting models of the metalloboranes, it will be used here. There are in fact only a few published theoretical treatments of metalloboranes but for those that exist we shall be mainly interested in the qualitative orbital description of the interaction between borane and metal fragments relative to that in similar organometallic systems. In some cases the results of UV–photoelectron spectroscopy have provided experimental information on the differences between metalloboranes and organometallic compounds.

III

BORANES AS LIGANDS

A. *Analogs of Hydrocarbon π Ligands*

The key that unlocked the field of metallocarborane chemistry was the theoretical and practical recognition that a five-atom open face of a nido icosahedral carborane could bind a transition metal fragment in a manner resembling that of η^5-$C_5H_5^-$ (*2, 29*). However, such fragments, e.g., η^5-$C_5H_5^-$ and $C_2B_9H_{11}^{2-}$, are not isoelectronic. One of the exciting recent developments has been the preparation of metalloboranes with boranes that are strictly isoelectronic to the ligands in metal–hydrocarbon π-complexes. These compounds, which range in complexity from two to five boron atoms, and their organic counterparts are sketched in Fig. 6. Inspection of the figure shows that all the compounds contain a single transition metal, the borane ligands have hydrogen bridges whereas the hydrocarbons do not, and the formal charges of the ligands can be different. In addition, just as η^5-$C_5H_5^-$ finds an analog in $C_2B_9H_{11}^{2-}$, so too may analogs of the smaller hydrocarbon ligands be found in polyhedral systems. That is, the $B-B$ bond in B_6H_{10} functions as an olefin analog towards a transition metal (*192, 193*) as does $B_2H_5^-$. The metal–boron interaction in a platinathiadecaborane is also described as similar to that in a π-allyl metal complex (*217*). Thus although the subsequent discussion focuses on the smaller borane ligands shown in Fig. 6, some larger polyhedral systems are also considered relevant since they include fragments structurally analogous to and exhibiting bonding modes reminiscent of the smaller boranes. In the following, each example in Fig. 6 is discussed in terms of geometric and electronic structure. Where information is available, reaction chemistry and ^{11}B NMR behavior are also summarized.

1. *Geometric and Electronic Structure*

The principal changes in the structure of a hydrocarbon ligand upon being bound in a π fashion to a transition metal are well documented and will not be repeated in detail here (*218*). The main purpose of this section is to review the observed changes in a borane ligand upon being bound to

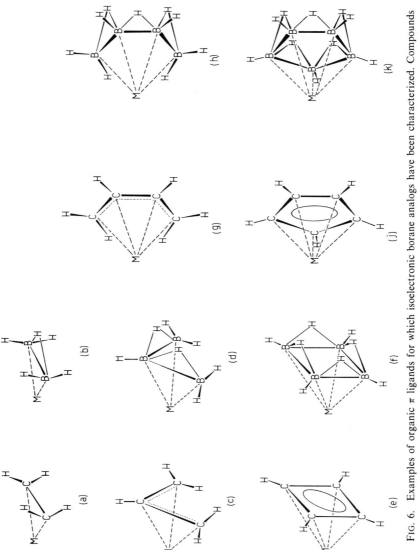

Fig. 6. Examples of organic π ligands for which isoelectronic borane analogs have been characterized. Compounds with references are listed in Table I. (a, b) C_2H_4 versus $B_2H_5^-$, (c, d) $C_3H_5^+$ versus B_3H_7, (e, f) C_4H_4 versus B_4H_8, (g, h) C_4H_6 versus $B_4H_9^-$, (j, k) $C_5H_5^-$ versus $B_5H_{10}^-$.

a transition metal relative to an isoelectronic hydrocarbon ligand. As will be seen, suitable information is not yet available on all systems, but some trends have been established.

a. $C_2H_4/B_2H_5^-$. The changes in the structure of ethylene on being bound to a transition metal are twofold. First, there is a significant increase in the C—C bond distance and second, upon coordination the olefin substituents bend away from the metal. Both changes are associated with σ charge transfer from the ligand to the metal and π charge transfer from the metal to the ligand. The rehybridization of the olefin has been associated with a requirement of maximum π overlap of the ligand orbitals with appropriate metal d functions, i.e., good back bonding (*219*). The explanation of these effects constitutes the Dewar–Chatt–Duncanson model for hydrocarbon π bonding, a concept that has had a large influence on the development of organometallic chemistry. In addition, the definite conformational preferences exhibited by a bound olefin with respect to the other metal ligands can be understood in terms of the symmetry of the ligand acceptor and the metal donor orbitals (*220*).

The $B_2H_5^-$ ligand, which is isoelectronic with C_2H_4, has been found in two complexes, one of which, $Cp(CO)_2FeB_2H_5$, has been structurally characterized (Table I and Fig. 6) (*112–114*). There is little difference between the B—B distance in this complex and the only known free ligand model, B_2H_6. In addition, in contrast to C_2H_4, there is only a small bending of the terminal hydrogens away from the metal. The complex $Cp(CO)_2FeB_2H_5$ has also been investigated by UV–photoelectron (PE) spectroscopy, and nonparameterized Fenske–Hall calculations for the complex as well as the free ligand have been carried out (*114*). The principal metal–boron interactions occur in the two orbitals sketched in Fig. 7 which may be seen to correlate with the highest occupied molecular orbital (HOMO) and lowest unoccupied molecular orbital (LUMO) of the ligand $B_2H_5^-$. Qualitatively, this overall bonding interaction is analogous to that which is now accepted for C_2H_4 and which is also illustrated in Fig. 7. There are, however, some notable differences. First, the B—B distance in $Cp(CO)_2FeB_2H_5$ is much longer (~ 0.4 Å) than the C—C distance in an olefin complex. The bending back of the hydrogens of C_2H_4 in bonding to a metal has been interpreted as being necessary for enhancing the back-bonding interaction (at the expense of the of σ interaction) by increasing the metal–carbon $d\pi$–$p\pi$ orbital overlap (*219*). That is, the carbons are too close together for effective $d\pi$–$p\pi$ overlap and bending the hydrogens away from the metal forces a rehybridization such that the lobes of both the HOMO and LUMO of C_2H_4 point away from the metal. Thus it would seem that the smaller bending back of the hydrogens in $B_2H_5^-$ when bound can be attributed to the longer

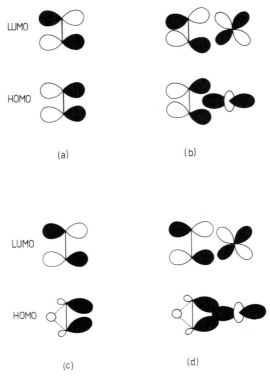

FIG. 7. Comparison of the HOMO–LUMO structures of (a) C_2H_4 and (c) $B_2H_5^-$ and the major metal–carbon and metal–boron interaction orbitals for the complexes (b) $(CO)_4FeC_2H_4$ and (d) $Cp(CO)_2FeB_2H_5$.

B—B distance in that effective back bonding can be achieved without rehybridization of the borons. There is another qualitative difference that shows up most strongly in the HOMOs of the free ligands. The presence of a bridging proton in the plane of the "π" bond of $B_2H_5^-$ rehybridizes the B2p components of this bond so that they point toward the metal position (Fig. 7). In going to the metal complex, this bent B—B bonding orbital becomes the boron–iron σ interaction MO that corresponds to the σ interaction orbital in $(CO)_4FeC_2H_4$. In contrast to C_2H_4 where the bending back of the hydrogens will weaken the σ interaction, the much smaller bending exhibited by $B_2H_5^-$ in binding to the metal along with the effect of the bridging hydrogen suggests a stronger σ interaction in this case. There is no major difference in the LUMOs in the free ligands as that of $B_2H_5^-$ contains no bridging hydrogen character because of symmetry (Fig. 7); however, as noted, bending back of the hydrogens is probably not required for good $d\pi-p\pi$ overlap for $B_2H_5^-$ because of the larger B—B distance.

Another view of the "olefinic" interaction with a metal in the borane case may be obtained from a consideration of the changes in a B—B bond in a polyhedral framework on being bound to a metal. Theoretical studies on B_6H_{10} have indicated the double bond character of the unbridged basal B—B bond in this polyhedral borane (221). Experimental work has demonstrated the ability of this bond to function as a site of Lewis basicity both toward a proton and a transition metal (192, 193). In going from free B_6H_{10} to the $Cl_2Pt(B_6H_{10})_2$ complex, the initially short basal B—B bond length increases from 1.60 Å to about 1.8 Å (193, 194). This change is very similar to that observed for C_2H_4 upon complexing and may suggest that B_2H_6 is not a good model for uncomplexed $B_2H_5^-$. It must be remembered, however, that the B—B bond in B_6H_{10} is not bridged by hydrogen and is part of an extended cage network.

The $(CO)_4FeB_6H_{10}$ complex has been the subject of an extended Hückel quantum chemical study (171). This work suggests that a major source of the metal–boron interaction is to be found in ligand to metal donation from the HOMO of B_6H_{10}, which is mainly localized in the unique basal B—B bond, to a low-lying empty orbital of the $Fe(CO)_4$ fragment. In B_6H_{10} the HOMO points toward the binding site of the metal just as it does for $B_2H_5^-$. In this sense the interactions of $B_2H_5^-$ and B_6H_{10} with a metal are similar. However, despite the fact that earlier theoretical calculations on B_6H_{10} indicate that there is a low-lying empty orbital analogous to the π^* orbital of ethylene (221), no mention of any back-bonding interaction appears in the theoretical discussion of $(CO)_4FeB_6H_{10}$.

The $B_5H_8^-$ ion also contains a B—B bond that can act as an electron donor in the same sense as the B—B bond in B_6H_{10}. Indeed several complexes with the metal in a μ-bridging position have been isolated (Table I) and the existing structural determinations suggest an interaction similar to that described above for $(CO)_4FeB_6H_{10}$. However, an analysis of the bonding on the extended Hückel level, is reported to show a difference in the bonding of μ-$(H_3P)_2CuB_5H_8$ and μ-$H_3SiB_5H_8$ (170). As the analysis of the bonding in the latter compound is similar to that in $(CO)_4FeB_6H_{10}$, it would appear that the copper derivative is different from $(CO)_4FeB_6H_{10}$ in a detailed sense. In terms of counting electrons on the metal, the iron derivative is an 18-electron compound whereas the copper derivative is a 16-electron compound and it has been suggested that 18-electron μ-$B_5H_8^-$ complexes are unstable with respect to rearrangement to an exoderivative (50, 166). Thus the reported difference in the bonding may well lie more with the acceptor properties of the metal rather than the donor properties of the B—B bond in B_6H_{10} and $B_5H_8^-$.

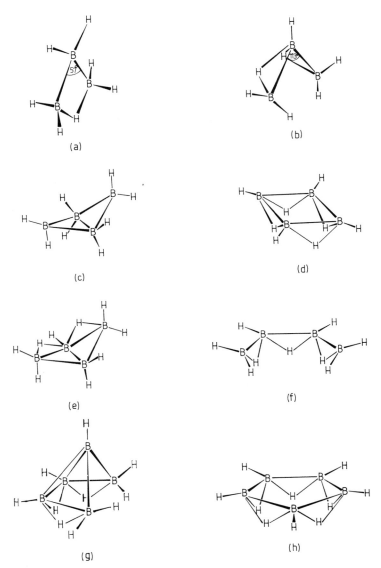

FIG. 8. Comparison of the calculated, proposed, or observed structures of borane ligands in unbound and bound states: (a, b) B_3H_7 (*223, 133*), (c, d) B_4H_8 (*223, 148*), (e, f) $B_4H_9^-$ (*224, 153*), and (g, h) $B_5H_{10}^-$ (*172, 173*).

b. $C_3H_5^+/B_3H_7$. The allyl ligand and its interactions with transition metals has been thoroughly studied both structurally and theoretically (*222*). In a transition metal complex this ligand is bent, often with equal C—C distances and equal C—M distances. Free B_3H_7 is not a stable species as its lowest energy structure (Fig. 8) contains a low energy unfilled orbital analogous to that in BH_3 (*223*). Like BH_3 it is known in the form of an adduct with a Lewis base (*2*). B_3H_7 exhibits the structure shown in Fig. 8 in the platinum derivative (Table I). Thus on binding to a metal there is a large increase in the BBB angle and considerable rearrangement of both bridging and terminal hydrogens. Qualitatively, the information suggests a much larger structural change in the borane relative to the hydrocarbon on metal complexation.

A detailed, but qualitative, extended Hückel MO comparison of borallyl, B_3H_7, and $C_3H_5^+$ has been carried out (*129*). The HOMO–LUMO structure of the two ligands are very similar both in terms of energetics and symmetry properties. Thus both have similar requirements as far as bonding to the transition metal fragment is concerned. As is the case with $B_2H_5^-$ and C_2H_4 the major difference between $C_3H_5^+$ and B_3H_7 lies in the spatial orientation of the HOMO and LUMO with respect to the metal site. The bridging hydrogens in B_3H_7 cause the HOMO to a large extent and the LUMO to a smaller extent to tilt toward the site of the metal in the complex (Fig. 9). As the B—B distances in complexed B_3H_7 are ~0.5 Å longer than the C—C distances in $C_3H_5^+$ this effect tends to strengthen the σ interaction. Calculations on $(CO)_2NiC_3H_5^+$ and $(CO)_2NiB_3H_7$ demonstrate definite similarities between the metal–carbon and metal–boron interactions; however, the results suggest that there is significantly more scrambling of metal and π-ligand orbitals in the metalloborane compared to the organometallic compound.

c. C_4H_4/B_4H_8. Cyclobutadiene and its complexes have been exhaustively studied (*225*). Despite some controversy on the subject, it appears that the ground state of C_4H_4 is rectangular (*226*). In going to a metal complex, e.g., $(CO)_3FeC_4H_4$, the ligand adopts a geometry such that the carbons lie at the corners of a square and the hydrogens are bent slightly away from the metal. The B_4H_8 moiety is known only as an adduct of a Lewis base, but the calculated most stable structure of B_4H_8 is indicated in Fig. 8. In coordinating to a metal (Fig. 4), B_4H_8 goes from a diamond-shaped fragment to a square with all borons equivalent, e.g., $(CO)_3FeB_4H_8$. Thus the available information suggests a fairly substantial change in geometry between free and bound ligands.

Of the metalloboranes shown in Fig. 6, $1\text{-}(CO)_3FeB_4H_8$ is the compound that has been most thoroughly characterized in terms of electronic struc-

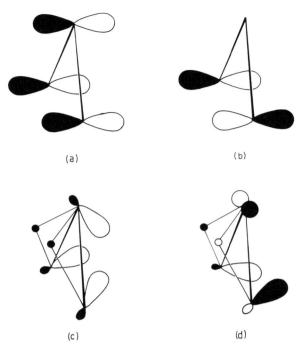

(a)   (b)

(c)   (d)

FIG. 9.   Comparison of the HOMO–LUMO structure of (a, b) $C_3H_5^+$ and (c, d) B_3H_7 (*129*).

ture. It has been studied with three different quantum chemical methods as well as by PE spectroscopy (*149–151*). The calculations and spectroscopic results are in agreement as to the proper MO description of the electronic structure; however, they differ as to the analysis of the metal–boron interaction. Two methods suggest the interaction is represented by three MOs in the complex, a nearly degenerate pair (of *e* symmetry if the C_{3v} symmetry of the Fe(CO)$_3$ fragment is neglected) and a lower energy, a_1 symmetry, orbital (Fig. 10) (*149, 150, 227*). The interpretation of the extended Hückel results (*151*), but not the Fenske–Hall calculations (*227*), tends to minimize the importance of the a_1 orbital attributing it to the ligand alone. Despite this, it is clear that the description is very similar to that proposed for (CO)$_3$FeC$_4$H$_4$ in spite of some possible quantitative differences. It is interesting to note that a recent fragment analysis of the bonding in the latter compound suggests that (CO)$_3$FeC$_4$H$_4$ is best described as a FeC$_4$H$_4$ fragment perturbed by three CO ligands rather than Fe(CO)$_3$ perturbed by C$_4$H$_4$ (*228*). If correct, this suggests that the cluster view (see Section IV) is an appropriate one. The PE results (*149*) suggest

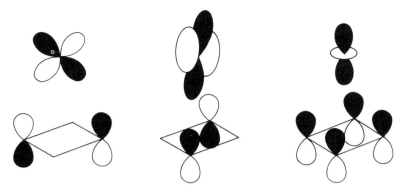

FIG. 10.   Schematic drawing of the principal metal–boron interaction orbitals in $(CO)_3FeB_4H_8$ (*149*).

the isoelectronic ferraborane is even more cluster-like, although a similar fragment analysis has not yet been done. Independent of which view, cluster or metal–ligand, is judged to be the better description, the similarity of the bonding in $(CO)_3FeB_4H_8$ to the organometallic analog is well established.

d. $C_4H_6/B_4H_9^-$. This is the system that is most clearly defined in an experimental sense as both free ligands and both complexes have been structurally characterized (*153, 224*). The changes in 1,3-butadiene on complexation to a transition metal have been revealed by several studies (*229*). It is clear that the geometry of the ligand changes such that the C—C bond lengths are more closely related to the geometry of the first excited state (or the geometry of the anion) than to the ground state. The absolute changes, however, are fairly small. The static structure of $B_4H_9^-$ as defined by low temperature ^{11}B NMR is indicated in Fig. 8. On being bound to the $Ir(CO)(PR_3)_2$ fragment (Fig. 6), the tetraboron "butterfly" opens up to form a planar B_4 unit very similar to the basal B_4 unit in B_5H_{11}. Although there are significant structural changes in both cases, the largest change takes place in the case of the isoelectronic borane ligand. Despite the fact that the free ligand, metal complex, free organic ligand, and organic π-complex have all been well characterized, no analysis of the bonding in the metalloborane has appeared. Clearly this would be useful for comparison.

e. $C_5H_5^-/B_5H_{10}^-$. The behavior of the cyclopentadienyl ion toward metals has been studied since the early days of organometallic chemistry yet its bonding with metals and the ring structure in its bound form still provoke theoretical and experimental work (230). The isoelectronic borane analog of ferrocene, *pentaboraferrocene*, has been recently prepared and the proposed structure is shown in Fig. 6 (172–173). As the skeletal structure of $B_5H_{10}^-$ as a free ligand presumably is that of B_5H_{11} (Fig. 8), there is a substantial structural change in going from the stable form of the free ligand to the bound form. Once again no theoretical work is available on this compound, and therefore, a more detailed comparison of the bonding is not possible.

In summary, there appear to be substantially larger changes in going from the free borane to the metalloborane than from the free hydrocarbon to metal–hydrocarbon π-complex. Generally speaking this suggests a greater perturbation of the borane on metal binding than the hydrocarbon. In those cases that have been studied theoretically there appears to be a definite qualitative similarity in the frontier orbital behavior of borane and isoelectronic hydrocarbon although the quantitative details of the interaction are different.

A general observation about the larger hydrocarbon and counterpart borane ligands is that whereas each of the former exhibits essentially the same geometry both when free and when bound to a metal, the latter undergo significant structural changes upon complexation. For example, the differences noted for C_4H_6 versus $B_4H_9^-$ arise primarily because the ground state geometry of $B_4H_9^-$ is analogous to the higher energy bicyclobutane conformer of C_4H_6. Should bicyclobutane be allowed to interact with, e.g., $[Fe(CO)_3]$, it is possible that the same structural changes evident in the coordination of $B_4H_9^-$ would occur in the hydrocarbon case. In addition, note that the direct borane analog of C_4H_6 is the neutral system B_4H_{10}, rather than the anion derived from B_4H_{10} by loss of H^+. As pointed out above, the role of the proton is an important one.

2. *Comparative Chemical Reactivity*

The rich reactivity of organometallic compounds has fueled and sustained the growth and activity of this area of chemistry. The reactivity of these metalloborane analogs of hydrocarbon π-complexes remains to be fully exploited. Perhaps the most complete comparative study is one in which the photochemical reactivity of $1\text{-}(CO)_3FeB_4H_8$ with respect to alkynes was compared to that of $(CO)_3FeC_4H_4$ (152). Under photochemical activation the latter adds alkynes and effects the insertion of the C_2 fragment into the cyclobutadiene ring leading to substituted benzene as a final product (Fig. 11) (231). The ferraborane under photochemical conditions also effects the insertion of alkynes into the B_4 fragment but with substantial differences. The major product, produced in about 60% yield, results from

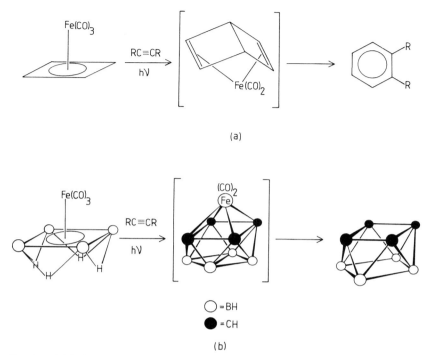

(a)

(b)

○ = BH

● = CH

Fig. 11. (a) Schematic representation of the photochemical reaction of $(CO)_3FeC_4H_4$ with an alkyne (231). (b) Schematic representation of the photochemical reaction of $(CO)_3FeB_4H_8$ with an alkyne (152).

the insertion of two moles of alkyne rather than one (Fig. 11). In addition, evidence suggesting the insertion of three and four moles of alkyne was found. The observations are interpreted in terms of an active intermediate that in the case of the ferraborane has a longer lifetime than the proposed intermediate in the analogous organometallic reaction. As a working hypothesis, a tautomeric equilibrium was suggested to account for the longer lifetime of the metalloborane. Some support for this view results from the fact that an unstable intermediate containing ferraborane and two moles of alkyne was isolated and partially characterized. It may well be that metalloborane analogs of hydrocarbon π-complexes will serve as routes to heteroborane systems not accessible by other means. If so, these compounds will become as useful to borane cage chemistry as organometallic compounds are to organic chemistry.

Despite chemistry reminiscent of organometallic compounds, boranes retain reactivity characteristic of boranes. For example, $(CO)_3FeB_4H_8$ may be deprotonated with KH to form $(CO)_3FeB_4H_7^-$ as may other metallo-boranes containing $B—H—B$ bridges (*26, 43*).

3. ^{11}B Chemical Shifts

One of the great advantages of working in the area of boron cluster chemistry before the age of multinuclear NMR spectrometers was the ready availability of ^{11}B NMR (*232*). Besides providing valuable structural information, the ^{11}B chemical shift contains information on the nature of

TABLE II

^{11}B CHEMICAL SHIFTS OF SELECTED METALLOBORANES

Class	Compound[a]	$\delta^{11}B$[b]
1	$Be(BH_4)_2$	-32.2
1	$2\text{-}(CO)_5MnB_5H_8$	$-48.6(1); 0.2(2); -10.3(3\text{-}5)$
1	$2\text{-}(CO)_5ReB_5H_8$	$-49(1); -11.5(2); -10.6(3\text{-}5)$
1	$2\text{-}[Cp(CO)_2Fe]B_5H_8$	$-48.5(1); 7.9(2); -10.6(3, 5); -14.3(4)$
2	$(CO)_6Fe_2B_2H_6$	-24
2	$(CO)_4MnB_3H_8$	$-42.2(1, 3); 1.5(2)$
2	$(CO)_3MnB_3H_8$	-47.2
2	$(CO)_3MnB_8H_{13}$	$-57.3(2); -52.0(3, 8); -7.8(1); -0.3(5,$ $6 \text{ or } 4, 7); 7.5(4, 7 \text{ or } 5, 6)$
3	$Cp(CO)_2FeB_2H_5$	-6.5
3	$(R_3P)_2CuB_5H_8$	$-47.8(1); -15.0(2\text{-}5)$
3	$(R_3P)_2BrPtB_5H_8$	$-48.3(1); -14.3(2\text{-}5)$
3	$(CO)_4FeB_6H_{10}$	$-54.4(1); 0.2(2, 3); 4.9(4, 6); 11.4(4)$
4	$(R_3P)_2(CO)HIrB_3H_7$	$-15(1, 3); 3(2)$
4	$1,2\text{-}[(CO)_3Fe]_2B_3H_7$	$4.2(3, 5); 12.1(4)$
4	$1\text{-}[(CO)_3Fe]B_4H_8$	-4.7
4	$1\text{-}[CpCo]B_4H_8$	-4.4
4	$2\text{-}[CpCo]B_4H_8$	$-15.9(2\text{-}5); 6.2(1)$
4	$(R_3P)_2CuB_4H_9$	$-55.2(1); 3.2(2\text{-}4)$
4	$1\text{-}[CpFe]B_5H_{10}$	5.1
4	$2\text{-}[CpFe]B_5H_{10}$	$-53.0(1); 8.2(4, 5); 44.4(3, 6)$
4	$2\text{-}[(CO)_3Mn]B_5H_{10}$	$-53.7(1); 11.3(4, 5); 31.8(3, 6)$
4	$2\text{-}[(CO)_3Fe]B_5H_9$	$-47.0(1); 2.3(4 \text{ or } 5); 13.5(5 \text{ or } 4);$ $37.7(6); 55.5(3)$
4	$2\text{-}[R_3P)_2(CO)IrB_5H_9$	$-28(1); 10(4, 5); 42(3, 6)$
4	$2\text{-}[(CO)_3Co]B_5H_8$	$-45.0(1); 6.8(4, 5); 57.0(3, 6)$

[a] References given in Table I.

[b] ^{11}B shifts in ppm given with $\delta BF_3 \cdot O(C_2H_5)_2 = 0$ and positive values downfield. The number in parentheses is the boron position with the apical position designated as 1.

the bonding environment of the boron observed. In particular the changes in the chemical shifts of the boron atoms of the borane ligand on being bound to a metal fragment constitute a measure of the perturbation of the borane on metal–boron bond formation. In principle these chemical shift differences contain information on the redistribution of electron density but, because of ambiguities in interpretation, they are used here only as an indicator of the magnitude of the perturbation caused by the metal fragment. A selection of ^{11}B chemical shifts for metalloboranes and boranes is given in Table II. For the compounds considered in this section, we compare differences between the free ligand spectra and the metalloborane. These differences, which reflect the magnitude of the overall structural change on bonding, are given in Table III. Free ligand spectra are known for only two of the six ligands. For the "olefinic" interaction of B_6H_{10} with $Fe(CO)_4$ in which the geometric changes in the borane are modest a net upfield shift on coordination is observed. For a similar interaction of $B_5H_8^-$ with CuL_2 there is a slight upfield shift of the basal borons and slight downfield shift of the apical boron. In both cases, the ^{11}B data suggest relatively minor perturbation of the basic boron skeletal bonding. For the other four compounds, only comparison with models of the free ligands is presently possible. Using B_2H_6 as a model for $B_2H_5^-$ suggests an upfield shift on interaction with iron; however, the magnitude of this shift would be reduced for $B_2H_5^-$ as deprotonation itself causes an upfield shift. $B_3H_7 \cdot$ THF as a model for B_3H_7 suggests no shift; however, as the spectrum of free B_3H_7 would probably be downfield of that of $B_3H_7 \cdot THF$, the net shift on coordination of B_3H_7 is probably upfield. Likewise 1-COB_4H_8 can be

TABLE III

Comparison of ^{11}B Chemical Shifts Hydrocarbon π-Ligand Analogs

Ligand[a]	Metallo-borane[a]	Free ligand or[b] (model)	$\Delta\delta_{M-F}$[c]
$B_2H_5^-$	−6.5	17.5(B_2H_6)	−24
B_6H_{10}	−54.4, 4.3(av)	−50.9, 16.1	−3, −11
$B_5H_8^-$	−47.8, −15.0	−52.8, −12.5	5, −3
B_3H_7	−15; 3	−8.4($B_3H_7 \cdot$THF)	∼0
	−9(av)		
B_4H_8	−4.7	−58.7; −1.5, −2.1(B_4H_8CO)	54, −3, −3
$B_4H_9^-$		−52.2, −9.7	
$B_5H_{10}^-$	5.1		

[a] See Tables I and II.

[b] See Refs. 224 and 232.

[c] Chemical shift difference (ppm) metalloborane − free ligand; positive values downfield.

considered a model for B_4H_8. Ignoring the boron at which substitution takes place and with the same considerations as for B_3H_7, there is also a suggestion of a net upfield shift on coordination. Although not definitive, these comparisons suggest modest upfield shifts on coordination with metals. The ^{13}C coordination shift for olefins on being bound to a transition metal is to higher field and is in the range of -25 to -90 ppm (233). Using an empirical relationship between ^{11}B and ^{13}C chemical shifts developed for carbocations and boranes (234), one would predict an upfield shift for the metal–carbon compounds of -8 to -30 ppm if they are indeed analogs of the metal–carbon compounds. The relatively small ^{11}B shifts on coordination suggest a fair amount of localization of the metal–boron interaction. In a metal–ligand complex, retention of much free ligand character in the bound state is expected and, thus assuming the ^{11}B shift is reasonably sensitive to the nature of the bonding, the chemical shift behavior supports the idea of considering these compounds as true analogs of hydrocarbon–metal complexes.

B. *Ligand Behavior Unique to Boranes*

The borane fragments described in the preceding section clearly are related to hydrocarbons in terms of interactions with transition metals. On the other hand, boranes have a well characterized tendency to form bonds to metals via bridging hydrogens—a tendency only rarely glimpsed in hydrocarbons. This characteristic is most fully documented in the metal tetrahydroborates (33). The ability to form $M-H-B$ bonds is also demonstrated by higher polyboranes. The purpose of this section is to review this mode of bonding.

1. *Geometric and Electronic Structure*

a. BH_4^-. Complexes containing tetrahydroborate have been the subject of comprehensive reviews in which structures and bonding models have been explored (33, 37). The BH_4^- ion displays mono-, bi-, and tridentate coordination (Fig. 12) and forms complexes with the majority of metals in the periodic table (Table I). An analogy of the bonding of BH_4^- and η^3-allyl has been drawn on the basis of empirical observations, i.e., transition metal complexes that differ only by the exchange of BH_4^- for allyl (33). In addition there is both experimental and theoretical evidence for only a small energy difference between the bi- and tridentate modes of bonding. One type of metal–boron interaction that is of interest with respect

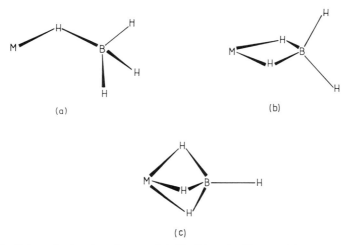

FIG. 12. Schematic drawing of mono- (a), bi- (b), and tridentate (c) BH_4^- bonding to a transition metal.

to the last section is the monodentate mode of bonding, of which one example has been established (Table I) (*92, 93*). In the case of $(CO)_5MnBH_4$, for which such a structure is also considered plausible (Fig. 13), (*87*) BH_4^- replaces CH_3^- in the isoelectronic organometallic compound, the Mn—C bond being replaced by a Mn—H—B bond. The tetrahydroborate is, however, very unstable. Note that for higher boranes direct analogs of σ hydrocarbon ligands exist; for example in 2-$(CO)_5MnB_5H_8$ the Mn—B bond is comparable to the Mn—C bond in $(CO)_5MnCH_3$.

b. $B_2H_6^{2-}$. This hexahydrodiborate has been identified in three compounds, two of which have been characterized by single crystal X-ray diffraction studies (*115, 116, 119*). These compounds whose structures are sketched in Fig. 14 all possess one or more ethane-like $B_2H_6^{2-}$ ligands bridging a dimetal fragment. The metal–boron interaction is via M—H—B bridges; four in the case of $(CO)_6Fe_2B_2H_6$, eight for $[CpNbB_2H_6]_2$, and six for $(CO)_{10}HMn_3B_2H_6$. This multidentate versatility is reminiscent of the situation with BH_4^-, i.e., each terminal BH_3 of $B_2H_6^{2-}$ is either bi- or tridentate. Except for the Mn derivative these compounds are rather new and it remains to be seen whether the size of this class will approach that of either the B_1 or B_3 derivatives.

The electronic structure of $(CO)_6Fe_2B_2H_6$ has been studied by means of theoretical calculations as well as PE spectroscopy (*117, 118*). The ethane-like geometric structure of the borane ligand can be considered

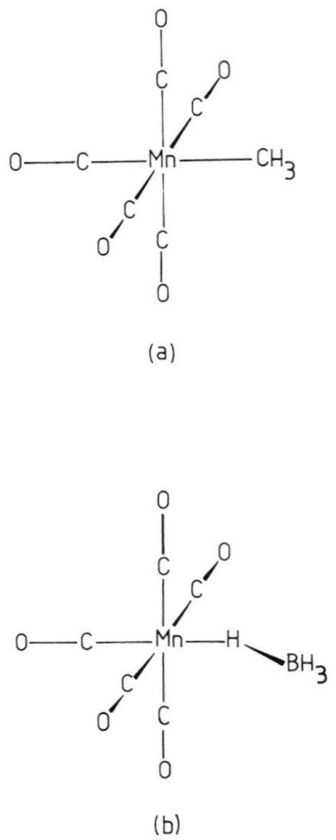

(a)

(b)

FIG. 13. Comparison of the structure of $(CO)_4MnCH_3$ (a) with that proposed for $(CO)_4MnBH_4$ (b) (33).

evidence for a negatively charged ligand, and two rather different calcu-
lational techniques used support this conclusion. In addition, both the PE
results and the calculations suggest a high-lying filled orbital of large boron
character that is associated with the B—B bond. The energy and properties
of this orbital are similar to the B—B bond in B_6H_{10} and it may well be
that this feature of the bonding is indicative of similar reactivity relative
to Lewis acids.

c. $B_3H_8^-$. The octahydrotriborate ion, like BH_4^-, is found in both ionic
salts as well as covalent transition metal compounds. A fair variety of
compounds is known in which the $B_3H_8^-$ ion coordinates to a metal via two
adjacent M—H—B interactions (Fig. 5); however, there are also some
variations that will be presented below. The bidentate $B_3H_8^-$ complexes
have been likened to B_4H_{10} (Fig. 15) in which the metal fragment replaces

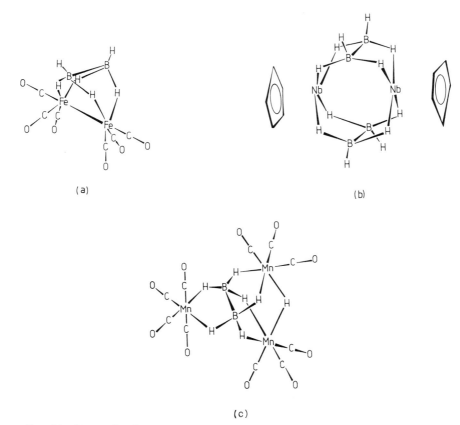

(a)

(b)

(c)

FIG. 14. Proposed or known structures of (a) $(CO)_6Fe_2B_2H_6$ (*116*), (b) $Cp_2Nb_2(B_2H_6)_2$ (*115*), and (c) $(CO)_{10}Mn_3H(B_2H_6)$ (*119*).

FIG. 15. Structure of B_4H_{10} (*1*).

a "wing-tip" BH_2^+ in the borane butterfly (Fig. 5) (235). It should be noted that $B_3H_8^-$ can be formally generated from B_3H_7 by the addition of H^-.

The electronic structure of a typical metal complex, $(CO)_4MnB_3H_8$, has been examined by the extended Hückel method. The free ligand itself is characterized by a large HOMO–LUMO gap (236). This contrasts with the situation of borallyl, B_3H_7 (see above), where the smaller HOMO–LUMO gap is typical of that found in hydrocarbon π systems. In forming the metal complex the major interactions are found to take place between the symmetric HOMO of the ligand with a symmetric empty orbital of the metal fragment and an antisymmetric filled ligand orbital with an antisymmetric empty metal orbital (Fig. 16) (129). The latter interaction rationalizes the orientation of the ligand with respect to the metal fragment in the same way that the orientation of olefins in complexes is rationalized. That is, for effective overlap the BBB plane of the ligand must be tilted with respect to the equatorial plane of the metal fragment and the B—B bond must lie in a plane parallel to but below the equatorial plane of the metal fragment. As this antisymmetric ligand orbital has a significant contribution from the terminal hydrogens adjacent to the B—B bond, the

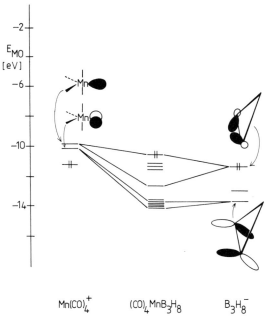

$$Mn(CO)_4^+ \qquad (CO)_4MnB_3H_8 \qquad B_3H_8^-$$

FIG. 16. Diagrammatic representation of the primary metal–boron interaction in $(CO)_4MnB_3H_8$ (129).

participation of these hydrogens in forming B—H—Mn bridges is reasonable.

As $B_3H_8^-$ exists as the free ligand it is of interest to examine the structural changes taking place on the formation of a metal complex. The most probable structure of $B_3H_8^-$ is shown in Fig. 17 where it is seen that the structure is basically that of cyclopropane in which two of the C—C bonds have been protonated. In fact the MO structure of $B_3H_8^-$ exhibits a distinct relationship to that of cyclopropane. In being bound to a metal the symmetry of $B_3H_8^-$ is reduced from C_{2v} to C_s and the antisymmetric orbital (Fig. 16) is polarized toward the terminal hydrogens on the same side of the BBB plane as the bridging hydrogens thereby enhancing the M—H—B interaction (Fig. 5). In a sense the structural change that takes place tends to focus the ligand orbitals toward metal binding but, in fact, the actual structural change is relatively small. Thus geometrically speaking the $B_3H_8^-$ moiety retains much of its ligand character in the metal complex. With the exception of the two primary donor orbitals the complexed B_3H_8 moiety also retains much of its electronic character and the calculated Mulliken charge on the bound ligand is found to be −0.8 in the extended Hückel treatment (129). Finally, one notes that the weighted average ^{11}B chemical shift of $(CO)_4MnB_3H_8$ ($\delta = -28$) is essentially equal to that of the free and fluxional ligand (Table II) (232). The suggestion is that there are no gross changes in the character of $B_3H_8^-$ on being bound to a transition metal which is, in effect, a definition of ligand behavior. There are, however, interesting changes in the fluxional behavior of the octahydrotriborate ion on being bound to a transition metal fragment (51). The bidentate $(CO)_4MnB_3H_8$ compound is static; however, the tridentate $(CO)_3MnB_3H_8$ (see below) is "selectively nonrigid" even at low temperatures in that the Mn—H—B hydrogens are static while the other hydrogens are made equiv-

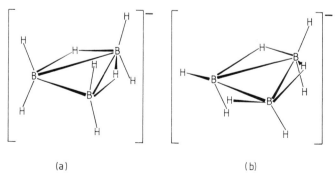

(a)   (b)

FIG. 17.   Representation of the $B_3H_8^-$ structure in unbound state (a) and on being bound to a transition metal (b).

alent by an internal exchange that is rapid on the NMR time scale. The fluxional behavior is very sensitive to ostensibly small changes, e.g., the substitution of Br for one of the terminal hydrogens in $(CO)_4MnB_3H_8$ produces a molecule that is static at low temperatures but selectively non-rigid at room temperature.

As $B_3H_8^-$ transition metal chemistry develops, other modes of bonding are being revealed. For example, the pyrolysis or photolysis of $(CO)_4MnB_3H_8$ with continuous removal of released CO yields $(CO)_3MnB_3H_8$ in which the $B_3H_8^-$ ligand functions as a tridentate ligand to the metal (Fig. 18) (51, 128). The entire process is reversible [Eq. (22)] which suggests a fairly small

$$(CO)_3MnB_3H_8 \overset{\Delta}{\rightleftharpoons} (CO)_3MnB_3H_8 + CO \qquad (22)$$

energy difference, a situation reminiscent of the bi- and tridentate behavior of BH_4^- (see above). A higher homolog of this tridentate $B_3H_8^-$ metallo-

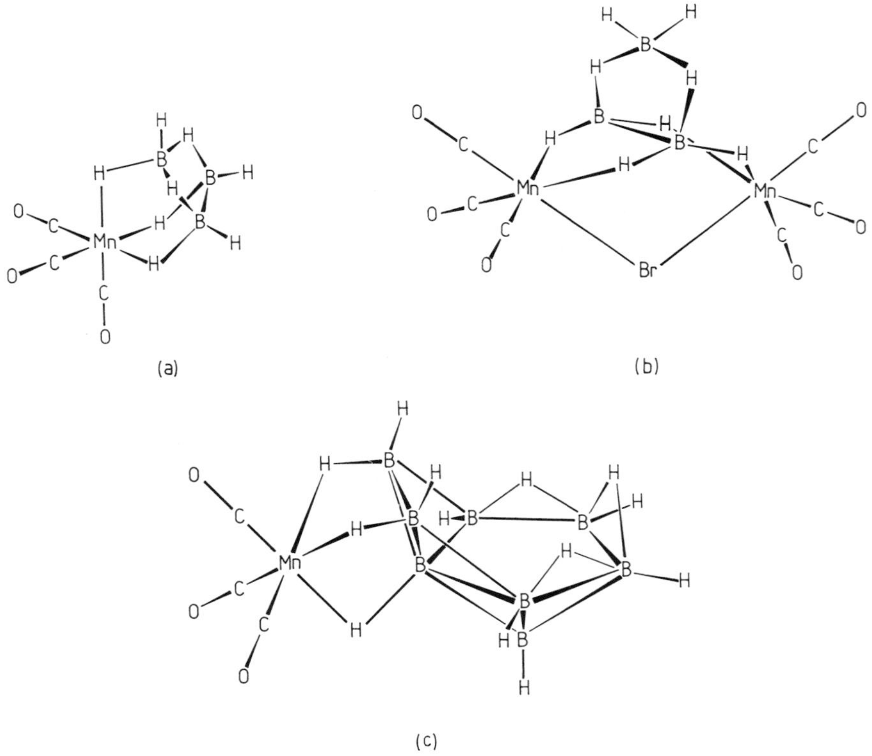

(a)

(b)

(c)

FIG. 18. Comparison of the structures of (a) $(CO)_3MnB_3H_8$ (128); (b) $(CO)_6BrMn_2B_3H_8$ (140); and (c) $(CO)_3MnB_8H_{13}$ (197).

borane is found in $(CO)_3MnB_8H_{13}$ (*197*), in which the B_8H_{13} fragment acts as a tridentate ligand to $Mn(CO)_3$ via three $Mn-H-B$ bridges (Fig. 18). This compound has been described as a polyhedral borane with a face-bridging $Mn(CO)_3$ fragment. As such, the metal fragment is an exopolyhedral substituent. $B_3H_8^-$ has been found to act as a bridge between two metal atoms in $(CO)_6BrMn_2B_3H_8$ (*140*) (Fig. 18), a compound that may be considered analogous to $(CO)_{10}HMn_3B_2H_6$ (*119*) (Fig. 14) by formally replacing BH_2 with $Mn(CO)_4$. It will also be noted that as far as the metal–boron bonding is concerned this compound is similar to both $(CO)_6Fe_2B_2H_6$ (*116*) and $[CpNbB_2H_6]_2$ (*115*) (Fig. 14).

With the exception of the eight-boron compound mentioned above there are no examples of this type of what might be described as hydroborate ligand behavior for boranes with more than three borons. Yet, as will be discussed more fully in the next section, there are a number of compounds in which the metal–boron interaction takes place partly through $M-H-B$ bridges. One compound that demonstrates a particularly interesting hybrid metal boron interaction is $(R_3P)_2CuB_5H_8Fe(CO)_3$ (Fig. 19) (*187*). The iron atom is considered as part of the six-atom cage skeleton and will be discussed in Section IV. The copper is bonded to the B_5Fe unit through an interaction that is related to that observed for both $(CO)_4FeB_6H_{10}$ or $(R_3P)_2CuB_5H_8$ and $(R_3P)_2CuB_3H_8$. The formation of the $Cu-H-B$ bond has been suggested as helping copper to achieve an 18-electron configuration and that the exohydrogens are more available for such an interaction in the pentagonal base of a six-atom nido cage than they are in the square base of a five-atom nido cage.

2. ^{11}B Chemical Shifts

As in Section III, A, the changes in the chemical shifts of the boron atoms of the borane ligand on being bound to the metal fragment constitute one measure of the extent of the perturbation of the borane by the metal.

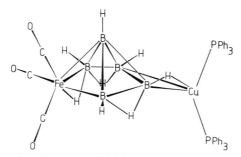

FIG. 19.   Structure of $(CO)_3FeB_5H_8Cu(PR_3)_2$ (*187*).

Pertinent data for the metal compounds are given in Table II. Data exist for only two ligands, BH_4^- and $B_3H_8^-$ (*232*). For the former only small changes in the chemical shift are observed on metal binding. For the latter, if one compares the weighted average of the shifts in $(CO)_4MnB_3H_8$ to the observed value in the fluxional $B_3H_8^-$ ligand, again there is little change observed. Both these observations support the view that the borane fragments considered in this section are properly considered as ligands.

IV

BORANES AS CLUSTER FRAGMENTS

As was pointed out in the introductory remarks, an analogy between boranes and metal clusters has been drawn that allows information on the former to be used to provide a first-order picture of the behavior of the latter (*10*). Metalloboranes, of course, lie in the middle and have been assigned a key role in relating structures of borane clusters to those metal clusters, and, as discussed above, to metal–hydrocarbon π-complexes (*28*). This section is used to summarize the cluster view of the metalloboranes and the analogous hydrocarbon π-complexes discussed above, as well as to present information on metalloboranes that can only be reasonably described as clusters. This section also demonstrates that the cluster view is broader in scope than a metal–ligand picture in that it encompasses in a single model description the analogs of hydrocarbon π-complexes and the hydroborate complexes as well as metalloboranes not easily described by any metal–ligand bonding scheme. As such, the cluster view is a unifying concept that adds to the earlier considerations rather that superseding them. Finally a few examples of metalloboranes are given demonstrating that a cluster view based solely on the behavior of metal-free polyhedral boranes is itself limited in that the observed structures are not necessarily those of the analogous boranes.

Before proceeding, it is necessary briefly to summarize the polyhedral skeletal electron pair theory (PSEPT) as it applies to metalloboranes. Extensive discussions of this theory have appeared elsewhere and only the bare outline is given here (*10*). This theory first assumes that the skeletal bonding is separable from the bonding of the exopolyhedral cluster ligands. If the cluster is made up of main group atoms of the first row with H as the exopolyhedral "ligand," this assumption allows three orbitals per cluster atom to be assigned to skeletal bonding. A molecular orbital treatment of the cluster bonding then suggests that a cluster structure having atoms at the vertices of a deltahedron (polyhedron with only triangular faces) will be stable if $\mathbf{p} + 1$, where $\mathbf{p}$ is the number of vertices of the deltahedron, electron pairs are available. If, as is usually the case, each cluster atom of the closed polyhedron skeleton utilizes one electron in forming an exopolyhedral bond, it contributes its valence electrons minus one to cluster bonding. Thus, for example, the BH fragment contributes two electrons to cluster bonding and the octahedral B_6H_6 cluster is predicted to be stable as the dianion. For obvious reasons this type of cluster is designated *closo*. If the number of skeletal bonding pairs exceeds $\mathbf{p} + 1$, the cluster adopts a geometry based on a deltahedron with the number of vertices greater than $\mathbf{p}$ and equal to one less than the number of available skeletal electron pairs. As in this case there are insufficient cluster atoms to occupy all the vertices of the deltahedron, the cluster must be open. Clusters with $\mathbf{p} + 2$ electron pairs are designated *nido* (one vertex unoccupied) and those with $\mathbf{p} + 3$ electron pairs *arachno* (two vertices unoccupied) (Fig. 20). It has been demonstrated that this treatment is not restricted to boranes and that a wide variety of other atoms and groups can replace BH to form heteroboranes obeying the same electron counting rules, e.g., in $B_6H_6^{2-}$ replacing two BH^- with CH yields the neutral carborane $C_2B_4H_6$.

It has been found that certain transition metal fragments have three valence orbitals of suitable symmetry and spatial orientation to function in cluster bonding in the same manner as, for example, BH. Thus $Fe(CO)_3$ is said to be isolobal and pseudoisoelectronic with BH in the sense that it contributes three orbitals and two electrons to cluster bonding (*10, 11*). From this point of view the analog of $(CO)_3FeC_4H_4$, $(CO)_3FeB_4H_8$, is seen to be analogous to B_5H_9 in the sense that the apical BH of the latter has been replaced with an $Fe(CO)_3$ fragment. Likewise, the metalloborane serves formally to relate B_5H_9 and $C_4H_4Fe(CO)_3$. These interrelations have

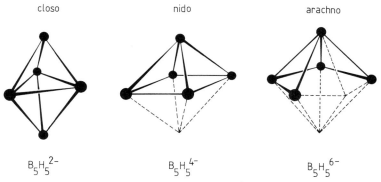

FIG. 20. The relationship between skeletal electron count and cluster geometry for hypothetical five-atom borane anions.

been discussed and the family ties of these formerly unrelated classes of compounds firmly established (28, 34).

A. *Metal–Boron Bonding Uniquely Cluster–Like*

Wade's approach is a unifying concept but one can ask whether, in fact, it is unique or even necessary, i.e., is there any real evidence for cluster behavior in metalloboranes? The answer is yes, and we detail below three ways in which metalloboranes exhibit distinctly cluster-like properties.

1. *Geometric Isomers of Hydrocarbon π-Complex Analogs*

π-Hydrocarbon metallo derivatives and the analogous metalloboranes can be considered as clusters based on skeletal electron counting rules. $(CO)_3FeC_4H_4$ (7 electron pairs, nido) exists as the 1-isomer. The analogous metalloborane, (Fig. 21) $CpCoB_4H_8$ exists as both the 1- and 2-isomers (57, 59) and there is evidence that the isoelectronic compound $(CO)_3FeB_4H_8$ does as well (41, 237). In ferrocene, $CpFeCp$, the iron atom shares the apical site of two adjacent pentagonal pyramids and the compound is there-

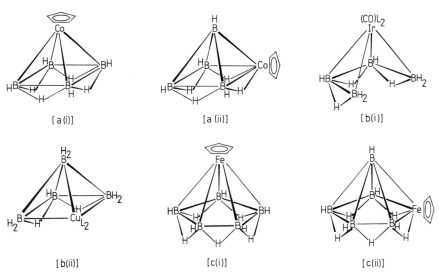

FIG. 21. Comparison of proposed or known structures of metalloborane geometric isomers: (a) $1\text{-}CpCoB_4H_8$ versus $2\text{-}CpCoB_4H_8$ (57, 59); (b) $1\text{-}(R_3P)_2(CO)IrB_4H_9$ versus $2\text{-}(R_3P)_2CuB_4H_9$ (153, 154); and (c) $1\text{-}CpFeB_5H_{10}$ versus $2\text{-}CpFeB_5H_{10}$ (172, 173).

fore viewed as involving two nido skeletons fused in the 1-position. However, $CpFeB_5H_{10}$ is found as both 1- and 2-isomers (*172, 173*) and $B_5H_{10}FeB_5H_{10}$ has been isolated as the 2,2'-isomer (*167*). Finally, only the 1-isomer of the arachno butadiene metal complex is known but there is good evidence for the existence of both 1- and 2-isomers of the analogous $B_4H_9^-$ ligand ($1-[(R_3P)_2(CO)Ir]B_4H_9$ and $2-[(R_3P)_2Cu]B_4H_9$; Table 1) (*153, 154*). This of course does not imply that other isomers of hydrocarbon π-complexes cannot exist but it does demonstrate that there is a greater tendency for the analogous metalloboranes to form isomers that are obviously cluster-like. The interesting question that has been raised earlier (*27*) is whether or not organometallic analogs of the 2-isomers might be found if they were looked for.

In the case of both $CpCoB_4H_8$ and $CpFeB_5H_{10}$ it is the 2-isomer that is initially formed in the preparative procedure. The 2-isomer is then thermally converted to the 1-isomer demonstrating unequivocally that the 1-isomer is more stable in a thermodynamic sense. Thus for the metalloboranes the kinetic product is the one isolated and there is a significant barrier separating it from the thermodynamically more stable 1-isomer. It is of course unknown whether or not the analogous hydrocarbon 2-isomer may be formed, or indeed is formed but with only a very small energy barrier separating it from the 1-isomer.

Besides direct analogs of hydrocarbon π-ligand complexes there are metalloboranes that are difficult to explain except as derivatives of a borane cluster in which a cluster component has been replaced by an isolobal transition metal fragment. An isoelectronic series of metalloboranes is illustrated in Fig. 22 for which no organometallic analogs exist but which

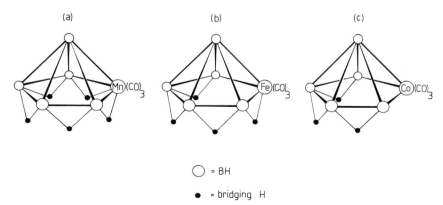

$\bigcirc$ = BH

● = bridging H

FIG. 22. Comparison of the proposed structures of (a) $(CO)_3MnB_5H_{10}$ (*164*); (b) $(CO)_3FeB_5H_9$ (*168, 169*); and (c) $(CO)_3CoB_5H_8$ (*167, 177*).

are easily described as metal derivatives of a parent hexaborane(10) in which a BH_2 unit (one H bridging) has been replaced with $Co(CO)_3$, $HF(CO)_3$, or $H_2Mn(CO)_3$.

Finally, although carboranes have been excluded from this review, it should be noted that there is only one example of a $C_5H_5^-$ ring serving as the central ligand in a "triple-decker" sandwich complex (238) whereas there are many examples of heteroborane rings doing such service (27, 239). As the two metals and central ligand of these complexes constitute a cluster, this observation serves as additional evidence for an intrinsic ability of boron to stabilize forms of bonding associated with clusters.

2. Multinuclear Metal Systems

The second type of metalloborane that requires a cluster description as opposed to a metal–ligand complex description is found in systems that contain more than one metal atom. For example, the series B_5H_9, $(CO)_3FeB_4H_8$, and $(CO)_6Fe_2B_3H_7$ shown in Fig. 23 demonstrates this point. The first member, B_5H_9, can only be reasonably represented as a cluster. The second, $(CO)_3FeB_4H_8$, can be represented equally well as either a metal–ligand complex [isoelectronic with $(CO)_3FeC_4H_4$] or a metal–boron cluster; [$Fe(CO)_3$ is isolobal and pseudoisoelectronic with BH]. The third, $(CO)_6Fe_2B_3H_7$, contains the B_3H_7 ligand which may be formally considered to be the same as the borallyl B_3H_7, found in $L_2PtB_3H_7$ (Table I). In fact, by doing so one can assign 18 electrons to each iron. However, the actual structure of the complex clearly demonstrates that the B_3H_7 fragment in the diiron metalloborane has neither the structure of borallyl nor that of

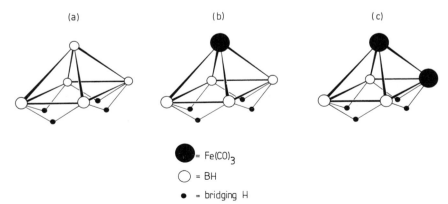

FIG. 23.   Comparison of the proposed or known structures of (a) B_5H_9 (1); (b) $(CO)_3FeB_4H_8$ (148); and (c) $(CO)_6Fe_2B_3H_7$ (142).

the B_3H_7 fragment in $B_3H_8^-$ (*142*). It is much more appropriate, and useful, to describe this compound as a derivative of B_5H_9 in which a basal and an apical BH fragment are replaced by isolobal and pseudoisoelectronic $Fe(CO)_3$ fragments. This is not to say that there are no organometallic analogs of compounds like $(CO)_6Fe_2B_3H_7$; in fact, there are several. For example, the dimetal compound shown in Fig. 24 formed from the reaction of an alkyne with a dimetal bridging alkylidyne ligand is a direct analog of $(CO)_6Fe_2B_3H_7$ (*240*). The C_3 moiety is bonded to tungsten via the two terminal carbons and to iron via all three carbons. It has been described as an $Fe(CO)_3$ group bonded to a tungstacyclobutadiene ring; however, it is perhaps more easily described as a seven-electron pair, nido cluster [each RC donates three electrons, $Cp(CO)_2W$ donates three electrons, and $(CO)_3Fe$ donates two electrons to cluster bonding]. A dimetal compound that can be viewed as an analog of $L_2(CO)IrB_4H_9$ is also shown in Fig. 24 (*241*). There is some advantage to describing this compound as an eight-electron pair, arachno cluster [each RC donates three electrons, $Cp(CO)Ru$ donates three electrons, CpRu donates one electron, CO bridging donates

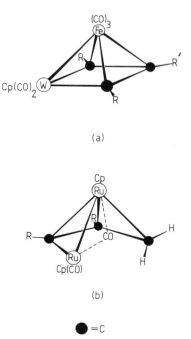

(a)

(b)

● = C

FIG. 24.   (a) Structure of an organometallic analog of $(CO)_6Fe_2B_3H_7$ (*240*) and (b) an organometallic analog of $L_2(CO)IrB_4H_9$ (*241*).

two electrons, and the endo hydrogen donates one electron to cluster bonding] rather than by means of the three equivalent representations presented in the original work. Compounds such as these suggest there are also organometallic systems that are more easily, if not more appropriately, considered as clusters rather than as metal–ligand complexes.

As $(CO)_6Fe_2B_3H_7$ is a key compound in probing cluster behavior, two detailed studies of the nature of the bonding in this compound have been carried out. In one (118), the extended Hückel method, the nonparameterized Fenske–Hall method, and UV–photoelectron spectroscopy have been utilized to probe the changes in the bonding through the series depicted in Fig. 23. One of the most interesting observations is that the valence electron density distribution of $(CO)_6Fe_2B_3H_7$ mimics in a qualitative way the charge distribution in B_5H_9. In the latter compound the apical boron carries a significant negative charge and the molecule has a substantial dipole moment. Although the calculated Mulliken charge of the apical iron in $(CO)_6Fe_2B_3H_7$ is positive, it is much less positive than that of the basal iron. This relative charge distribution, which is supported by the spectroscopic results, demonstrates the appropriateness of considering metalloboranes such as $(CO)_6Fe_2B_3H_7$ as clusters. The second study (129) compares in a qualitative way the similarities and differences between the frontier orbitals of B_3H_7 as found in $(CO)_6Fe_2B_3H_7$ with those in $L_2PtB_3H_7$, (borallyl) as well as in $(CO)_4MnB_3H_8$. The latter two compounds contain triboron fragments that display true ligand properties (See Section III,A). The frontier orbital properties of the latter two triborane ligands lead to reasonable descriptions of the mononuclear metalloboranes as metal–ligand complexes. However, the frontier orbitals of the triborane unit in $(CO)_6Fe_2B_3H_7$ are complex, being a hybrid, as it were, of those of borallyl and $B_3H_8^-$ and no simple metal–ligand description of the diiron complex results from this treatment.

Figure 25 gives the structures of examples of di- and tri-nuclear cobalt metalloboranes (57, 144–146, 155). These are further examples that establish the necessity of a cluster description. Both of these compounds exhibit structures easily rationalized in terms of clusters. The CpCo fragment is isolobal and pseudoisoelectronic with BH, thus according to Wade's rules (PSEPT) both have seven electron pairs associated with six-vertex cluster bonding and should be closo, i.e., octahedral, as they obviously are. On the other hand the only way to describe $Cp_3Co_3B_3H_5$ in terms of a metal–ligand complex is to consider a $B_3H_3^{3-}$ ligand bound to a $Cp_3Co_3H_2^{3+}$ fragment. This does not seem particularly realistic.

Just as there are metal clusters that exhibit structures for which there are no analogous boranes, so too there are metalloboranes that constitute exceptions to PSEPT (Wade's rules). Three examples are shown in Fig.

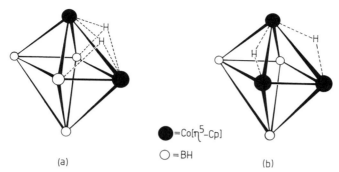

FIG. 25. Structures of multinuclear metalloboranes conforming to the PSEPT formalism: (a) $Cp_2Co_2B_4H_6$ (*57, 146, 155*) and (b) $Cp_3Co_3B_3H_5$ (*144, 145*).

26 all of which contain several metal atoms. The first compound, $Cp_3Co_3B_4H_4$ (*156, 157*), is unknown as far as polyhedral boranes are concerned in that it contains a BH fragment capping the triangular Co_3 face of a Co_3B_3 cluster. Such structural features are known in metal clusters, e.g., $Os_7(CO)_{21}$ (*242*), and can be accommodated in the PSEPT approach by noting that the presence of the capping fragment does not add to the number of cluster MOs required by the principal cluster (*243*). That is, $Cp_3Co_3B_4H_4$ [and $Os_7(CO)_{21}$] has seven cluster pairs that are only sufficient to meet the requirements of a closo, six-atom cluster. Hence, the compound adopts a capped octahedral structure that only requires seven

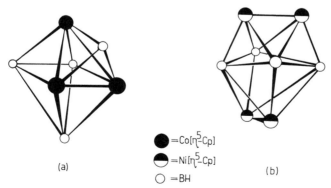

FIG. 26. Structures of multinuclear metalloboranes not conforming to the PSEPT formalism: (a) $Cp_3Co_3B_4H_4$ (*156a,b*) and (b) $Cp_4M_4B_4H_4$, M = Ni. For M = Co the positions of atoms M and B in the figure are interchanged. (*158, 160, 161*).

electron pairs to accommodate seven cluster atoms. Again, this behavior is cluster-like and the metal–ligand picture is inappropriate.

The second two examples in Fig. 26 have structures that are not accounted for under the PSEPT approach. Both $Cp_4Ni_4B_4H_4$ and $Cp_4Co_4B_4H_4$ have similar structures, being formally closo dodecahedra with idealized D_{2d} symmetry (158, 160, 161). The cobalt compound has eight cluster pairs, one short of the number required by PSEPT for an eight-atom closo system. The nickel compound on the other hand has ten pairs, one more than required for a closo structure. In this regard these two compounds are examples of metalloboranes that constitute exceptions to the PSEPT approach and exhibit some of the variable behavior of metal cluster systems.

The electronic factors leading to the observed structures of $Cp_4M_4B_4H_4$ have been discussed on the basis of extended Hückel calculations and a perceptive fragment analysis (159). The key experimental observation is that there is a real difference between the cobalt and nickel structures. Both are closo dodecahedral structures but in the cobalt derivative the metal atoms occupy the high connectivity sites whereas in the nickel derivative they are found in the low connectivity sites. If the dodecahedron is viewed as formed from the fusion of an elongated tetrahedron and a flattened tetrahedron (Fig. 27) the former represents the low connectivity sites and the latter the high connectivity sites. Separate considerations show that the elongated tetrahedron is able to accommodate ten skeletal electron pairs while the flattened tetrahedron is suited for eight skeletal electron pairs. It is then noted that the valence orbitals of the metal fragment and the BH fragment differ in a significant fashion in that the energies of the former are lower than those of BH. Therefore when the two tetrahedral fragments are fused to form a dodecahedron the filled orbitals generated are more metal than boron-like. The net conclusion is that when the metal

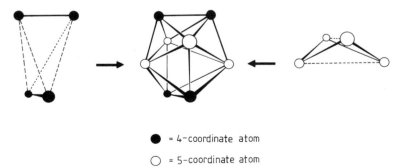

● = 4-coordinate atom

○ = 5-coordinate atom

FIG. 27. The formation of a D_{2d} dodecahedron from an elongated tetrahedron (black circles) and a flattened tetrahedron (open circles).

atoms occupy the high connectivity sites in $Cp_4M_4B_4H_4$ the optimal skeletal electron count for the eight-atom cluster is controlled by the requirements of the flattened tetrahedron, i.e., eight pairs, and when the metal atoms occupy the low connectivity sites the optimal electron count is controlled by the requirements of the elongated tetrahedron, i.e., ten pairs. Hence, this analysis rationalizes the observed behavior of both $Cp_4Ni_4B_4H_4$ and $Cp_4Co_4B_4H_4$. It must be kept in mind, however, that there are several skeletal structures for an eight-atom cluster system that do not differ appreciably in stability. Factors other than electronic appear to be important in controlling the actual structure observed in any given instance (244). In fact, an alternative explanation of $Cp_4Ni_4B_4H_4$ and $Cp_4Co_4B_4H_4$ based on special properties of dodecahedral boranes has been proposed (245). It is noted that the HOMO and LUMO in $B_8H_8^{2-}$ are nondegenerate whereas for $B_nH_n^{2-}$ for $n = 5, 6, 7, 10,$ or 12 the HOMO–LUMO pair is degenerate. Thus in principle two electrons can be subtracted or added to $B_8H_8^{2-}$ to form B_8H_8 ($Cp_4Co_4B_4H_4$) or $B_8H_8^{4-}$ ($Cp_4Ni_4B_4H_4$) without generating an open-shell system that would be expected to be subject to Jahn–Teller distortion as would be the case for $B_nH_n^{2-}$ for $n = 5, 6, 7, 10,$ or 12. The preference of CpCo and CpNi for high and low coordination sites, respectively, is rationalized on the basis of the isolobal analogy, i.e., CpCo $\cong$ BH and CpNi $\cong$ CH, and the relative connectivity preferences of the main group fragments. Yet another explanation has recently appeared (246).

3. Interstitial Compounds

The preparation and characterization of metal clusters containing atoms encapsulated by the metal cluster has revealed some fascinating materials that find no analogs in borane chemistry (14). The chemical manipulation of carbide clusters to expose the carbide carbon to reaction allows the first systematic exploration of a bound form of carbon proposed to be similar to that found in bulk metals (247, 248). Recently the preparation and structural characterization of a novel iron–boron cluster containing four iron atoms and one boron has been reported (111). The compound, Fig. 28, consists of a $HFe_4(CO)_{12}$ butterfly fragment bridged across the wing tips by a BH_2 ligand. Both the geometric structure and the ^{11}B chemical shift ($\delta = 106$) suggest that this compound should be viewed as an iron boride rather than the "normal" ferraborane discussed above. This compound is related to tetrairon carbides, e.g., $CFe_4(CO)_{13}$ (248a) and $HFe_4(CH)(CO)_{12}$ (248b), and serves as a structural model for the interaction of CH_2 with a four-atom cluster. Although the electron counting rules (PSEPT) can easily account for the structure of $HFe_4(BH_2)(CO)_{12}$

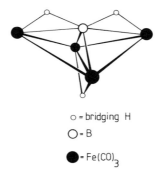

o = bridging H

O = B

● = Fe(CO)$_3$

FIG. 28.   The skeletal structure of HFe$_4$(BH$_2$)(CO)$_{12}$ (*III*).

[the boron is considered interstitial and the four-atom cluster arachno; two electrons for each Fe(CO)$_3$, three from boron, and one each from the bridging hydrogens], there is no borane analog of this compound. Hence its existence reflects cluster behavior restricted to the transition metals.

B. *The Transition from Complex to Cluster*

There are metalloboranes that can justifiably be described as metal–ligand complexes and there are metalloboranes that can only usefully be described as clusters. The series of compounds containing the B$_5$H$_8$ fragment shown in Figs. 5, 19, and 22 demonstrates the transition in the metal–boron interaction from one that is purely metal–ligand in character to one that is closely cluster-like. In Cp(CO)$_2$FeB$_5$H$_8$ the metal–boron interaction is simply a two-center two-electron σ bond. In L$_2$CuB$_5$H$_8$ the borane acts as a two-electron π donor to the metal or, alternatively, the metal fragment replaces a bridging proton in B$_5$H$_9$. Note that the metal is located well below the plane defined by the basal borons of the five-atom borane. In (CO)$_3$FeB$_5$H$_8$CuL$_2$ the boron–copper interaction is more complex. The copper is much closer to the basal boron plane but not within interaction distance of the apical boron. By adopting this geometry it is suggested that the borane is able to supply more than two electrons to the copper (see Section III,B,1). Finally, in (CO)$_3$CoB$_5$H$_8$ the metal clearly interacts with three borons and the metal is considered to be part of a six-atom cluster.

The ^{11}B NMR behavior (Table II) of the compounds containing the B$_5$H$_8$ fragment suggests an increasing perturbation of the borane fragment in going from Cp(CO)$_2$FeB$_5$H$_8$ to (CO)$_3$CoB$_5$H$_8$. In the first metalloborane only the ^{11}B shift of the boron to which the metal is attached is significantly perturbed. For the second compound the chemical shifts are virtually the same as in B_5H_9. For the third compound the chemical shifts have not yet been reported and in any case the interpretation would be confused by the iron atom in the cluster. Finally, in the fourth compound the shifts of all the basal borons are considerably perturbed relative to those for B_5H_9.

Apparently the structures of metalloboranes containing the B_5H_8 fragment also define intermediates on a real reaction pathway (50). As noted in Fig. 29, the formation of the σ-bonded metalloborane as well as the metalloborane in which the metal atom forms part of the cluster network can be considered to arise from the μ-bridged ("olefinic") $B_5H_8^-$ metal complex. The evidence for the scheme is circumstantial but persuasive. First, μ-bridged complexes of some metals have been isolated. For $M = R_3Ge$ or R_3Si the μ-isomer can be isolated and then converted into the 2-isomer (249). Finally, an unstable and uncharacterized intermediate has been reported in the formation of the "metallohexaborane" $L_2(CO)IrB_5H_8$ from $B_5H_8^-$ and $Ir(CO)ClL_2$ (50). This intermediate is suggested to be the μ-isomer.

Lastly, this series of compounds illustrates a type of valence tautomerism of the B_5H_8 fragment. Valence tautomerism in transition metal hydrocarbon complexes is well known, e.g., σ and π allyl, but the versatility displayed by B_5H_8 in this series in accommodating different metal requirements is unparalleled in organometallic chemistry.

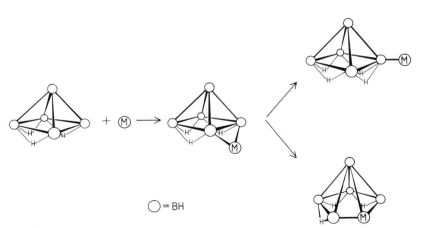

FIG. 29. Scheme for the conversion of a five-atom borane to three types of five-atom metalloboranes (50, 166).

V

METALLOBORANES AS CHEMICAL INTERMEDIATES

The usefulness of metalloboranes in synthetic procedures has not been established but there are indications that these compounds may well eventually be valued intermediates. In this section we sketch out reactions that are potentially useful in this regard.

Boranes as ligands (Section III) appear to be relatively weakly bound and under appropriate conditions are displaced by other Lewis bases. For example, the utilization of BH_4^- as a leaving group led to the preparation of the first reported carbonyl derivative of Zr [Eq. (23)] (77).

$$[Cp_2Zr(BH_4)]_2 \xrightarrow[CO]{Me_3N} Cp_2Zr(CO)_2 \qquad (23)$$

The displacement of the borallyl ligand led to the preparation of the first example of a tetrakis(trialkylphosphine)platinum(0) complex [Eq. (24)] ($132, 133$)

$$(R_3P)_2PtB_3H_7 + 2PR_3 \rightarrow Pt(PR_3)_4 \qquad (24)$$

and the displacement of $B_3H_8^-$ by phosphine has also been used to prepare derivatives difficult to synthesize by other routes [Eq. (25)] (235).

$$(CO)_4MoB_3H_8^- + 2PH_3 \rightarrow (CO)_4Mo(PH_3)_2 + B_3H_8^- \qquad (25)$$

The reaction of tetrahydroborates with transition metal complexes produces metal tetrahydroborates that in some cases decompose to transition metal hydrides [Eq. (26)] (90).

$$Ir(H)Cl_2[P-(t\text{-}Bu)_2Me]_2 \rightarrow Ir(H)_2(BH_4)[P-(t\text{-}Bu)_2Me]_2 \rightarrow Ir(H)_5[P-(t\text{-}Bu)_2Me]_2 \qquad (26)$$

Lewis bases may also be used to displace BH_3 from metal tetrahydroborates to produce metal hydrides (250).

The metal atom of some metalloboranes serves as a site for the insertion of alkynes into the borane fragment [Eqs. (27) and (28) and Section III,A,2] ($146, 152$).

$$2\text{-}CpCoB_4H_8 \xrightarrow{HC\equiv CH} 1\text{-}CpCoC_2B_3H_7 \qquad (27)$$

$$1\text{-}(CO)_3FeB_4H_8 + CH_3C\equiv CCH_3 \xrightarrow{h\nu} (CH_3)_4C_4B_4H_4 \qquad (28)$$

These reactions suggest ways of preparing heteroborane clusters that heretofore have been unavailable. Metalloboranes may also prove useful in fusing or coupling cages. Although this technique has been used successfully in coupling carboranes, so far the mercury(II) derivative of $B_5H_8^-$ has yielded only traces of coupled product on heating [Eq. (29)] (186).

$$B_5H_8HgB_5H_8 \xrightarrow{\Delta} (B_5H_8)_2 \qquad (29)$$

Finally, Eq. (22) demonstrates that a borane ligand can function in a reversible multidentate manner. Thus $(CO)_3MnB_3H_8$ (Fig. 18) is a compound that reversibly "reveals" an unsaturated metal site to potential ligands [Eq. (30)] where the brackets indicate B_3H_8 coordination as in Fig. 5 (*51*, *127*, *128*). As metal unsaturation is intimately

$$(CO)_3MnB_3H_8 \rightleftarrows [(CO)_3B_3H_8Mn] \overset{+L}{\underset{-L}{\rightleftharpoons}} (CO)_3LMnB_3H_8 \tag{30}$$

involved in catalytic cycles, metalloboranes may well be found to be active in this regard, making them comparable to metallocarboranes already known to function in this way (*251*).

VI

CONCLUSIONS

In this review we have tried to analyze the various faces that metalloboranes present to an observer. Clearly metalloboranes mimic very closely some aspects of metal–hydrocarbon π-complexes and much can be learned from the comparison of the two. But in the metal–hydroborates, boranes exhibit ligand behavior for which there are few hydrocarbon analogs. On the other hand, in contrast to metal–hydrocarbon complexes, the electropositive character of boron apparently permits borane fragments to simulate metals in terms of cluster behavior. This leads to classes of metalloboranes, the metal–boron interactions within which can only be realistically described as cluster-like. Thus metalloboranes serve as a real bridge between metal–hydrocarbon π-complexes and metal clusters. But they have the potential to be more than a simple conceptual tool. The chemistry of these materials, which is gradually being revealed, suggests some dimensions in reactivity beyond those of either metal–hydrocarbon complexes or metal clusters.

ACKNOWLEDGMENTS

We thank Professors S. G. Shore, John E. Bercaw, D. M. P. Mingos, T. R. Spalding, and K. Wade for information in advance of publication and the National Science Foundation for the generous support of our work in this area.

REFERENCES

1. W. N. Lipscomb, "Boron Hydrides." Benjamin, New York, 1963.
2. E. L. Muetterties, ed., "Boron Hydride Chemistry." Academic Press, New York, 1975.
3. H. C. Brown, "Hydroboration." Benjamin, New York, 1962.
4. K. Wade, "Electron Deficient Compounds." Nelson, London, 1971.

5. T. Onak, "Organoborane Chemistry." Academic Press, New York, 1975.
6. D. F. Gaines, M. W. Jorgenson, and M. A. Kulzick, *Chem. Commun.* p. 381 (1979).
7. P. Chini and B. T. Heaton, *Top. Curr. Chem.* **71,** 1 (1977).
8. F. Demartin, M. Manassero, M. Sansoni, L. Garlaschelli, C. Raimondi, S. Martinengo, and F. Canziani, *Chem. Commun.* p. 528 (1981).
9. P. L. Bellon, F. Cariati, M. Manassero, L. Naldini, and M. Sansoni, *Chem. Commun.* p. 1423 (1971).
10. K. Wade, *Adv. Inorg. Chem. Radiochem.* **18,** 1 (1976).
11. D. M. P. Mingos, *Nature (London) Phys. Sci.* **236,** 99 (1972).
12. R. N. Grimes, *Ann. N.Y. Acad. Sci.* **239,** 180 (1974).
13. R. W. Rudolph, *Acc. Chem. Res.* **9,** 446 (1976).
14. B. F. G. Johnson, ed., "Transition Metal Clusters." Wiley, New York, 1980.
15. P. Chini, *Inorg. Chim. Acta* **2,** 31 (1968).
16. E. L. Muetterties, T. N. Rhodin, E. Band, C. F. Brucker, and W. R. Pretzer, *Chem. Rev.* **79,** 91 (1979).
17. K. Wade, *Chem. Commun.* p. 792 (1971).
18. D. M. P. Mingos, *J. Chem. Soc., Dalton Trans.* p. 133 (1974).
19. J. W. Lauher, *J. Organomet. Chem.* **213,** 25 (1981).
20. G. Ciani and A. Sironi, *J. Organomet. Chem.* **197,** 233 (1980).
21. K. E. Inkrott and S. G. Shore, *Inorg. Chem.* **10,** 2817 (1979).
22. G. Schmid, *Angew. Chem., Int. Ed. Engl.* **9,** 819 (1970).
23. D. Seyferth, G. Raab, and O. S. Grimme, *J. Org. Chem.* **26,** 3034 (1961).
24. H. Nöth and G. Schmid, *Angew. Chem., Int. Ed. Engl.* **2,** 623 (1963).
25. R. K. Nanda and M. G. H. Wallbridge, *Inorg. Chem.* **3,** 1798 (1964).
26. N. N. Greenwood, *Pure Appl. Chem.* **49,** 791 (1977).
27. R. N. Grimes, *Acc. Chem. Res.* **11,** 420 (1978).
28. M. E. O'Neill and K. Wade, *in* "Metal Interactions with Boron Clusters" (R. N. Grimes, ed.), Chap. 1, p. 1. Plenum, New York, 1982.
29. M. F. Hawthorne, *J. Organomet. Chem.* **100,** 97 (1975).
30. M. F. Hawthorne and R. L. Pilling, *J. Am. Chem. Soc.* **87,** 3987 (1965).
31. N. N. Greenwood and I. M. Ward, *Chem. Soc. Rev.* **3,** 231 (1974).
32. R. N. Grimes, *Coord. Chem. Rev.* **28,** 47 (1979).
33. T. J. Marks and J. R. Kolb, *Chem. Rev.* **77,** 263 (1977).
34. R. W. Parry and G. Kodama, eds., "Boron Chemistry-4," Pergamon, Oxford, 1980.
35. R. N. Grimes, ed., "Metal Interactions with Boron Clusters." Plenum, New York, 1982.
36. R. N. Grimes, *in* "Organometallic Reactions and Syntheses" (E. I. Becker and M. Tsutsui, eds.), p. 63. Plenum, New York, 1977.
37. B. D. James and M. G. H. Wallbridge, *Prog. Inorg. Chem.* **11,** 99 (1970).
38. E. L. Muetterties, *Pure Appl. Chem.* **29,** 585 (1972).
39. *Gmelin, New Suppl. Seri.* **19,** Part 3, 185 (1975).
40. R. Snaith and K. Wade, *MTP Int. Rev. Sci.: Inorg. Chem., Ser. Two* p. 1 (1975).
41. D. F. Gaines, M. B. Fischer, S. J. Hildebrandt, J. A. Ulman, and J. W. Lott, *Adv. Chem. Ser.* **150,** 311 (1976).
42. L. J. Todd, *Adv. Chem. Ser.* **150,** 302 (1976).
43. S. G. Shore, *Pure Appl. Chem.* **49,** 717 (1977).
44. S. Heřmánek, J. Plešek, B. Štíbr, and Z. Janoušek, *Proc. ICCC, 19th, 1978* p. 35 (1978).
45. N. N. Greenwood, *Proc. ICCC, 19th, 1978* p. 71 (1978).
46. K. Niedenzu, *J. Organomet. Chem.* **147,** 63 (1978).
47. D. S. Matteson, *J. Organomet. Chem.* **98,** 295 (1975).
48. D. S. Matteson, *J. Organomet. Chem.* **147,** 17 (1978).
49. D. S. Matteson, *J. Organomet. Chem.* **180,** 19 (1979).

50. N. N. Greenwood, J. D. Kennedy, W. S. McDonald, D. Reed, and J. Staves, *J. Chem. Soc., Dalton Trans.* p. 117 (1979).
51. D. F. Gaines and S. J. Hildebrandt, *Inorg. Chem.* **17**, 795 (1978).
52. H. D. Johnson, R. A. Geanangel, and S. G. Shore, *Inorg. Chem.* **9**, 908 (1970).
53. V. T. Brice and S. G. Shore, *Chem. Commun.* p. 1312 (1970).
54. V. T. Brice and S. G. Shore, *J. Chem. Soc., Dalton Trans.* p. 334 (1975).
55. G. G. Outterson, Jr., V. T. Brice, and S. G. Shore, *Inorg. Chem.* **15**, 1456 (1976).
56. N. N. Greenwood, J. A. K. Howard, and W. S. McDonald, *J. Chem. Soc., Dalton Trans.* p. 37 (1977).
57. V. R. Miller and R. N. Grimes, *J. Am. Chem. Soc.* **95**, 5079 (1973).
58. V. R. Miller, R. Weiss, and R. N. Grimes, *J. Am. Chem. Soc.* **99**, 5646 (1977).
59. L. G. Sneddon and D. Voet, *Chem. Commun.* p. 118 (1976).
60. A. Almenningen, G. Gundersen, and A. Haaland, *Chem. Commun.* p. 557 (1967).
61. A. Almenningen, G. Gundersen, and A. Haaland, *Acta Chem. Scand.* **22**, 859 (1968).
62. T. H. Cook and G. L. Morgan, *J. Am. Chem. Soc.* **91**, 774 (1969).
63. D. R. Armstrong and P. G. Perkins, *Chem. Commun.* p. 352 (1968).
64. D. F. Gaines, J. L. Walsh, and D. F. Hillenbrand, *Chem. Commun.* p. 224 (1977).
65. G. Gunderson, L. Hedberg, and K. Hedberg, *J. Chem. Phys.* **59**, 3777 (1973).
66. J. Bicerano and W. N. Lipscomb, *Inorg. Chem.* **18**, 1565 (1979).
67. D. F. Gaines and J. L. Walsh, *Chem. Commun.* p. 482 (1976).
68. D. F. Gaines and J. L. Walsh, *Inorg. Chem.* **17**, 1238 (1978).
69. D. F. Gaines, J. L. Walsh, and J. C. Calabrese, *Inorg. Chem.* **17**, 1242 (1978).
70. P. C. Maybury and J. E. Ahnell, *Inorg. Chem.* **6**, 1286 (1967).
71. A. Almenningen, G. Gundersen, and A. Haaland, *Acta Chem. Scand.* **22**, 328 (1968).
72. A. C. Bond and F. L. Himpsl, Jr., *J. Am. Chem. Soc.* **99**, 6906 (1977).
73. H. Nöth and R. Rurländer, *Inorg. Chem.* **20**, 1063 (1981).
74. A. J. Downs, R. G. Egdell, A. F. Orchard, and P. D. P. Thomas, *J. Chem. Soc., Dalton Trans.* p. 1755 (1978).
75. A. J. Downs and P. D. P. Thomas, *Chem. Commun.* p. 825 (1976).
76. G. I. Mamaeva, I. Hargittai, and V. P. Spiridonov, *Inorg. Chim. Acta* **25**, L123 (1977).
77. G. Fachinetti, G. Fochi, and C. Floriani, *Chem. Commun.* p. 230 (1967).
78. V. V. Volkov, K. G. Myakishev, and G. I. Bagryanstev, *Russ. J. Inorg. Chem.* **15**, 996 (1970).
79. T. J. Marks and L. A. Shimp, *J. Am. Chem. Soc.* **94**, 1542 (1972).
80. P. H. Bird and M. R. Churchill, *Chem. Commun.* p. 403 (1967).
81. B. E. Smith, H. F. Shurvell, and B. D. James, *J. Chem. Soc., Dalton Trans.* p. 710, (1978).
82. R. K. Nanda and M. G. H. Wallbridge, *Inorg. Chem.* **3**, 1798 (1964).
83. R. V. Ammon, B. Kanellakopulos, G. Schmid, and R. D. Fischer, *J. Organomet. Chem.* **25**, Cl (1970).
84. T. J. Marks and W. J. Kennelly, *J. Am. Chem. Soc.* **97**, 1439 (1975).
85. H. Nöth and M. Seitz, *Chem. Commun.* p. 1004 (1976).
86. S. W. Kirtley, M. A. Andrews, R. Bau, C. W. Grynkewich, T. J. Marks, D. L. Tipton, and B. R. Whittlesey, *J. Am. Chem. Soc.* **99**, 7154 (1977).
87. P. H. Bird and M. G. H. Wallbridge, *Chem. Commun.* p. 687 (1968).
88. G. W. Parshall, *J. Am. Chem. Soc.* **86**, 361 (1964).
89. H. D. Empsall, E. Mentzer, and B. L. Shaw, *Chem. Commun.* p. 861 (1975).
90. H. D. Empsall, E. M. Hyde, E. Mentzer, B. L. Shaw, and M. F. Uttley, *J. Chem. Soc., Dalton Trans.* p. 2069 (1976).
91. T. Saito, M. Nakajima, A. Kobayashi, and Y. Sasaki, *J. Chem. Soc., Dalton Trans.* p. 482 (1978).

92. J. L. Atwood, R. D. Rogers, C. Kutal, and P. A. Grutsch, *Chem. Commun.* p. 593 (1977); C. Kutal, P. Grutsch, J. L. Atwood, and R. D. Roger, *Inorg. Chem.* **17**, 3558 (1978); F. Takusagawa, A. Fumagalli, T. P. Koetzle, S. G. Shore, T. Schmitkons, A. F. Fratini, K. W. Morse, C. Wei, and R. Bau, *J. Am. Chem. Soc.* **103**, 5165 (1981).
93. J. C. Bommer and K. W. Morse, *Chem. Commun.* p. 137 (1977).
94. S. J. Lippard and K. M. Melmed, *Inorg. Chem.* **6**, 2223 (1967).
95. M. Grace, H. Beau, and C. H. Bushweller, *Chem. Commun.* p. 701 (1970).
96. C. H. Bushweller, H. Beau, M. Grace, W. J. Dewkett, and H. S. Bilofsky, *J. Am. Chem. Soc.* **93**, 2145 (1971).
97. V. I. Mikheeva, N. N. Mal'tseva, and L. S. Alekseeva, *Russ. J. Inorg. Chem.* **13**, 682 (1968).
98. H. Nöth, E. Wiberg, and L. P. Winter, *Z. Anorg. Chem.* **370**, 209 (1969).
99. T. A. Keiderling, W. T. Wozniak, R. S. Gay, D. Jurkowitz, E. R. Bernstein, S. J. Lippard, and T. G. Spiro, *Inorg. Chem.* **14**, 576 (1975).
100. N. Davies, C. A. Smith, and M. G. H. Wallbridge, *J. Chem. Soc. A* p. 2601 (1969).
101. R. R. Rietz, A. Zalkin, D. H. Templeton, N. M. Edelstein, and L. K. Templeton, *Inorg. Chem.* **17**, 653 (1978).
102. A. Zalkin, R. R. Rietz, D. H. Templeton, and N. M. Edelstein, *Inorg. Chem.* **17**, 661 (1978).
103. R. R. Rietz, N. M. Edelstein, H. W. Ruben, D. H. Templeton, and A. Zalkin, *Inorg. Chem.* **17**, 658 (1978).
104. P. B. Armentrout and J. L. Beauchamp, *Inorg. Chem.* **18**, 1349 (1979).
105. P. Zanella, G. DePadi, G. Bombieri, G. Zanotti, and R. Rossi, *J. Organomet. Chem.* **142**, C21 (1977).
106. T. J. Marks and J. R. Kolb, *J. Am. Chem. Soc.* **97**, 27 (1975).
107. R. H. Banks, N. M. Edelstein, B. Spencer, D. H. Templeton, and A. Zalkin, *J. Am. Chem. Soc.* **102**, 620 (1980).
108. G. Schmid, V. Bätzel, G. Etzrodt, and R. Pfeil, *J. Organomet. Chem.* **86**, 257 (1975).
109. F. Klanberg, W. B. Askew, and J. L. Guggenberger, *Inorg. Chem.* **7**, 2265 (1968).
110. C. R. Eady, B. F. G. Johnson, and J. Lewis, *J. Chem. Soc., Dalton Trans.* p. 477 (1977).
111. K. S. Wong, W. R. Scheidt, and T. P. Fehlner, *J. Am. Chem. Soc.* **104**, 1111 (1982).
112. G. Medford and S. G. Shore, *J. Am. Chem. Soc.* **100**, 3953 (1978).
113. J. S. Plotkin and S. G. Shore, *J. Organomet. Chem.* **182**, C15 (1979).
114. S. G. Shore, J. S. Plotkin, J. C. Huffman, G. J. Long, T. P. Fehlner, and R. L. DeKock, *Abstr. Pap., 181st Natl. Meet., Am. Chem. Soc.* INORG 104 (1981).
115. J. E. Bercaw, private communication.
116. E. L. Andersen and T. P. Fehlner, *J. Am. Chem. Soc.* **100**, 4606 (1978).
117. E. L. Andersen and T. P. Fehlner, *Inorg. Chem.* **18**, 2325 (1979).
118. E. L. Andersen, R. L. DeKock, and T. P. Fehlner, *Inorg. Chem.* **20**, 3291 (1981).
119. H. D. Kaesz, W. Fellman, G. R. Wilkes, and L. F. Dahl, *J. Am. Chem. Soc.* **87**, 2756 (1965).
120. D. F. Gaines and J. H. Morris, *Chem. Commun.* p. 626 (1975).
121. J. C. Calabrese, D. F. Gaines, S. J. Hildebrandt, and J. H. Morris, *J. Am. Chem. Soc.* **98**, 5489 (1976).
122. D. F. Gaines, J. L. Walsh, J. H. Morris, and D. F. Hillenbrand, *Inorg. Chem.* **17**, 1516 (1978).
123. J. Borlin and D. F. Gaines, *J. Am. Chem. Soc.* **94**, 1367 (1972).
124. F. Klanberg, E. L. Muetterties, and L. J. Guggenberger, *Inorg. Chem.* **7**, 2272 (1968).
125. F. Klanberg and L. J. Guggenberger, *Chem. Commun.* p. 1293 (1967).
126. L. J. Guggenberger, *Inorg. Chem.* **9**, 367 (1970).

127. D. F. Gaines and S. J. Hildebrandt, *J. Am. Chem. Soc.* **96,** 5574 (1974).
128. S. J. Hildebrandt, D. F. Gaines, and J. C. Calabrese, *Inorg. Chem.* **17,** 790 (1978).
129. C. E. Housecroft and T. P. Fehlner, *Inorg. Chem.* **21,** 1739 (1982).
130. M. W. Chen, J. C. Calabrese, D. F. Gaines, and D. F. Hillenbrand, *J. Am. Chem. Soc.* **102,** 4928 (1980).
131. N. N. Greenwood, J. D. Kennedy, and D. Reed, *J. Chem. Soc., Dalton Trans.* p. 196 (1980).
132. A. R. Kane and E. L. Muetterties, *J. Am. Chem. Soc.* **93,** 1041 (1971).
133. L. J. Guggenberger, A. R. Kane, and E. L. Muetterties, *J. Am. Chem. Soc.* **94,** 5665 (1972).
134. S. J. Lippard and D. Ucko, *Chem. Commun.* p. 983 (1967).
135. S. J. Lippard and D. A. Ucko, *Inorg. Chem.* **7,** 1051 (1968).
136. H. Beall, C. H. Bushweller, and M. Grace, *Inorg. Nucl. Chem. Lett.* **7,** 641 (1971).
137. C. H. Bushweller, H. Beall, and W. J. Dewkett, *Inorg. Chem.* **15,** 1739 (1976).
138a. R. K. Hertz, R. Goetze, and S. G. Shore, *Inorg. Chem.* **18,** 2813, (1979).
138b. J. T. Gill and S. J. Lippard, *Inorg. Chem.* **14,** 751 (1975).
139. B. G. Cooksey, J. D. Gorham, J. H. Morris, and L. Kane, *J. Chem. Soc., Dalton Trans.* p. 141 (1978).
140. M. W. Chen, D. F. Gaines, and L. G. Hoard, *Inorg. Chem.* **19,** 2989 (1980).
141. E. L. Andersen, K. J. Haller, and T. P. Fehlner, *J. Am. Chem. Soc.* **101,** 4390 (1979).
142. K. J. Haller, E. L. Andersen, and T. P. Fehlner, *Inorg. Chem.* **20,** 309 (1981).
143. G. J. Zimmerman, L. W. Hall, and L. G. Sneddon, *Inorg. Chem.* **19,** 3642 (1980).
144. V. R. Miller, R. Weiss, and R. N. Grimes, *J. Am. Chem. Soc.* **99,** 5646 (1977).
145. J. R. Pipal and R. N. Grimes, *Inorg. Chem.* **16,** 3255 (1977).
146. R. Weiss, J. R. Bowser, and R. N. Grimes, *Inorg. Chem.* **17,** 1522 (1978).
147. F. L. Himpsl, Jr. and A. C. Bond, *J. Am. Chem. Soc.* **103,** 1098 (1981).
148. N. N. Greenwood, C. G. Savory, R. N. Grimes, L. G. Sneddon, A. Davison, and S. S. Wreford, *Chem. Commun.* p. 718 (1974).
149. J. A. Ulman, E. L. Andersen, and T. P. Fehlner, *J. Am. Chem. Soc.* **100,** 456 (1978).
150. D. R. Salahub, *Chem. Commun.* p. 385 (1978).
151. P. Brint and T. R. Spalding, *Inorg. Nucl. Chem. Lett.* **15,** 355 (1979).
152. T. P. Fehlner, *J. Am. Chem. Soc.* **102,** 3424 (1980).
153. S. K. Boocock, M. J. Toft, S. G. Shore, J. C. Huffman, and K. Folting, *Abstr. Pap., 182nd Natl. Meet., Am. Chem. Soc.* INORG 149 (1981).
154. K. E. Inkrott and S. G. Shore, *Chem. Commun.* p. 866 (1978).
155. J. R. Pipal and R. N. Grimes, *Inorg. Chem.* **18,** 252 (1979).
156a. V. R. Miller and R. N. Grimes, *J. Am. Chem. Soc.* **98,** 1600 (1976).
156b. J. R. Pipal and R. N. Grimes, *Inorg. Chem.* **16,** 3255 (1977).
157a. T. L. Venable and R. N. Grimes, *Inorg. Chem.* **21,** 887 (1982).
157b. T. L. Venable, E. Sinn, and R. N. Grimes, *Inorg. Chem.* **21,** 904 (1982).
157c. T. L. Venable, E. Sinn, and R. N. Grimes, *Inorg. Chem.* **21,** 895 (1982).
158. J. R. Pipal and R. N. Grimes, *Inorg. Chem.* **18,** 257 (1979).
159. D. N. Cox, D. M. P. Mingos, and R. Hoffmann, *J. Chem. Soc., Dalton Trans.* p. 1788 (1981).
160. J. R. Bowser and R. N. Grimes, *J. Am. Chem. Soc.* **100,** 4623 (1978).
161. J. R. Bowser, A. Bonny, J. R. Pipal, and R. N. Grimes, *J. Am. Chem. Soc.* **101,** 6229 (1979).
162. D. F. Gaines, K. M. Coleson, and J. C. Calabrese, *J. Am. Chem. Soc.* **101,** 3979 (1979).
163. D. F. Gaines, K. M. Coleson, and J. C. Calabrese, *Inorg. Chem.* **20,** 2185 (1981).
164. M. B. Fischer and D. F. Gaines, *Inorg. Chem.* **18,** 3200 (1979).

165. D. F. Gaines and T. V. Iorns, *Inorg. Chem.* **7,** 1041 (1968).
166. N. N. Greenwood, J. D. Kennedy, C. G. Savory, J. Staves, and K. R. Trigwell, *J. Chem. Soc., Dalton Trans.* p. 237 (1978).
167. S. G. Shore, J. Ragaini, T. Schmitkons, L. Barton, G. Medford, and J. Plotkin, *Abstr. IMEBORON, 4th, 1979* p. 36 (1979).
168. T. P. Fehlner, J. Ragaini, M. Mangion, and S. G. Shore, *J. Am. Chem. Soc.* **98,** 7085 (1976).
169. S. G. Shore, J. D. Ragaini, R. L. Smith, C. E. Cottrell, and T. P. Fehlner, *Inorg. Chem.* **18,** 670 (1979).
170. P. Brint and T. R. Spalding, *J. Chem. Soc., Dalton Trans.* p. 1236 (1980).
171. P. Brint, W. K. Pelin, and T. R. Spalding, *J. Chem. Soc., Dalton Trans.* p. 546 (1981).
172. R. Weiss and R. N. Grimes, *Inorg. Chem.* **18,** 3291 (1979).
173. R. Weiss and R. N. Grimes, *J. Am. Chem. Soc.* **99,** 8087 (1977).
174. J. A. Ulman and T. P. Fehlner, *Chem. Commun.* p. 632 (1976).
175. W. K. Pelin, T. R. Spalding, and R. P. Brint, private communication.
176. R. W. Wilczynski and L. G. Sneddon, *Inorg. Chem.* **18,** 864 (1979).
177. S. G. Shore, private communication.
178. M. R. Churchill, J. J. Hackbarth, A. Davison, D. D. Traficante, and S. S. Wreford, *J. Am. Chem. Soc.* **96,** 4041 (1974).
179. M. R. Churchill and J. J. Hackbarth, *Inorg. Chem.* **14,** 2047 (1975).
180. N. N. Greenwood and J. Staves, *J. Chem. Soc., Dalton Trans.* p. 1788 (1977).
181. A. Davison, D. D. Traficante, and S. S. Wreford, *J. Am. Chem. Soc.* **96,** 2802 (1974).
182. A. B. Burg and F. B. Mishra, *J. Organomet. Chem.* **24,** C33 (1970).
183. N. N. Greenwood, J. D. Kennedy, and J. Staves, *J. Chem. Soc., Dalton Trans.* p. 1146 (1978).
184. N. N. Greenwood and J. Staves, *J. Chem. Soc., Dalton Trans.* p. 1144 (1978).
185. N. N. Greenwood and J. Staves, *J. Chem. Soc., Dalton Trans.* p. 1786 (1977).
186. N. S. Hosmane and R. N. Grimes, *Inorg. Chem.* **18,** 2886 (1979).
187. M. Mangion, J. D. Ragaini, T. A. Schmitkons, and S. G. Shore, *J. Am. Chem. Soc.* **101,** 754 (1979).
188. L. W. Hall, G. J. Zimmerman, and L. G. Sneddon, *Chem. Commun.* p. 45 (1977).
189. D. L. Denton, W. R. Clayton, M. Mangion, S. G. Shore, and E. A. Meyers, *Inorg. Chem.* **15,** 541 (1976).
190. R. J. Remmel, D. L. Denton, J. B. Leach, M. A. Toft, and S. G. Shore, *Inorg. Chem.* **20,** 1270 (1981).
191. S. G. Shore, *Proc. Int. Conf. Coord. Chem., 19th, 1978* Vol. 1, p. 86 (1978).
192. A. Davison, D. D. Traficante, and S. S. Wreford, *Chem. Commun.* p. 1155 (1972).
193. A. Davison, D. D. Traficante, and S. S. Wreford, *J. Am. Chem. Soc.* **96,** 2802 (1974).
194. J. P. Brennan, R. Schaeffer, A. Davison, and S. S. Wreford, *Chem. Commun.* p. 354 (1973).
195. N. N. Greenwood, M. J. Hails, J. D. Kennedy, and W. S. McDonald, *Chem. Commun.* p. 37 (1980).
196. M. Mangion, W. R. Clayton, O. Hollander, and S. G. Shore, *Inorg. Chem.* **16,** 2110 (1977).
197. J. C. Calabrese, M. B. Fischer, D. F. Gaines, and J. W. Lott, *J. Am. Chem. Soc.* **96,** 6318 (1974).
198. S. K. Boocock, N. N. Greenwood, M. J. Hails, J. D. Kennedy, and W. S. McDonald, *J. Chem. Soc., Dalton Trans.* p. 1415 (1981).
199. J. W. Lott, D. F. Gaines, H. Shenhaw, and R. Schaeffer, *J. Am. Chem. Soc.* **95,** 3042 (1973).
200. G. J. Zimmerman, L. W. Hall, and L. G. Sneddon, *Inorg. Chem.* **19,** 3642 (1980).

201. V. R. Miller, R. Weiss, and R. N. Grimes, *J. Am. Chem. Soc.* **99**, 5646 (1977).
202. J. R. Pipal and R. N. Grimes, *Inorg. Chem.* **16**, 3251 (1977).
203. R. N. Leyden and M. F. Hawthorne, *Chem. Commun.* p. 310 (1975).
204. N. N. Greenwood and J. A. McGinnety, *J. Chem. Soc. A* p. 1090 (1966).
205. N. N. Greenwood, B. S. Thomas, and D. W. Waite, *J. Chem. Soc., Dalton Trans.* p. 299 (1975).
206. N. N. Greenwood and J. A. K. Howard, *J. Chem. Soc., Dalton Trans.* p. 177 (1976).
207. F. Sato, T. Yamamoto, J. R. Wilkinson, and L. J. Todd, *J. Organomet. Chem.* **86**, 243 (1975).
208. A. R. Siedle, *J. Organomet. Chem.* **97**, C4 (1975).
209. F. Klanberg, P. A. Wegner, G. W. Parshall, and E. L. Muetterties, *Inorg. Chem.* **7**, 2072 (1968).
210. L. J. Guggenberger, *J. Am. Chem. Soc.* **94**, 114 (1972).
211. T. E. Paxson and M. F. Hawthorne, *Inorg. Chem.* **14**, 1604 (1975).
212. N. N. Greenwood, J. A. McGinnety, and J. D. Owen, *J. Chem. Soc. A* p. 809 (1971); N. N. Greenwood and N. F. Traver, *J. Chem. Soc. A* p. 3257 (1971).
213. D. F. Gaines and G. A. Steehler, *Chem. Commun.* **122**, (1982).
214. B. P. Sullivan, R. N. Leyden, and M. F. Hawthorne, *J. Am. Chem. Soc.* **97**, 456 (1975); R. N. Leyden, B. P. Sullivan, R. T. Baker, and M. F. Hawthorne, *J. Am. Chem. Soc.* **100**, 3758 (1978).
215a. R. L. Sneath, J. L. Little, A. R. Burke, and L. J. Todd, *Chem. Commun.* p. 693 (1970).
215b. S. K. Boocock, N. N. Greenwood, J. D. Kennedy, W. S. McDonald, and J. Staves, *J. Chem. Soc. Dalton* p. 2573 (1981).
216. K. S. Wong and T. P. Fehlner, *J. Am. Chem. Soc.* **103**, 966 (1981).
217. T. K. Hilty, D. A. Thompson, W. M. Butler, and R. W. Rudolph, *Inorg. Chem.* **18**, 2642 (1979).
218. S. D. Ittel and J. A. Ibers, *Adv. Organomet. Chem.* **14**, 33 (1976).
219. D. M. P. Mingos, *Adv. Organomet. Chem.* **15**, 1 (1977).
220. T. A. Albright, R. Hoffman, J. C. Thibeault, and D. L. Thorn, *J. Am. Chem. Soc.* **101**, 3801 (1979).
221. I. R. Epstein, J. A. Tossell, E. Switkes, R. M. Stevens, and W. N. Lipscomb, *Inorg. Chem.* **10**, 171 (1971).
222. B. E. R. Schilling, R. Hoffmann, and J. W. Faller, *J. Am. Chem. Soc.* **101**, 592 (1979).
223. I. M. Pepperberg, T. A. Halgren, and W. N. Lipscomb, *Inorg. Chem.* **16**, 363 (1977).
224. S. G. Shore, *in* "Boron Hydride Chemistry" (E. L. Muetterties, ed.), p. 120. Academic Press, New York, 1975.
225. R. Pettit, *J. Organomet. Chem.* **100**, 205 (1975).
226. L. J. Schaad, B. A. Hess, Jr., and C. S. Ewig, *J. Am. Chem. Soc.* **101**, 2181 (1979).
227. R. L. DeKock and T. P. Fehlner, unpublished data.
228. B. E. Bursten and R. F. Fenske, *Inorg. Chem.* **18**, 1760 (1979).
229. M. R. Churchill and R. Mason, *Adv. Organomet. Chem.* **5**, 93 (1967).
230. P. J. Fitzpatrick, Y. LePage, J. Sedman, and I. S. Butler, *Inorg. Chem.* **20**, 2852 (1981).
231. J. S. Ward and R. Pettit, *J. Am. Chem. Soc.* **93**, 262 (1971).
232. G. R. Eaton and W. N. Lipscomb, "NMR Studies of Boron Hydrides and Related Compounds." Benjamin, New York, 1969.
233. P. Jolly and R. Mynott, *Adv. Organomet. Chem.* **19**, 257 (1981).
234. R. E. Williams and L. D. Field, *in* "Boron Chemistry-4" (R. W. Parry and G. Kodama, eds.), p. 131. Pergamon, Oxford, 1980.
235. P. A. Wegner, *in* "Boron Hydride Chemistry" (E. L. Muetterties, ed.), p. 431. Academic Press, New York, 1975.

236. I. M. Pepperberg, D. A. Dixon, W. N. Lipscomb, and T. A. Halgren, *Inorg. Chem.* **17,** 587 (1978).
237. T. P. Fehlner, unpublished data.
238. A. Salzer and H. Werner, *Angew. Chem., Int. Ed. Engl.* **11,** 930 (1972).
239. W. Siebert, *Adv. Organomet. Chem.* **18,** 301 (1980).
240. J. C. Jeffry, K. A. Mead, H. Razay, F. G. A. Stone, M. J. Went, and P. Woodward, *Chem. Commun.* p. 867 (1981).
241. A. F. Dyke, J. E. Guerchais, S. A. R. Knox, J. Roué, R. L. Short, T. E. Taylor, and P. Woodward, *Chem. Commun.* p. 537 (1981).
242. C. R. Eady, B. F. G. Johnson, J. Lewis, R. Mason, P. B. Hitchcock, and K. M. Thomas, *Chem. Commun.* p. 385 (1977).
243. D. M. P. Mingos and M. I. Forsyth, *J. Chem. Soc., Dalton Trans.* p. 610 (1977).
244. E. L. Muetterties, E. L. Hoel, C. G. Salentine, and M. F. Hawthorne, *Inorg. Chem.* **14,** 950 (1975).
245. M. E. O'Neill and K. Wade, *Inorg. Chem.* **21,** 461 (1982).
246. R. B. King, *Polyhedron* **1,** 132 (1982).
247. M. Tachikawa and E. L. Muetterties, *J. Am. Chem. Soc.* **100,** 4606 (1978).
248a. J. S. Bradley, G. B. Anell, M. E. Leonowicz, and E. W. Hill, *J. Am. Chem. Soc.* **103,** 4968 (1981).
248b. M. A. Beno, J. M. Williams, M. Tachikawa, and E. L. Muetterties, *J. Am. Chem. Soc.* **103,** 1485 (1981).
249. D. F. Gaines and T. V. Iorns, *J. Am. Chem. Soc.* **90,** 6617 (1968).
250. B. D. James, R. K. Nanda, and M. G. H. Wallbridge, *Inorg. Chem.* **6,** 1979 (1967).
251. M. S. Delaney, C. B. Knobler, and M. F. Hawthorne, *Inorg. Chem.* **20,** 3124 (1981).

ADVANCES IN ORGANOMETALLIC CHEMISTRY, VOL. 21

Mechanistic Pathways for Ligand Substitution Processes in Metal Carbonyls

DONALD J. DARENSBOURG*

Department of Chemistry
Tulane University
New Orleans, Louisiana

I

INTRODUCTION

During the past decade soluble transition metal catalysts have become increasingly more important in a wide range of reactions; included in these are the "oxo process" or hydroformylation, olefin metathesis or isomerization, hydrogenation, polymerization, and oligomerization (1, 2). In all of these processes coordination of the substrate to the transition metal is an essential step. However, before coordination of the substrate can occur, a vacant site must be available in the coordination sphere of the metal (3). In many cases, the vacant site or free coordination site is created in solution by dissociation of a bonded ligand either thermally or photochemically promoted. In this connection knowledge of ligand substitution reactions on low-valent saturated metal centers, whether these be mononuclear metal complexes or metal clusters, remains a basic requirement for the proper design and understanding of homogeneous catalytic processes.

---

* Present address: Department of Chemistry, Texas A & M University, College Station, Texas 77843.

Kinetic studies of ligand substitution reactions of metal carbonyl derivatives have played a particularly important role in the elucidation of the details of reactions in general involving the displacement of neutral ligands (e.g., CO, phosphines, olefins, and amines) from metal centers in low oxidation states. The generalized reaction for ligand substitution entailing derivatives of metal carbonyls may be represented by Eq. (1). Basolo (4) has recently published a very interesting chronological survey of the mechanisms of carbon monoxide replacement reactions in metal carbonyls. In addition, ligand substitution reactions in metal carbonyls have been the subject of excellent reviews (e.g., see Refs. 5–9). Hence, the object of this particular article is not to provide a comprehensive coverage of all aspects of the subject. Instead, this article will focus on the basic mechanistic chemistry of ligand substitution reactions of octahedral metal carbonyl complexes with special emphasis being placed on stereochemical considerations. Many of the recent developments in this area have come about because of the availability of stable isotopes of carbon and oxygen at significantly reduced prices, along with rapid synergistic advances in instrumentation (in particular, Fourier transform nuclear magnetic resonance spectroscopy).

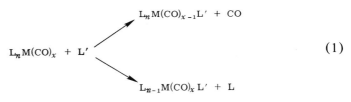

$$L_nM(CO)_x + L' \begin{cases} L_nM(CO)_{x-1}L' + CO \\ L_{n-1}M(CO)_x L' + L \end{cases} \quad (1)$$

II

CLASSIFICATION OF LIGAND SUBSTITUTION MECHANISMS

Langford and Gray (10) have classified the three possible intimate mechanisms for ligand substitution processes as (a) dissociative (D), (b) associative (A), and (c) interchange (I). The latter designation may be further subdivided into I_d and I_a, depending on the extent to which both the entering and leaving groups participate in the transition state.

A. *The Dissociative Pathway*

The dissociative mechanism [Eqs. (2) and (3)] that affords an intermediate of reduced coordination number, is by far the most common pathway for substitution reactions involving 18-electron complexes. The rate constant expression depicted in Eq. (4) reduces to $k_{obs} = k_1$ for the condition that is usually satisfied, $k_{-1}[A] \ll k_0[B]$. That is, reactions are generally carried out in solutions where the concentration of the incoming ligand (B) exceeds that of the leaving ligand (A). Both ligand competition and flash photolysis studies have shown that the 16-electron intermediates are highly reactive, and hence exhibit little selectivity in achieving a coordinatively saturated species ($k_{-1} \approx k_0$). This is illustrated in Table I which contains a compilation of ligand-competition studies, where k_{-1}/k_0 ratios have been determined for various pentacoordinate intermediates reacting with a range of incoming ligands (11–15). Similar ligand-competition ratios indicative of nondiscriminating transients have been observed for the tricoordinate intermediate [Ni(CO)$_3$] (16), the tetracoordinate intermediates [Fe(CO)$_3$C(OC$_2$H$_5$)CH$_3$] (17) and [Fe(CO)$_4$] (18), and intermediates arising from CO dissociation in Co$_4$ (19) and Ru$_3$ (20) carbonyl clusters.

$$M(L)_nA \underset{k_{-1}}{\overset{k_1}{\rightleftarrows}} M(L)_n + A \tag{2}$$

$$M(L)_n + B \underset{\text{rapid}}{\overset{k_0}{\longrightarrow}} M(L)_nB \tag{3}$$

TABLE I

COMPETITION RATIOS FOR THE COMBINATION OF FIVE-COORDINATE INTERMEDIATES
WITH VARIOUS LEWIS BASES

Intermediate	Departing ligand (A)	Entering ligand (B)	k_{-1}/k_0	Reference
Mo(CO)$_5$[a]	Piperidine	Ph$_3$As	1.00	11
		Ph$_3$P	0.85	11
		(MeO)$_3$P	0.74	11
Mo(CO)$_4$PPh$_3$[a]	Piperidine	CO	3.24	12
		Ph$_3$As	2.13	12
		(n-Bu)$_3$P	1.47	12
		Ph$_3$Sb	0.90	12
		Ph$_3$P	0.68	12
		C$_2$H$_5$C(CH$_2$O)$_3$P	0.42	12
V(CO)$_5^-$ [b]	PPh$_3$	(PhO)$_3$P	0.78	13
(Phen)Cr(CO)$_3$[c]	PPh$_3$	(n-Bu)$_3$P	1.0	14
		(PhO)$_3$P	1.0	14

[a] Hexane solvent.
[b] Tetrahydrofuran solvent.
[c] Dichloroethane solvent.

$$k_{obs} = \frac{k_1 k_0 [B]}{k_{-1}[A] + k_0 [B]} \qquad (4)$$

Dobson (14) has recently obtained parallel ligand-competition results and pulsed laser flash photolysis data for the reaction of the five-coordinate intermediate, (phen)Cr(CO)$_3$, with phosphine and phosphite ligands [Eq. (5)]. On the basis of "competition ratios" at several temperatures of 1.0 and identical bimolecular rate constants, $3.1 \times 10^6 \; M^{-1} \; sec^{-1}$ (versus the diffusion-control rate constant of $7.5 \times 10^9 \; M^{-1} \; sec^{-1}$), he concludes that the (phen)Cr(CO)$_3$ intermediates produced thermally and photochemically are the same and that by Hammond's postulate (21), the (phen)Cr(CO)$_3$ intermediate and the transition state leading to its formation are very similar in nature. It also follows that the products of ligand dissociation, the pentacoordinate 16-electron intermediates, in related systems such as those provided in Table I are the best models for the transition-state structures. More will be said about the structural and fluxional features of these pentacoordinate intermediates in a later section.

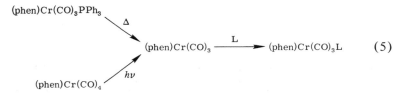

$$\qquad (5)$$

In general the rates of carbon monoxide dissociation from group VIB metal carbonyl compounds follow the order Mo > Cr > W, and similarly for the group VIIB metal carbonyl compounds where Mn > Re. Table II contains representative data, along with activation parameters, illustrating this metal dependence. The enthalpy and entropy of activations listed in Table II are consistent with metal–ligand bond-breaking processes. It is also evident from the rate data provided in Table II that metal–CO bond lability is extremely sensitive to the nature of the ligands in the coordination sphere of the metal. Substitution of the carbon monoxide ligand by "hard" Lewis bases (28) greatly accelerates the rates of M—CO bond cleavage. The origin of this phenomenon will be discussed in detail in a later section of this article.

There are several systematic rate studies for dissociation of ligands other than carbon monoxide in group VIB metal carbonyl derivatives (29–35). One such investigation where ligand steric effects are minimal has been carried out for the reaction described in Eq. (6) (34, 36). The rate parameters determined for this substitution reaction (collected in Table III) are

TABLE II

RATE PARAMETERS FOR CO DISSOCIATION IN GROUP VIB
AND VIIB CARBONYL COMPOUNDS[a]

Compound	Solvent	k_1 (sec^{-1})	ΔH^* (kcal mol^{-1})	ΔS^* (eu)	Reference
$Cr(CO)_6$	Decalin	1×10^{-12}	40.2	·22.6	22
$Cr(CO)_5PR_2R'$	Octane	1.5×10^{-10}	32.3	6.4	23[b]
$Cr(CO)_5Cl^-$	Diglyme	1.5×10^{-4}	—	—	24
$Mo(CO)_6$	Decalin	5×10^{-10}	31.7	6.7	22
$Mo(CO)_5PR_2R'$	Octane	2×10^{-8}	28.7	3.4	25[b]
$Mo(CO)_5Cl^-$	Diglyme	$>10^{-3}$	—	—	24
$W(CO)_6$	Decalin	1×10^{-14}	39.9	13.8	22
$W(CO)_5PR_2R'$	Nonane	4×10^{-13}	37.0	9.6	23[b]
$W(CO)_5Cl^-$	Diglyme	5×10^{-5}	—	—	24
$Mn_2(CO)_{10}$[c]	p-Xylene	1×10^{-11}	37.0	—	26
$Mn(CO)_5Br$[c]	Hexane	2.8×10^{-5}	29.2	—	27
$Re_2(CO)_{10}$[d]	Decalin	$<10^{-13}$	—	—	26
$Re(CO)_5Br$[d]	Hexane	5×10^{-7}	—	—	27

[a] Taken from tables in Ref. 9. Group VIb metal carbonyl rate data at 30°C. Original data are to be found in the references cited in the above table.

[b] $PR_2R' = Ph_2PCH_2CH_2PPh_2$.

[c] Data at 23°C.

[d] Data at 30°C.

congruous with rate-determining dissociation of the ligand L and reflects the Cr—L binding strengths. This latter conclusion is based on the fact that the common intermediate in these reactions, $Cr(CO)_5$, closely resembles the transition state (*vide supra*) as depicted in Fig. 1. The order of ligand lability from Eq. (6) was found to be py > AsPh$_3$ > CO $\approx$ PPh$_3$

TABLE III

RATE PARAMETERS FOR DISSOCIATION OF L IN $Cr(CO)_5L$ COMPOUNDS AT 130°C

L	$k \times 10^6$ (sec^{-1})	ΔH^* (kcal mol^{-1})	ΔS^* (eu)	Reference
Pyridine	1.34×10^6	25.4 ± 1.1	3.1 ± 3.5	36
Ph$_3$As	8600	35.3 ± 0.4	22.2 ± 1.2	34
CO[a]	106	40.2 ± 0.6	22.6 ± 1.5	22
Ph$_3$P	99.7	36.3 ± 1.9	12.5 ± 4.8	34
(PhO)$_3$P	15.6	31.9 ± 1.1	-2.0 ± 2.6	34
(CH$_3$O)$_3$P	0.548	—	—	34

[a] Reaction temperature 129.2°C.

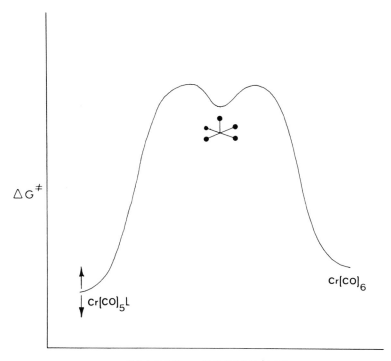

REACTION COORDINATE

FIG.'1. Plot of free energy versus reaction coordinate for the reaction of $Cr(CO)_5L$ + CO → $Cr(CO)_6$ + L.

> $P(OC_6H_5)_3$ > $P(OCH_3)_3$ > $P(n\text{-}Bu)_3$, with $Cr(CO)_5P(n\text{-}Bu)_3$ existing in equilibrium with $Cr(CO)_6$ (K_{eq} = 4.2 × 10^{-5} at 130°C). A similar less dramatic trend in Cr—L bond strengths was noted from an investigation of ligand dissociation from $(phen)Cr(CO)_3L$ [Eq. (7)] that provided an order of Cr—L bond strengths:

$$Cr(CO)_5L + CO \rightarrow Cr(CO)_6 + L \tag{6}$$

$$(phen)Cr(CO)_3L + CO \rightarrow (phen)Cr(CO)_4 + L \tag{7}$$

$P(n\text{-}Bu)_3$ > $P(OEt)_3$ > CO > $P(OPh)_3$ > PPh_3. These kinetically determined metal–ligand binding strengths are qualitatively observed in a variety of low-valent mononuclear and cluster metal carbonyl derivatives where the stronger metal–phosphorus bonds involve good sigma donor or good π-acceptor ligands. X-Ray structural data for M—P bond distances are supportive of these conclusions (37).

Ligand steric requirements, as semiquantitatively defined by their cone angles (*38*) (in general obtained from molecular models), can often be of major importance in determining the rates of ligand dissociation in complexes where steric congestion is large. We have for some time maintained an interest in steric effects on the rates of ligand substitution processes in metal carbonyl derivatives (*32, 33*). For example, investigations involving *cis*-$Mo(CO)_4L_2$ derivatives [Eq. (8)] show steric interactions to be of paramount importance in determining rates of phosphorus ligand dissociation (see Table IV). Indeed X-ray structural data indicate significant steric crowding when L is a bulky phosphorus ligand (*40*). This is illustrated in an ORTEP drawing of the *cis*-$Mo(CO)_4[PPh_3]_2$ compound in Fig. 2, where the P—Mo—P angle was found to be 104.6°. This represents a gross distortion of the idealized octahedral geometry, presumably caused by interligand steric interactions. In contrast, in the *cis*-$Mo(CO)_4[PPh_2Me]_2$ derivative, where the two PPh_2Me ligands are able to orient themselves so as to minimize steric interactions (see Fig. 3), the P—Mo—P angle was determined to be 92.5°. As listed in Table IV, the rates of dissociative loss of PPh_3 and PPh_2Me from the corresponding *cis*-$Mo(CO)_4L_2$ compounds are 3.16×10^{-3} sec^{-1} and 1.33×10^{-5} sec^{-1}, respectively. Although the Mo—P distances of 2.555(10) Å in *cis*-$Mo(CO)_4[PPh_2Me]_2$ and 2.577(2) Å in *cis*-$Mo(CO)_4[PPh_3]_2$ indicate some of the decrease in lability in the PPh_2Me derivative to be electronic in nature, most of this decrease is believed to be caused by steric considerations. Nevertheless, electronic ef-

TABLE IV

RATES OF LIGAND (L) REPLACEMENT BY CO AT 70°C IN CO-SATURATED TETRACHLOROETHYLENE FOR *cis*-$Mo(CO)_4L_2$ DERIVATIVES

L	Cone angle (deg)[a]	Rate (sec^{-1})	ΔH^* (kcal mol^{-1})	ΔS^* (eu)
$PN_3(CH_2)_6$	102[b]	$<1.0 \times 10^{-6}$		
PMe_2Ph	122	$<1.0 \times 10^{-6}$		
$PMePh_2$	136	1.33×10^{-5}		
PPh_3	145	3.16×10^{-3}	29.7(5)	14.2(14)
$PPhCy_2$	162[c]	6.40×10^{-2}	30.2(16)	21.7(39)
$P(OCH_2)_3CEt$	101	$<1.0 \times 10^{-6}$		
$P(OPh)_3$	128	$<1.0 \times 10^{-5}$		
$P(O\text{-}o\text{-tol})_3$	141	1.60×10^{-4}	31.9(11)	14.4(35)

[a] Taken from Ref. *38* unless otherwise noted.

[b] R. J. DeLerno, L. M. Trefonas, M. Y. Darensbourg, and R. J. Majeste, *Inorg. Chem.* **15**, 816 (1976).

[c] Taken from Ref. *32*.

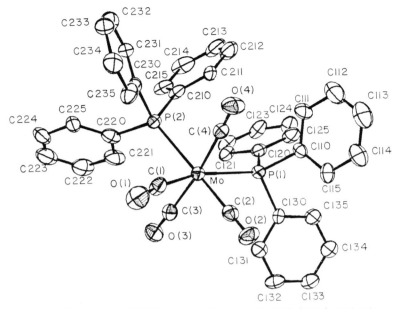

FIG. 2. Perspective ORTEP drawing of the *cis*-Mo(CO)₄(PPh₃)₂ molecule.

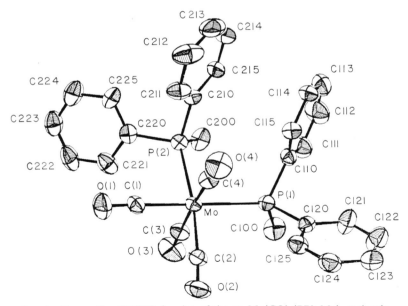

FIG. 3. Perspective ORTEP drawing of the *cis*-Mo(CO)₄(PPh₂Me)₂ molecule.

fects are superimposed on steric interactions in determining the rates of phosphorus ligand dissociation in cis-Mo(CO)$_4$L$_2$ species are seen in Table IV where phosphites of similar cone angle to phosphines (but having greater π-acceptor abilities) dissociate more slowly. The origin of the steric accelerations is often of a transition-state nature where the relief of interligand steric repulsions upon ligand dissociation allows the remaining metal–ligand bonds to attain values closer to their electronic equilibrium positions.

$$cis\text{-Mo(CO)}_4\text{L}_2 + \text{CO} \rightarrow \text{Mo(CO)}_5\text{L} + \text{L} \qquad (8)$$

Similarly, comparative kinetic studies of either cis-CO labilization or phosphine dissociation in W(CO)$_5$PMe$_3$ and W(CO)$_5$P(t-Bu)$_3$ derivatives demonstrate both processes to be greatly enhanced in the sterically demanding P(t-Bu)$_3$ species (41). These observations are consistent with the solid-state structures of these complexes, where the W—P bond distance is some 0.170(6) Å longer in the P(t-Bu)$_3$ derivative and the P—W—C$_{eq}$ angles average 96.6° (42, 43). The Tolman cone angles for these two phosphine ligands which are of similar basicities, PMe$_3$ and P(t-Bu)$_3$, are 118 and 182°, respectively (38). It should be noted parenthetically that in employing cone angles as defined by Tolman (38), it is important to consider ligand intermeshing as illustrated by Clark (44).

A kinetic study of ligand dissociation from a variety of $trans$-Cr(CO)$_4$L$_2$ derivatives has been reported that was designed to minimize interligand steric interactions and maximize electronic effects (35). The activation parameters for the process described in Eq. (9) (see Table V) are indicative of a rate-determining dissociation of the ligand L to afford the 16-electron Cr(CO)$_4$L intermediate ($vide\ infra$). The order of ligand dissociation, AsPh$_3$ > PPh$_3$ > P(n-Bu)$_3$ > P(OPh)$_3$ > CO > P(OCH$_3$)$_3$, is similar to that observed by these researchers for ligand dissociation in Cr(CO)$_5$L

TABLE V

ACTIVATION PARAMETERS FOR THE DISSOCIATION
OF L FROM $trans$-Cr(CO)$_4$L$_2$ DERIVATIVES[a]

L	ΔH^* (kcal mol^{-1})	ΔS^* (eu)
P(OPh)$_3$	37.6(10)	18.6(26)
P(n-Bu)$_3$	42.5(18)	33.2(45)
P(OCH$_3$)$_3$	43.4(6)	25.5(16)
PPh$_3$[b]	31.3(12)	21.2(36)

[a] Taken from Ref. 35.

[b] J. D. Atwood and M. J. Wovkulich, personal communication.

(*34*). The exception in the ordering of ligand dissociation in the two studies [Eqs. (6) and (9)], $P(n\text{-Bu})_3$, was explained on the basis of a ground-state bond weakening in the $trans\text{-}Cr(CO)_4[P(n\text{-Bu})_3]_2$ species due to excessive buildup of σ-electron density along the $P{-}Cr{-}P$ axis. The π-bonding capability of the trans disposed ligands (L) was thought to be important in determining the rates of ligand dissociation in these bis chromium derivatives.

$$trans\text{-}Cr(CO)_4L_2 + CO \rightarrow Cr(CO)_5L + L \qquad (9)$$

Pi-Bonding effects have also been used to account for the increase in PPh_3 lability in $V(CO)_5PPh_3^-$ in the presence of sodium cations (*46*). Ion pairing between Na^+ and the axial CO ligand in $V(CO)_5L^-$ derivatives has been established via infrared spectroscopy in tetrahydrofuran solvent (*46, 47*). Hence, the faster rate of PPh_3 dissociation was suggested to be due to a loss of $V \rightarrow P$ π-back bonding in the presence of trans $CO \cdot \cdot \cdot Na^+$, a ligand that competes better for metal d electrons than does unperturbed CO. On the other hand in the $V(CO)_5P(n\text{-Bu})_3^-$ species, where π bonding between phosphorus and vanadium is expected to be minimal, no significant rate enhancement was observed in the presence of Na^+ ions although a great deal of ion pairing existed. As provided in Table I, the anionic five-coordinate $V(CO)_5^-$ intermediate in these ligand dissociation reactions exhibited little discrimination in its reactivity toward incoming Lewis bases, a feature in common with its neutral analogs. Parallel ion pairing effects are responsible for the large rate acceleration observed for dissociative CO loss in the anionic $HFe(CO)_4^-$ species in the presence of the sodium cation as compared with the PPN^+ [bis(triphenylphosphine)iminium] cation (*48*).

The effect of charge on the metal on the rates of ligand dissociation is evident when comparing reactions of the monosubstituted, isoelectronic $V(CO)_5PPh_3^-$ and $Cr(CO)_5PPh_3$ derivatives (*34, 46*). Dissociation of the PPh_3 ligand occurs more readily from $V(CO)_5PPh_3^-$ as might be expected because of less donation by the ligand to the more negative metal center resulting in a weaker $M{-}PPh_3$ bond.

B. *The Associative Pathway*

The associative mechanism [Eq. (10)] involves addition of incoming ligand to the complex in the first step affording an intermediate of increased coordination number. Because of the strong tendency for organometallic compounds to adhere to the rule of not exceeding 18 electrons in the coordination sphere of the metal (*45*), the associative pathway is not in general a route open to 18-electron complexes for ligand substitution. Indeed, Tolman (*1*) has formulated a rule that essentially states that 18-electron complexes will not undergo ligand substitution via a 20-electron intermediate, but will instead react by dissociation involving a 16-electron intermediate. Basolo (*4*) has recently qualified this statement by formally stating the following rule: *Substitution reactions by 18-electron transition metal organometallic compounds may proceed by an associative mechanism provided the metal complex can delocalize a pair of electrons onto one of its ligands.* This dictum is employed in explaining the second order kinetics observed in the displacement of arenes in (arene)Mo(CO)$_3$ by phosphorus ligands [Eq. (12)] (*49*). In this instance the arene goes from $\eta^6 \rightarrow \eta^4$ in the transition state or intermediate, thus preserving the 18-electron configuration about the metal center [Eq. (12a)]. Reaction products of this type, (triene)Mo(CO)$_3$L, have actually been isolated from the reaction of (triene)Mo(CO)$_3$ (triene = bicyclo[6.1.0]nona-2,4,6-triene) with phosphites and carbon monoxide (*50, 51*). Moreover, the arene ligand has been structurally characterized in its η^4-bonding mode in the solid state (*52, 53*), and implication of η^4-arene–metal complexes as intermediates in the catalytic hydrogenation of arenes have been reported (*39, 54*).

$$M(L)_nA + B \xrightarrow{k_2} M(L)_n(A)B \tag{10}$$

$$\downarrow$$

$$M(L)_nB + A$$

$$\text{rate} = k_2[M(L)_nA][B] \tag{11}$$

$$(\text{arene})Mo(CO)_3 + L \rightarrow fac\text{-}Mo(CO)_3L_3 + \text{arene} \tag{12}$$

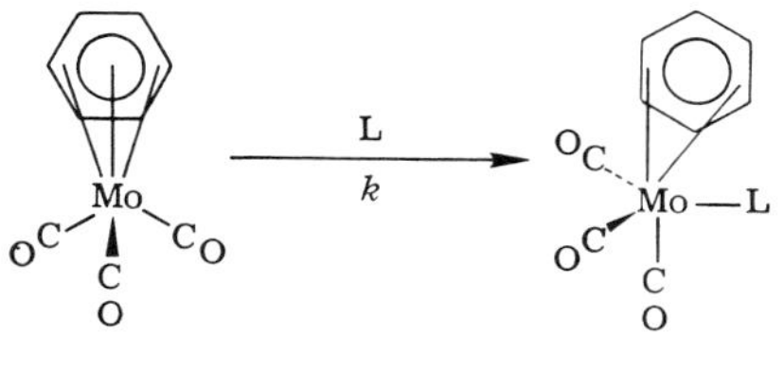

$$(12a)$$

Other instances of ligands capable of delocalizing a pair of electrons and, hence, allowing the substitution process to occur by means of an associative route include NO and cyclopentadiene (*55, 56*). For example, although Fe(CO)$_5$ undergoes carbon monoxide substitution by a dissociative mechanism (*57, 58*), ligand substitution in the isoelectronic Mn(CO)$_4$NO complex occurs by an associative process (*55*). This latter process presumably proceeds via a bent nitrosyl ligand in the transition state (i.e., $Mn \leftharpoonup N = \ddot{O}: \rightarrow M^+ \leftarrow N^{-\nearrow \ddot{O}:}$). Similarly, $Co(CO)_2(\eta^5\text{-}C_5H_5)$ exhibits second order kinetics in its CO substitution reactions, which is explained by the intermediacy of a $Co(\eta^3\text{-}C_5H_5)$ species (56). Pertinent to this proposal was the presentation of infrared evidence for $(\eta^3\text{-}C_5H_5)Co(CO)_3$ in a CO matrix at 12°K (59).

Evidence has recently been provided for solvent coordination in a migratory carbon monoxide insertion reaction (Scheme 1) in which the solvent is displaced by incoming phosphine by an associative pathway (60). This presumably is another instance of an $\eta^5\text{-}C_5H_5 \rightarrow \eta^3\text{-}C_5H_5$ transformation in the transition state (*vide infra*). Rate measurements in solvents of varying donicity (61) (THF, 2-MeTHF, 3-MeTHF, and 2,5-Me$_2$THF) but similar polarities were used to demonstrate that direct attack of the donor solvents at the metal center was occurring concurrently with alkyl migration.

The cyclopentadiene ligand is capable of delocalizing an additional pair of electrons with concomitant formation of an $\eta^1\text{-}C_5H_5$ species. An illustration of this is seen in the work of Casey and Jones (62) for the conversion of $\eta^5\text{-}C_5H_5Re(NO)(CO)CH_3$ and two equivalents of PMe$_3$ to $\eta^1\text{-}C_5H_5Re(NO)(CO)(CH_3)[PMe_3]_2$. This conversion was shown to follow second order kinetics, first order in both $\eta^5\text{-}C_5H_5Re(NO)(CO)CH_3$ and PMe$_3$, presumably through an intermediate possessing a "slipped" C_5H_5 ligand.

A recent kinetic investigation of a carbon monoxide substitution process involving a saturated metal center that proceeds by an associative mechanism is described in Eq. (13) (63). This represents yet another ligand

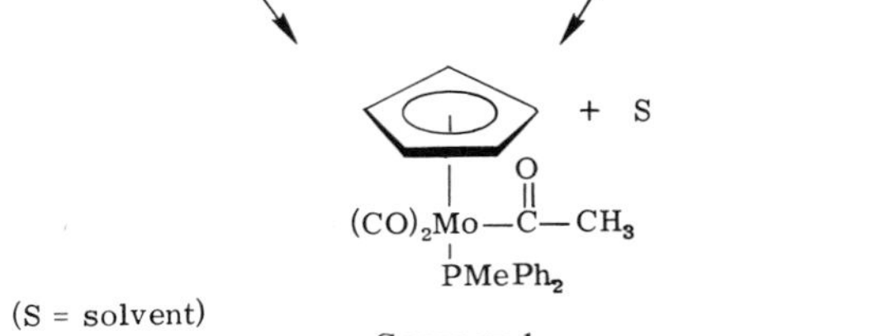

$$
\begin{array}{c}
\text{(CO)}_3\text{Mo—CH}_3 \quad + \; S \;
\xrightleftharpoons[k_{-1}]{k_1} \;
\text{(CO)}_2\text{Mo—C—CH}_3 \\
\end{array}
$$

(S = solvent)

SCHEME 1

system capable of delocalizing a pair of electrons onto itself [Eq. (14)]. Stabilization of this transition state by Lewis acids such as BF_3 was found greatly to facilitate nucleophilic attack at the iron center, as noted by a 10^6-fold increase in the rate of reaction (13) when $L = PPh_3$. The activation parameters support the associative nature of this reaction, e.g., for the incoming ligand $L = PMe_3$, $\Delta H^* = 6.9 \pm 2$ kcal mol^{-1} and $\Delta S^* = -31.4 \pm 0.7$ eu. It is informative to compare these parameters to those obtained in reaction processes where attack occurs not at the metal center but instead at the carbon atom of the carbon monoxide ligand in metal carbonyl derivatives. An example of such a process is given in Eq. (15), where $\Delta H^* = 9.4 \pm 0.9$ kcal mol^{-1} and $\Delta S^* = -44.0 \pm 3.2$ eu (64).

$$\text{(13)}$$

Ground state   Transition state

$$\text{(14)}$$

$$Ph_3PFe(CO)_4 + BzMgCl \rightarrow [Ph_3PFe(CO)_3C(O)Bz^-][MgCl^+] \qquad (15)$$

By way of contrast, carbon monoxide substitution in $Ph_3PFe(CO)_4$, where a mechanism for delocalizing a pair of electrons onto a ligand center does not exist, occurs by a rate-limiting CO dissociative pathway with $\Delta H^* = 42.5 \pm 1.2$ kcal mol^{-1} and $\Delta S^* = +18.4 \pm 2.8$ eu (65).

C. *The Interchange Pathway*

Many ligand substitution processes involving 18-electron transition metal organometallic compounds proceed through concurrent ligand-independent (*vide supra*) and ligand-dependent pathways. A classic illustration of this type behavior is exemplified in the substitution of carbon monoxide in group VIB $M(CO)_6$ species by phosphorus ligands [Eq. (16)] (22, 66). Notwithstanding Basolo's criterion for associative mechanisms, the ligand-dependent pathway for these type processes nevertheless has been widely rationalized in terms of an associative (S_N2) mechanism. However, contemporary thinking with regard to these processes is that they occur by concerted or interchange pathways designated by I_a or I_d. The I_a designation applies to processes where the transition state entails substantial bonding of both the incoming and leaving ligands, whereas in I_d processes the transition state involves only a weak bonding to both the incoming and leaving ligands. The distinguishing feature between the interchange pathway and those previously discussed (i.e., **D** and **A**) is "the absence of an intermediate in which the primary coordination number of the metal is modified" (*10*).

$$M(CO)_6 + L \rightarrow M(CO)_5L + CO \tag{16}$$

$$\text{rate} = (k_1 + k_2[L])[M(CO)_6] \tag{17}$$

The proposed mechanism for an interchange ligand substitution that proceeds to completion is given by Eqs. (18)–(20).

$$M(L)_nA + B \underset{}{\overset{K_1}{\rightleftarrows}} M(L)_nA \cdot B \text{ (rapid)} \tag{18}$$

$$M(L)_nA \cdot B \overset{k_2}{\rightarrow} M(L)_nB \cdot A \tag{19}$$

$$M(L)_nB \cdot A \overset{\text{fast}}{\rightarrow} M(L)_nB + A \tag{20}$$

The rate constant expressions [Eqs. (21) and (22)] are:

$$k_{obs} = \frac{k_2 K_1[B]}{1 + K_1[B]} \tag{21}$$

$$\frac{1}{k_{obs}} = \frac{1}{k_2} + \frac{1}{k_2 K_1[B]} \tag{22}$$

It should be stressed here that the I_a and **A** classifications are in general equivocal. In the absence of a kinetically detectable intermediate where the coordination sphere of the metal has been expanded, or at least activation parameters, it is difficult to distinguish between these two processes. For example, the associative (S_N2) displacement of solvent by phosphine in Scheme 1, which is accountable in terms of an $\eta^5 \rightarrow \eta^3$ transformation, can alternatively be described as occurring by means of an I_a pathway. Further, it is of importance to state explicitly that our working definition of the entering and departing ligands' interactions in the transition state, which allows for distinguishing between the I_a and I_d processes, involves the extent of bonding of those ligands with the metal center. This is to differentiate between interactions occurring elsewhere in the complex, e.g., hydrogen bonding between a ligand coordinated to the metal and an incoming ligand (*vide infra*) (*67*).

Kinetic differentiation of the I_d and **D** mechanisms has been proposed based on the fact that the rate constant expression for the interchange process [Eq. (22)] does not show a concentration dependence on the departing ligand (A) whereas the corresponding expression for the dissociative process [Eq. (4)] does (*68*). The fallacy with this reasoning is that for the commonplace occurrence, where the departing and entering ligands are similar in character, it is necessary to include the equilibrium process involving outer-sphere complex formation with the departing ligand (A) in the interchange process [Eq. (23)] (69). Inclusion of Eq. (23) in the interchange pathway described by Eqs. (18)–(20), results in the rate constant expression [Eq. (24)].

$$M(L)_nA + A \overset{K_2}{\rightleftarrows} M(L)_nA \cdot A \tag{23}$$

Equations (24) and (4) have equivalent algebraic forms, i.e., if $K_2[A]$ $\gg 1$ the slopes computed by the reciprocal forms of Eqs. (4) and (24) will vary identically with [A]. In other words, the overall rate can be equally affected by the competition of A and B for the highly reactive coordinatively unsaturated intermediate, $M(L)_n$, or by competition of A and B to form outer-sphere complexes with $M(L)_nA$. Equation (23) would not in general be of significance for the instance where the departing ligand (A) is carbon monoxide. Hence, the lone observation of a retardation of the rate of ligand substitution as a function of the departing ligand's concentration is insufficient evidence to differentiate kinetically between **D** and $\mathbf{I_d}$ pathways. Other necessary information would include activation parameters, solvent effects, and sensitivity of substitution rates to the nature of the entering ligand.

$$k_{obs} = \frac{k_2 K_1[B]}{1 + K_1[B] + K_2[A]} \tag{24}$$

Intimate mechanistic information for substitution reactions occurring at octahedral metal centers has been gained from studies of the replacement of the amine ligand in $M(CO)_5$–amine complexes, where M = Cr, Mo, W [Eq. (25)] (11, 67, 70–72). Because of the low nucleophilicity of carbon monoxide, its participation in the transition state when employed as the entering ligand in reactions of the type defined by Eq. (25) is minimal. Therefore, ligand substitution investigations of the reactions of $M(CO)_5$–amine with CO, in the absence of added amines, to afford $M(CO)_6$ readily provide the rate and activation parameters for M–amine dissociation. Both ligand-independent and ligand-dependent terms are observed in the rate constant expression for other entering ligands (L), such as phosphines or phosphites. This is analogous to the kinetic behavior noted for carbon monoxide substitution processes in the parent metal hexacarbonyls [Eq. (16)].

$$M(CO)_5amine + L \rightarrow M(CO)_5L + amine \tag{25}$$

The ligand-independent rate parameters obtained employing either CO or phosphines as incoming ligands were identical, thus signifying a common reaction pathway, i.e., a rate-determining dissociation of the amine providing $M(CO)_5$. Consistent with this proposal the first order rate parameters varied with the nature of the departing amine group, decreasing with increasing amine basicity. On the other hand, the ligand-dependent term exhibits a significant dependence on both the leaving and entering groups. Since in general these ligand-dependent pathways possess activation parameters that parallel those for the dissociative process over a wide range of substrates (see some examples in Table VI), a dissociative interchange mechanism (I_d) is inferred. The decrease in ΔH^* for the I_d pathways is due primarily to differing extents of bond breaking in the bond to the departing ligand (11).

For selected amine ligands that contain an NH grouping, spectroscopic evidence for the formation of a kinetically labile hydrogen-bonded adduct during the reaction of $M(CO)_5$amine with PR_3 in the presence of catalysts such as $(n\text{-Bu})_3P{=}O$ has been presented (67, 71). This process is described in Eqs. (26)–(29); where C represents the $(n\text{-Bu})_3P{=}O$ catalyst, B the incoming PR_3 ligand, and NHR_2 the departing amine group. Alternatively, B may directly add to the $(L)_nM$ intermediate as C removes the amine from the metal center. However, the rapid reversible association process [Eq. (26)] increases the effective concentration or activity of C at the reaction site and hence serves as an "entropy trap." The rate constant expressions for the general base catalyzed path are identical to those pre-

TABLE VI

ENTHALPIES OF ACTIVATION FOR FIRST AND SECOND ORDER REACTION
PROCESSES INVOLVING OCTAHEDRAL COMPLEXES

Reaction[a]	ΔH^* (kcal mol^{-1})		
	Ligand-independent	Ligand-dependent	Reference
$Cr(CO)_6 + P(n\text{-Bu})_3$	40.2	25.5	22
$Mo(CO)_6 + P(n\text{-Bu})_3$	31.7	21.7	22
$W(CO)_6 + P(n\text{-Bu})_3$	39.9	29.2	22
$Cr(CO)_5C(OMe)Me + PEt_3$	26.8	20.6	73
$Mo(CO)_5NHC_5H_{10} + PPh_3$	25.8	16.5	11
$Cr(CO)_6 + N_3^-$	—	18.3	74
$Mo(CO)_6 + N_3^-$	—	15.3	74
$W(CO)_6 + N_3^-$	—	12.8	74

[a] The reactions with phosphines involve simple ligand substitution to afford the phosphine derivatives, whereas the reaction with azide provides the $M(CO)_5NCO$ species.

viously described for the I_d process [Eqs. (21) and (22)], and are given by Eqs. (30) and (31) provided that the rate of formation of the intermediate $(L)_nMC$ is much lower than that of its subsequent reaction with B.

$$
\begin{array}{ccc}
H & & H\cdots C \\
| & & | \\
(L)_nMNR_2 + C & \overset{K_c}{\rightleftharpoons} & (L)_nMNR_2
\end{array}
\tag{26}
$$

$$
\begin{array}{c}
H\cdots C \\
| \\
(L)_nMNR_2 \overset{k_c}{\rightarrow} (L)_nMC + NHR_2
\end{array}
\tag{27}
$$

$$
(L)_nMC \rightleftharpoons (L)_nM + C \text{ (fast)}
\tag{28}
$$

$$
(L)_nM + B \rightarrow (L)_nMB \text{ (fast)}
\tag{29}
$$

$$
k_{obs} = \frac{k_c K_c[C]}{1 + K_c[C]}
\tag{30}
$$

$$
\frac{1}{k_{obs}} = \frac{1}{k_c} + \frac{1}{k_c K_c[C]}
\tag{31}
$$

For the D, I_d, and general base catalyzed (I_{bc}) pathways, all operating concurrently with kinetically negligible concentrations of both A and $M(L)_nC$ (Scheme 2), the rate constant expression is given by Eq. (32). Equation (32) cannot be cast into an analytically convenient inverted linear form; however, Eq. (31) can be used as a valid approximation under the usually prevailing conditions, i.e., $(k_1 + k'_2 K_1[B]) \ll k_c K_c[C]$ and $K_1[B] \ll 1$. This is illustrated in Fig. 4 for the reaction of $Mo(CO)_5NHC_5H_{10}$ with PPh_3 in the presence of $(n\text{-}Bu)_3P{=}O$.

$$
k_{obs} = \frac{k_1 + k'_2 K_1[B] + k_c K_c[C]}{1 + K_1[B] + K_c[C]}
\tag{32}
$$

Although the association between the $M(CO)_5NHR_2$ species and PR_3 was too weak for identification of an intermediate, the evidence in this instance is overwhelmingly in favor of formation of a transient hydrogen-bonded outer-sphere complex. This is particularly so since THF (and other

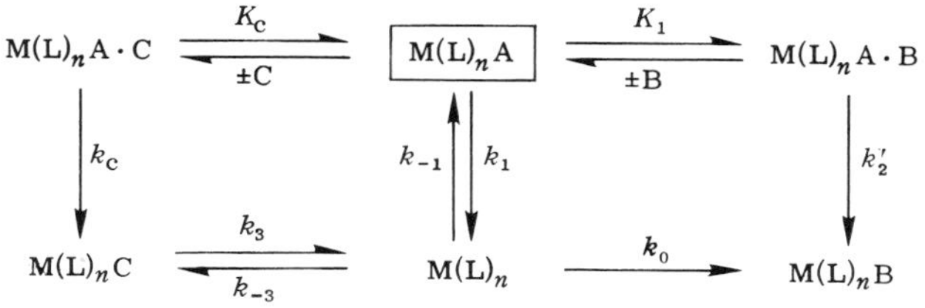

SCHEME 2

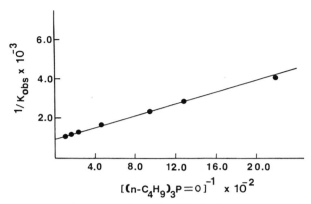

FIG. 4. Double-reciprocal plot according to Eq. (31) for the reaction of $Mo(CO)_5NHC_5H_{10}$ with PPh_3 in the presence of $(n\text{-Bu})_3P{=}O$ in hexane at 34.5°C.

hydrogen-bonding solvents) (11, 71) competes effectively with PR_3 and quenches the bimolecular ligand substitutional pathway entirely. Thus Eqs. (33)–(40) explain the delicate balance between I_d and D mechanisms when changing solvents where S is a solvent with an accessible lone pair (e.g., THF) and S′ is a π base (e.g., benzene). Solvents may thus act as catalysts or inhibitors in these substitution reactions through specific solvent effects. A stabilizing effect toward amine dissociation is also seen in cases where intramolecular hydrogen bonding between the departing amine group and an electronegative site within the complex is possible (75–77).

$$M(L)_nA + S \rightleftharpoons M(L)_nA \cdot S \qquad (33)$$

$$M(L)_nA \cdot S \rightarrow M(L)_nS \cdot A \qquad (34)$$

$$M(L)_nS \cdot A \rightarrow M(L)_nS + A \qquad (35)$$

$$M(L)_nS \rightarrow M(L)_n + S \qquad (36)$$

$$M(L)_n + B \rightarrow M(L)_nB \qquad (37)$$

or

$$M(L)_nA + S' \rightleftharpoons M(L)_nA \cdot S' \qquad (38)$$

$$M(L)_nA \rightleftharpoons M(L)_n + A \qquad (39)$$

$$M(L)_n + B \rightarrow M(L)_nB \qquad (40)$$

The limiting rate in the dissociative interchange mechanism is not radically different from those observed in the D pathway. Both ΔH^* (11) and ΔS^* (78) have been implicated as making the major contribution to this small rate enhancement for I_d versus D processes. The intermediacy of hydrogen-bonded species may be relevant to a large variety of amine substitutional processes, ranging from those involving coordination complexes to those of biological importance (69).

Although kinetically detectable intermediates were observed in the amine substitution reactions characterized above, the process is classified as an interchange pathway (as opposed to A) because the interaction for adduct formation does not directly involve the metal center. Other instances where analogous types of intermediates, involving significant interactions between the substrates and the incoming ligand away from the metal center, are observed include phosphine substitution with metal carbonyl carbenes (73) and metal carbonyl unsaturated hydrocarbon derivatives (79, 80). A caveat that should be noted in these cases is that it is not possible kinetically to distinguish whether or not these intermediates lie along the lowest energy reaction path for substitution (81). Criteria such as activation parameters and metal dependence, and their similarities with the dissociative processes, are often applied in resolving this issue.

Concerted processes, where a large degree of interaction takes place between the incoming ligand and the metal center (i.e., I_a processes), would be anticipated to display a metal dependence that did not necessarily parallel that of the dissociative process. This is illustrated in the reaction of the azide ion with the group VIB hexacarbonyls (see Table VI), where the rate varies $W > Mo > Cr$ and the transition state is proposed to resemble that shown in Eq. (41) (74). Similar rate behavior has been noted for OH^- addition to the carbon center in these metal hexacarbonyls, where the ultimate process involves carbon monoxide substitution by a hydride ligand (82).

$$M(CO)_6 + N_3^- \longrightarrow \left[(CO)_5M\overset{\overset{O}{\overset{\|}{C}}}{\underset{N\cdots N_2}{|}} \right]^{\ddagger} \overset{-N_2}{\longrightarrow} (CO)_5MNCO^- \quad (41)$$

III

ISOTOPIC LABELING STUDIES

The use of highly enriched ^{13}CO and $C^{18}O$ ($>90\%$) has contributed significantly to an understanding of the intimate mechanistic details of ligand substitution processes of metal carbonyl derivatives.

A. *Monitoring Techniques*

1. *Infrared*

A general method of following the course of labeled carbon monoxide in ligand substitution reactions is by observing frequency shifts in the $\nu(CO)$ vibrational modes as a consequence of isotopically labeled CO incorporation or loss in the substrate molecule. Similar shifts in the $\nu(CO)$ vibrational modes are observed upon either ^{13}CO or $C^{18}O$ substitution. The reliability of simplified restricted CO force field computations at deducing structural information has immeasurably aided in the utilization of this procedure (*83–85*). Nevertheless, this spectroscopic technique is often not straightforward in its application due to difficulties associated with quantitatively assessing the various isotopically substituted species present. This is primarily due to the fact that in metal carbonyl derivatives containing several CO ligands an array of species is possible in which the species possess electronically equivalent, but symmetry nonequivalent, CO ligands; hence giving rise to a large number of overlapping or degenerate infrared active bands. Notwithstanding, this technique's obvious advantage of requiring less sophisticated instrumentation, rapid data acquisition, and smaller sample size than the alternative NMR technique makes it often the method of choice.

2. *Nuclear Magnetic Resonance*

The other method commonly employed in tracing the fate of labeled carbon monoxide during ligand substitution processes is ^{13}C Fourier transform nuclear magnetic resonance spectroscopy. This may be accomplished either directly by using ^{13}CO or indirectly by employing $C^{18}O$ and noting the ^{18}O isotope shift on the ^{13}C NMR spectrum (*86, 87*). Additional recent publications have pointed out the potential of ^{17}O NMR in assaying the labeled carbon monoxide ligand (*88–90*). The most apparent advantage of these NMR approaches lies in their ability to quantify the extent of labeled CO in the substrate as well as accurately assigning its stereochemistry, generally without the necessity of computational methods.

B. *Viewing Ligand Substitution Processes*

An illustration of the sorts of information accessible from labeling studies is seen in photochemical and thermal substitution reactions of the group VIB metal tetracarbonylnorbornadiene derivatives (*85, 91*). These species are generally believed to typify intermediates in the photoassisted hydro-genation of dienes with the group VIB metal hexacarbonyls (*92–96*). Pho-tochemical substitution reactions of $M(CO)_4(NBD)$ (**1**) with ^{13}CO were shown via infrared spectroscopy, coupled with ^{13}C NMR, to occur with preferential loss of an axial CO ligand. In a subsequent step the stereo-selectively ^{13}CO labeled species (**2**) undergoes rearrangement upon thermal and/or photochemical activation (Scheme 3). The selective production of the axially di-^{13}CO substituted species **4** demonstrates that the stable form of the intermediate produced by means of loss of CO is a square pyramid and further that it does not intramolecularly scramble CO groups during its solution lifetime. This observation is in agreement with a growing body of evidence in support of 16 valence electrons, pentacoordinate metal car-bonyl derivatives exhibiting square-pyramidal, rather than trigonal-bipyr-amidal geometry, both in solution and in inert matrices (*84, 97–104*). In general this is an important conclusion as to the nature of five-coordinate intermediates in solution that can be addressed from labeling studies.

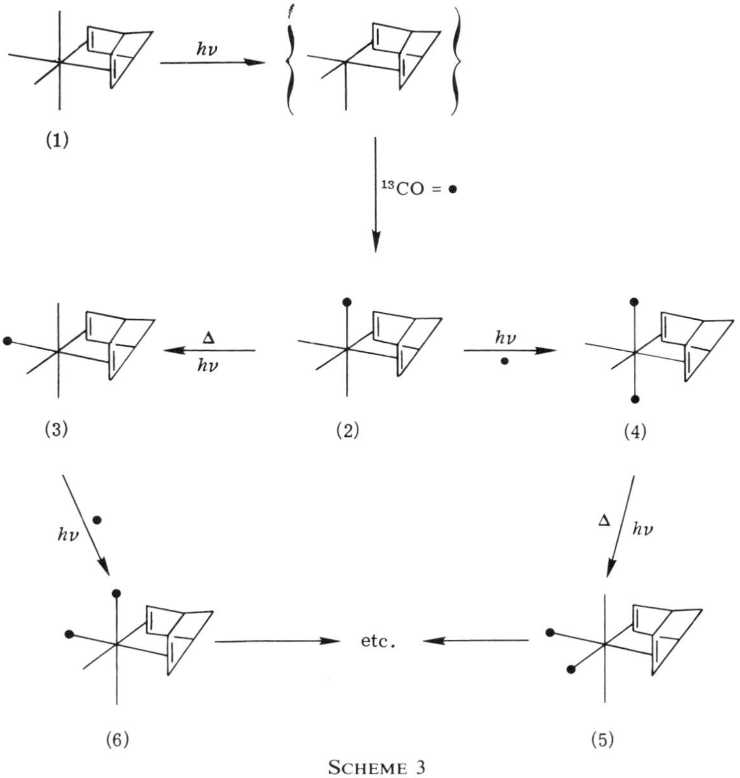

SCHEME 3

A mechanism that involves cleavage of one metal–olefin bond followed by a Berry permutation (*105*) of the four CO groups in the five-coordinate intermediate has been proposed for the rearrangement of the stereoselectively axially labeled species **2** in the tungsten derivative (Scheme 4). The activation parameters for this process were found to be $\Delta H^* = 18.4$ kcal mol^{-1} and $\Delta S^* = -12.3$ eu in saturated hydrocarbon solvent. Similar rate data were obtained for the di-^{13}CO axially substituted species **4** proceeding to the di-^{13}CO equatorially substituted species **5** with no concomitant production of a di-^{13}CO axial–equatorial labeled species **6** as would be dictated by a Berry pseudorotation mechanism. Since the rearrangement of CO groups involves partial rotation about the metal–olefin bond in the unsaturated intermediate, it would be anticipated that ring closure would have a smaller activation barrier than the pseudorotation process. This was verified by an investigation of the diene substitution reaction with phosphine which proceeds via a chelate ring-opening mechanism with a limiting rate observed at high phosphine concentrations. These kinetic measurements clearly indicate that the rate of ring opening is somewhat faster than intramolecular CO interchange.

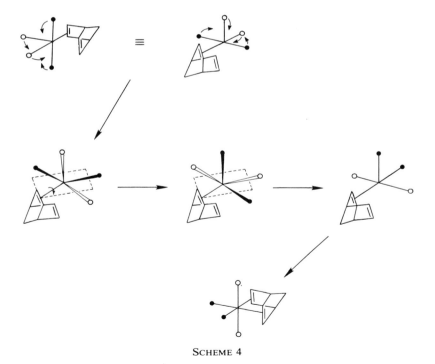

SCHEME 4

Although the stereoselectively ^{13}CO labeled tungsten derivative readily undergoes CO scrambling, the molybdenum analog was found to be much more inert with respect to CO rearrangement. This greater rigidity of the molybdenum carbonyl framework has been further demonstrated in several investigations. The reaction of the triene complex (bicyclo[6.1.0]nona-2,4,6-triene)molybdenum tricarbonyl 7 with ^{13}CO to afford stereoselectively the axially labeled ^{13}CO tetracarbonyl derivative 8 is consistent with the photochemical incorporation of ^{13}CO in $Mo(CO)_4(NBD)$ where ^{13}CO addition to a square pyramid intermediate occurs (Scheme 5) (106, 107). Reaction of species 8 with a large variety of ligands has provided an extremely efficient synthesis of stereoselectively ^{13}CO axially labeled molybdenum tetracarbonyl derivatives, e.g., cis-$Mo(CO)_4L_2$ (L = phosphine, phosphite, etc.), $Mo(CO)_4(NBD)$, and $Mo(CO)_4$(diamine). Similarly, substitution of the chelating ligands, NBD or diamine, provides an alternative route to axially ^{13}CO labeled cis-$Mo(CO)_4L_2$ species (108).

Substitution of the bidentate ligands, bicyclo[6.1.0]nona-2,4,6-triene, bicyclo[2.2.1]hepta-2,5-diene, and N,N,N',N'-tetramethyl-1,3-diamino-propane by Lewis based L proceeds via a mechanism involving a reversible ring-opening process [Eq. (42)]. Thus the lack of intramolecular carbonyl ligand rearrangement in the formation of molybdenum tetracarbonyl derivatives clearly demonstrates the rigidity of substituted, five-coordinate molybdenum carbonyl species (where the substituted ligand occupies an equatorial coordination site in the square-pyramidal structures) during

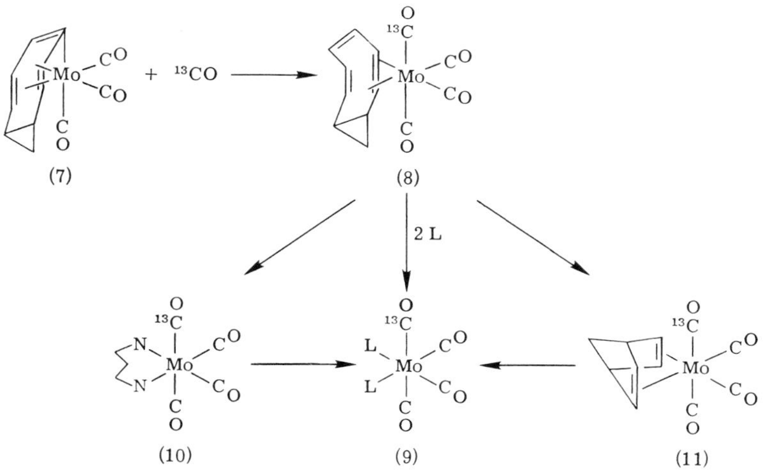

SCHEME 5

their solution lifetime at room temperature, i.e., the $||-||$-Mo(CO)$_4$, NN-Mo(CO)$_4$, and LMo(CO)$_4$ intermediates generated through the process described in Eq. (42). The average inverse lifetime of d^6 five-coordinate species in saturated hydrocarbon solvents is expected to be of the order of 10^4 sec^{-1} (*109–111*).

$$\left(\begin{smallmatrix} L' \\ \\ L' \end{smallmatrix} \right. \!\! M(CO)_4 \underset{k_{-1}}{\overset{k_1}{\rightleftharpoons}} \: :L'-L'-M(CO)_4 \overset{L}{\longrightarrow} \: :L'-L'-M(CO)_4L$$

$$\Big\downarrow \text{fast} \qquad (42)$$

$$\text{Product} \longleftarrow M(CO)_4L$$

The rigidity of the Mo(CO)$_4$ moiety has also been observed in Mo(CO)$_4$BH$_4^-$ (*112*). On the other hand, the carbonyl ligands have been reported to be quite fluxional in the *cis*-L$_2$Mo(CO)$_4$ (L$_2$ = substituted diazabutadienes) derivatives; however, this process is proposed to occur by a trigonal rotation mechanism involving no metal–ligand dissociation (*113*). In addition, the cis $\rightleftharpoons$ trans isomerization reactions of Mo(CO)$_4$[PR$_3$]$_2$ (R = Me, Et, *n*-Bu) [Eq. (43)] have been demonstrated to take place by an intramolecular nondissociative process on the basis of the lack of ^{13}CO incorporation into either *cis*- or *trans*-Mo(CO)$_4$[PR$_3$]$_2$ or formation of Mo(CO)$_4$(^{13}CO)PR$_3$ during the rearrangement (*114, 115*). The data summarized in Table VII for reaction (43) reveal the anticipated trend as a function of the size of the trialkylphosphine ligand. That is, the smaller PMe$_3$ ligand favors the cis stereochemistry slightly which minimizes the number of mutually trans CO groups; and concomitantly the rate of cis $\rightarrow$ trans isomerization is much slower than that for the larger PEt$_3$ and P(*n*-Bu)$_3$ species. These kinetic observations are directly compatible with the solid state structural results, where a greater distortion toward the trans

TABLE VII

COMPARATIVE RATE AND EQUILIBRIUM PARAMETERS FOR THE INTRAMOLECULAR
ISOMERIZATION OF Mo(CO)$_4$[PR$_3$]$_2$ DERIVATIVES AT 64.8°C

PR$_3$	cis $\rightarrow$ trans $k \times 10^4$ (sec^{-1})	$t_{1/2}$ (min)	K_{eq}	ΔG (kcal mol^{-1})
PMe$_3$ [a]	0.084	1375	0.62	0.32
PEt$_3$ [b]	2.10	55.0	7.0	−1.30
P(*n*-Bu)$_3$ [b]	1.67	69.2	5.3	−1.12

[a] Taken from Ref. *115*.
[b] Taken from Ref. *114*.

disposition of the phosphine ligands is seen in the ground state structures of the cis-$Mo(CO)_4[PR_3]_2$ (R = Et, n-Bu) derivatives (115). Consistent with these steric arguments, in the $Cr(CO)_4[PMe_3]_2$ derivative, where significantly greater distortions from ideal octahedral geometry are expected, the trans isomer is favored with $K_{eq} \sim 1.7$ at 32°C and the cis → trans isomerization rate has a $t_{1/2}$ value < 15 min (116). On the other hand $Cr(CO)_4[PEt_3]_2$ exists predominantly in the trans isomeric form. A trigonal rotation mechanism ($\Delta H^* = 24.5$ kcal mol^{-1}) was used to account for these isomerization processes. This is to be compared with processes involving phosphine dissociation in derivatives of the type cis-$Mo(CO)_4L_2$, where L is a bulky phosphine (e.g., PPh_3) (40), which occur with activation enthalpies of ~ 30 kcal mol^{-1} (33).

$$cis\text{-}Mo(CO)_4[PR_3]_2 \rightleftharpoons trans\text{-}Mo(CO)_4[PR_3]_2 \tag{43}$$

Rossi and Hoffman (117) have predicted that in d^6 square-pyramidal species weak σ donors should occupy an equatorial site, i.e., when considering only σ-bond strengths the expected trend for metal–ligand bond strengths is $M-L_{ax}$ stronger than $M-L_{eq}$. Thus ligands that are weak σ donors and non-π acceptors (saturated amines) (77) or weak π acceptors (olefins without electron withdrawing substituents, phosphines, etc.) should show a preference for an equatorial site in the square-pyramidal structure of d^6 species. This preference would be greater for the weak σ donor and non-π-acceptor ligands, diminishing as the unique ligand approaches the bonding characteristic of carbon monoxide.

The above arguments form the basis of Brown's site preference model for reactivity in these octahedral systems (9, 27, 118–121). For example, Atwood and Brown (27) have measured the rate constant for dissociative CO loss in $Mn(CO)_5Br$ [Eq. (44)] using infrared spectral techniques and found that the rate for cis CO dissociation was at least ten times that for trans CO dissociation. This observation is consistent with an earlier study of the exchange of ^{14}CO with $Mn(CO)_5X$ (X = Cl, Br, I) (122, 123), and clarifies further some misconceptions with regard to this process that have appeared in the literature (124, 125). These researchers were further able to establish that the five-coordinate intermediate, $Mn(CO)_4Br$, formed during the dissociation [Eq. (44)] is fluxional; i.e., randomization of the stereoselective label occurred on the same time scale as ligand substitution (27). We had some years earlier made an analogous observation for the $Mo(CO)_5$ intermediate procreated during amine replacement in $Mo(CO)_5(NHC_5H_{10})$ (126). The phenomenon of cis labilization, i.e., labilization of a position cis to a designated ligand as compared with when that ligand is CO, is believed to be rather prevalent in reactions involving dissociation of a ligand from six-coordinate low-valent metal carbonyl derivatives. The cis-labilization order derived by Atwood and Brown (9) was as follows: $CO < P(OPh)_3 < PPh_3 < I^- < Br^- < Cl^-$ (see Table II).

$$Mn(CO)_5Br + n^{13}CO \rightleftharpoons Mn(CO)_{5-n}(^{13}CO)_nBr + nCO \tag{44}$$

Oxygen donor bases have recently been shown to be very strongly labilizing ligands. This property of oxygen bases has been exploited in the syntheses of highly ^{13}CO enriched metal carbonyl derivatives via the skeletal sequence below [Eq. (45)], where $R = n$-Bu (127, 128). Monodentate bonded acetate and formate ligands have been observed to be extremely good CO-labilizing ligands in complexes as exemplified in Eq. (46) (129–132). Indeed this Co labilization occurs preferentially at sites cis to the oxygen bases. This was unequivocally demonstrated by ^{13}C NMR spectroscopy (129, 133–135). For example, the natural abundance ^{13}C NMR spectrum of $[PNP][W(CO)_5O_2CCH_3^-]$ exhibits two carbonyl resonances at 206.4 and 200.5 ppm in $CDCl_3$, which are assigned to the axial and equatorial carbon monoxide ligands, respectively. Upon incorporation of ^{13}CO into the $W(CO)_5O_2CCH_3^-$ anion at 0°C in tetrahydrofuran for 5 h, the sample was enriched to a total ^{13}CO content of 27.8% with the peak for the cis CO ligands at 200.5 ppm ($J_{W-C} = 131$ Hz) accounting for all the ^{13}CO uptake (see Fig. 5) (133, 134). In addition to the cis-labilization arguments of Brown, these processes involving monodentate acetate and formate ligands may be facilitated by the assistance of the free carboxylic oxygen atom in CO dissociation and/or adding stability to the five-coordinate intermediate or transition state. Other manifestations of the CO-labilizing ability of oxygen ligands are seen in decarbonylation of metal carbonyl compounds on alumina surfaces, a process of importance in heterogeneous catalysis (136).

$$\underset{\substack{|\\[-2pt]L}}{[M{+}CO]} \xrightarrow[+\,R_3P{=}O]{-\,L} \underset{\substack{||\\[-2pt]O\\[-2pt]|}}{\overset{PR_3}{[M{+}CO]}} \xrightarrow[+\,^{13}CO]{-\,CO} \underset{\substack{||\\[-2pt]O\\[-2pt]|}}{\overset{PR_3}{[M{+}^{13}CO]}} \xrightarrow[+\,^{13}CO]{-\,R_3P{=}O} \underset{\substack{|\\[-2pt]|}}{\overset{\substack{O\\[-2pt]||\\[-2pt]^{13}C}}{[M{+}^{13}CO]}} \tag{45}$$

$$W(CO)_5O_2CCH_3^- + n^{13}CO \rightleftharpoons W(CO)_{5-n}(^{13}CO)_nO_2CCH_3^- + n^{12}CO \tag{46}$$

The enhanced lability of the group VIB metal hexacarbonyls in the presence of hydroxide base, presumably due to a transient interaction of OH^- with the carbon atom of coordinated CO, is probably yet another instance of cis labilization (82, 137–139). This proposal is supported by the observation of oxygen-18 incorporation into the metal carbonyl compound under the reaction conditions for ligand substitution [Eq. (47)] (82, 138). A similar explanation has been offered for base catalyzed substitution reactions of $Fe(CO)_2(NO)_2$ [Eq. (48)] by Morris and Basolo (140). In

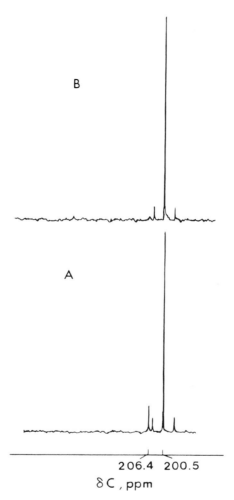

Fɪɢ. 5. ^{13}C NMR spectra of the W(CO)$_5$O$_2$CCH$_3^-$ anion in CDCl$_3$. (A) Natural abundance spectrum. (B) Spectrum of initial ^{13}CO exchange product indicating only equatorial ^{13}CO enrichment.

general base catalysis of ligand substitution reactions in metal carbonyls is anticipated to be of much importance in a variety of homogeneous and heterogeneous catalyzed processes.

$$M(CO)_6 + {}^{18}OH^- \rightleftharpoons M(CO)_5COOH^- \rightleftharpoons M(CO)_{5-n}(C^{18}O)_n + OH^- \qquad (47)$$

$$Fe(CO)_2(NO)_2 + L \rightarrow Fe(NO)_2(CO)L + CO \qquad (48)$$

The investigation of ligand substitution processes using isotopic replacement of the type reported on above is extremely informative. This is so because the free energy profile for the reactions [e.g., Eqs. (44) and (46)] is the same in either direction (microscopic reversibility), thus dictating that the stereochemistry of the products mirror precisely the stereochemistry of the dissociative process (125).

The greater thermodynamic stability of coordinatively unsaturated $M(CO)_4L$ species of d^6 metals where the substituent (L), a weak σ donor and non- to weak π-acceptor ligand, lies in the equatorial plane of the square-pyramidal structure, has provided the basis for much of the syntheses of stereoselectively ^{13}CO labelled mono- and dinuclear metal derivatives, e.g., Eq. (49) (33, 104, 141–143). Further, in the absence of a site preference ligand (L), for example in the intermediate $Mo(CO)_4(^{13}CO)$ afforded from amine dissociation in cis-$Mo(CO)_4(^{13}CO)NHC_5H_{10}$, reaction with ^{13}CO produces a statistical distribution of cis- and $trans$-$Mo(CO)_4(^{13}CO)_2$ (126). Presumably the carbonyl ligands completely scramble in the $Mo(CO)_4(^{13}CO)$ intermediate during its solution lifetime through the square-pyramidal equivalent of the Berry pseudorotation ($C_{4v} \rightarrow D_{3h} \rightarrow C_{4v}$). Similar observations have also been made in the matrix photochemistry of the group VIB hexacarbonyl species (99). At the other end of the spectrum, when L is a better σ-donor and π-acceptor ligand than CO, a site preference for the axial position in the $Mo(CO)_4L$ intermediate is expected. Thus the reaction of $trans$-$W(CO)_4(CS)I^-$ with Ag^+ in the presence of ^{13}CO yields the stereoselectively ^{13}CO-enriched $trans$-$W(CO)_4(CS)(^{13}CO)$ species (144). This latter labeled species played a key role in the very clever matrix isolation study of Martyn Poliakoff (145).

$$\tag{49}$$

In an investigation related to the subject of the solution stability of $M(CO)_4L$ species the stereoselectivity of the dissociative process (35) involving the reaction of $trans$-$Cr(CO)_4[PPh_3]_2$ with ^{13}C-labeled carbon monoxide to afford $Cr(CO)_5PPh_3$ has been examined (146). The data presented were consistent with a completely stereoselective process where the incoming ^{13}CO ligand occupies a site in the octahedral complex cis to the PPh_3 ligand [i.e., formation of cis-$Cr(CO)_4(^{13}CO)PPh_3$]. Hence rearrangement of the initially formed $Cr(CO)_4PPh_3$ intermediate of C_{4v} symmetry to that of C_s symmetry occurs faster than the bimolecular reaction of the intermediate with carbon monoxide, i.e., with a rate constant $>3 \times 10^4$ sec^{-1} (Scheme 6) (*109–111*). A subsequent reaction involving a stereomobility of the ligands in the six-coordinate derivative was observed to take place in the absence of ligand (CO or PPh_3) dissociation on a time scale similar to that for PPh_3 dissociation in the *trans*-$Cr(CO)_4(PPh_3)_2$ species. That is, the rate constant for PPh_3 dissociation in *trans*-$Cr(CO)_4(PPh_3)_2$ is 4.15×10^{-5} sec^{-1} at 40°C; whereas, the rate constant for CO scrambling in the six-coordinate $Cr(CO)_5PPh_3$ species is 3.13×10^{-5} sec^{-1} at 40°C. This latter rate constant was measured on a stereoselectively ^{13}CO enriched sample prepared by the previously described synthetic methodology [Eq. (50)].

$$cis\text{-}Cr(CO)_4[PPh_3]Cl^- + {}^{13}CO \xrightarrow[-50°C]{EtOH} cis\text{-}Cr(CO)_4({}^{13}CO)PPh_3 + Cl^- \qquad (50)$$

Presumably, the pathway for this nondissociative, intramolecular rearrangement process in d^6 octahedral complexes proceeds through either a trigonal–prismatic (*147–151*) or bicapped tetrahedron (*152*) intermediate or transition state. The relative energy barrier to intramolecular ligand rearrangement as a function of metal has been found to be Cr < W < Mo (*87*). Although this is somewhat unexpected based on steric constraints alone, where a greater ease of flexibility for a trigonal twist would be anticipated for the larger Mo and W species, it is evident that electronic factors are important in these processes. For example, the importance of the electronic environment about the metal center is revealed in the nondissociative ligand isomerization of *cis*-$Mo(CO)_4(PR_3)_2$ derivatives (R = Et, *n*-Bu) which readily occurs with $\Delta H^* = 24.5$ kcal mol^{-1} (*114*). On the other hand, CO stereomobility in the $Mo(CO)_5PR_3$ derivatives is apparently a more energetic process than ligand dissociation where $E_{act} > 30$ kcal mol^{-1} (*25*).

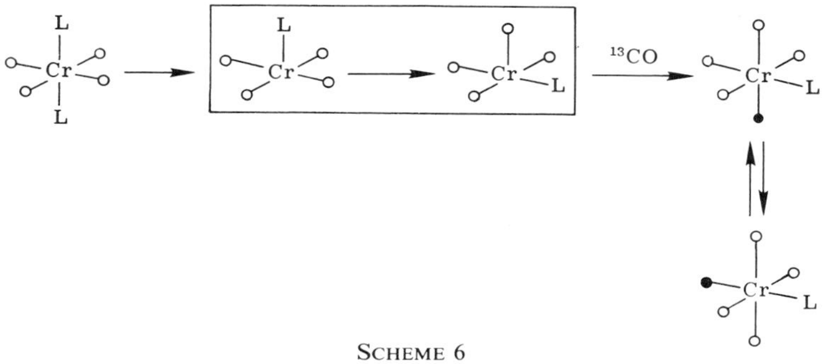

SCHEME 6

Thus far in our investigations of ligand stereomobility in $M(CO)_5L$ derivatives by nondissociative routes we have observed this occurrence in chromium and tungsten complexes containing a number of phosphorus donor ligands, including those where $L = PPh_3$, PPh_2Me, and $P(OMe)_3$. It is apparent that ΔH^* is less than that corresponding to ligand dissociation (i.e., <32 kcal mol^{-1}) and greater than 16 kcal mol^{-1}. This latter lower limit is suggested by the lack of line broadening in the CMR of the carbonyl resonances in the high temperature spectra of these derivatives.

A caveat that must be considered in CO-dissociative processes as described in Scheme 7, e.g., the reactions provided in Eqs. (44) and (46), is that at least part of the randomization of the introduced label ascribed to occur in the five-coordinate intermediate (boxed in Scheme 7) may be due to intramolecular scrambling in the once formed six-coordinate product (see Scheme 6). Further, in these processes where CO dissociation is slow relative to intramolecular ligand rearrangement in the six-coordinate species, no site preference for CO loss would be observable, although it would generally be anticipated based on empirical as well as theoretical considerations.

Dobson and co-workers (*153–155*) have employed synthetic techniques analogous to that depicted in Eq. (49) for the preparation of stereoselective ^{13}CO-labeled group VIB metal tetracarbonyl derivatives containing bidentate ligands. Subsequently, these derivatives were used in definitive experiments defining the site of metal–CO bond cleavage in dissociative CO substitution processes, along with a concomitant assessment of the fluxional behavior of the thus afforded five-coordinate intermediates. For example, in the axially ^{13}CO labeled 1,2-bis(diphenylphosphino)ethanetetracarbonylmolybdenum complex (**12**), these researchers proved that the site of bond breaking is axial only with the thus procreated intermediate $Mo(CO)_3(diphos)$ being partially fluxional ($\sim$83% CO randomization) at

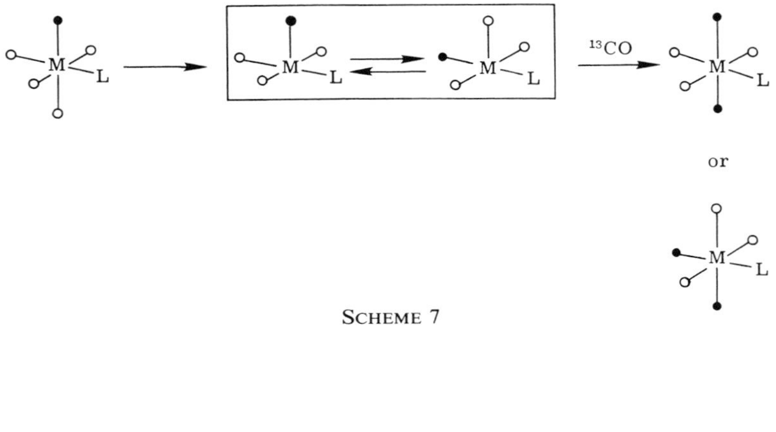

SCHEME 7

125°C (Eq. 51) (*153*). However, the predominant fluxional species is species **13** where the phosphine ligand occupies equatorial sites. On the other hand, at room temperature the $Mo(CO)_3(diphos)$ intermediate (**13**) is nonfluxional, consistent with other results previously discussed.

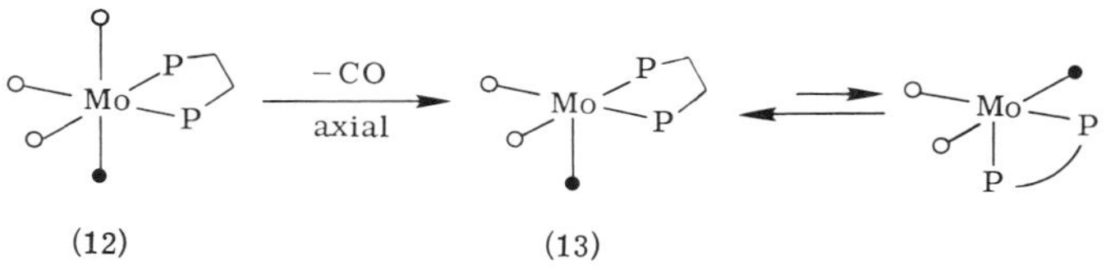

$$\text{(12)} \qquad\qquad\qquad \text{(13)} \qquad\qquad\qquad\qquad (51)$$

In parallel studies $(phen)M(CO)_4$ (M = Cr, Mo, W; and phen = *o*-phenanthroline) derivatives were found to undergo exclusive axial loss of CO with a statistical scrambling of carbonyls in the resulting five-coordinate intermediate being observed (*154, 155*).

IV

RADICAL PATHWAYS FOR LIGAND SUBSTITUTION REACTIONS

Many 17-electron radical species afforded by electrochemical oxidation of 18-electron substrates are suitably stable to have reversible cyclic voltammograms (e.g., see Ref. *156*, and references therein). Nevertheless, it has become quite apparent that one of the most important characteristics of many transition metal carbonyl radicals is their exceedingly high lability toward ligand substitution (*157*). Brown and co-workers have noted strong evidence that radicals such as $Mn(CO)_5\cdot$ (*158*), $Re(CO)_5\cdot$ (*159*), $[\eta^5\text{-}C_5H_5]Mo(CO)_3]\cdot$ (*160*), and $Co(CO)_4\cdot$ (*161*) all undergo CO substitution in the presence of phosphines or phosphites. A radical chain mechanism has been established for these processes as first proposed by Byers and Brown (*162*), and is represented in its general form in Eqs. (52)–(58).

$$(CO)_nMX + Q\cdot \rightarrow (CO)_nM\cdot + QX \qquad (52)$$

$$(CO)_nM\cdot \rightarrow (CO)_{n-1}M + CO \qquad (53)$$

$$(CO)_{n-1}M + L \rightarrow (CO)_{n-1}LM\cdot \qquad (54)$$

$$(CO)_{n-1}LM + (CO)_nMX \rightarrow (CO)_{n-1}LMX + (CO)_nM\cdot \qquad (55)$$

Chain termination steps:

$$2(CO)_nM\cdot \rightarrow (CO)_{2n}M_2 \qquad (56)$$

$$2(CO)_{n-1}LM\cdot \rightarrow (CO)_{2n-2}L_2M_2 \qquad (57)$$

$$(CO)_{n-1}LM \cdot + (CO)_n M \cdot \rightarrow (CO)_{2n-1}LM_2 \qquad (58)$$

Although both $Mn(CO)_5 \cdot$ and $Re(CO)_5 \cdot$ are believed to undergo CO loss by a dissociative thermal pathway, phosphine substitution appears to stabilize these radicals toward further dissociative loss of ligands. In these instances an associative pathway is evident, i.e., reaction (59) was observed to be first order in radical and CO (*163*). Consistent with an associative substitution process the bimolecular rate constant for reaction (59) is strongly dependent on the steric bulk of the L ligand, as the smaller $P(-n\text{-}Bu)_3$ ligand (cone angle = 132°) reacts about two orders of magnitude faster than the $P(-i\text{-}Bu)_3$ ligand (cone angle = 143°). Hence, as in an 18-electron substrate both dissociative and associative ligand substitution pathways are evident for radical species.

$$Mn(CO)_3L_2 \cdot + CO \rightarrow Mn(CO)_4L \cdot + L \qquad (59)$$

$$2Mn(CO)_4L \rightarrow Mn_2(CO)_8L_2 \qquad (60)$$

The kinetics of ligand substitution reactions of dinuclear group VIIB metal carbonyls have been interpreted in terms of a thermal cleavage of the metal–metal bond to afford radical species (*164*). This proposal however is presently a point of much contention, with a simple rate-determining CO dissociation in the dinuclear complex being alternatively suggested (*165–167*). Indeed metal crossover studies strongly support initial dissociation of CO in ligand substitution reactions of $MnRe(CO)_{10}$ and $Re_2(CO)_{10}$ (*168*). Further crossover experiments, in particular employing metal isotopic labeling studies, should settle such controversies.

Although detailed kinetic investigations of ligand substitution reactions on 17-electron species are lacking when compared with the corresponding processes involving an 18-electron substrate, this area will most certainly receive a great deal of attention during this decade. The intent of this very brief discussion of these processes was simply to point out that radicals are in general extremely labile when compared with their 18-electron counterparts.

V

CONCLUDING REMARKS

The general conclusions seem clear. Our level of understanding ligand substitution reactions at low-valent mononuclear metal centers as evinced by detailed kinetic studies is quite high. Over the past twelve years or so studies involving the use of isotopic labeling have revealed most of the principal features of the site of bond cleavage and fluxional behavior of the intermediates thus formed. Researchers have begun seriously to turn their attention to analogous investigations of metal carbonyl clusters (*169*). Carbon monoxide substitution reactions in these species bear several mechanistic similarities to the corresponding processes involving mononuclear metal carbonyls. That is, in both instances the kinetic behavior is often consistent with a two term rate law [Eq. (61)], where k_1 and k_2 are identified with dissociative and associative or interchange mechanisms, respectively. A feature generally available in metal clusters, which is less frequently open to 18-electron mononuclear species, is their ability to accommodate another ligand in the coordination sphere of the metal while still maintaining conformity with the 18-electron rule. This is achieved with the simultaneous breaking of a metal–metal bond in the cluster framework.

$$\text{rate} = k_{\text{obs}}[M_n(CO)_m] = (k_1 + k_2[L])[M_n(CO)_m] \qquad (61)$$

Unfortunately because of intramolecular ligand rearrangements in cluster substrate molecules, which commonly occur on a time scale faster than dissociative ligand loss, definitive answers to questions of site selectivity will often go unresolved. Notwithstanding, it is anticipated that with diligent efforts over the next decade the level of understanding of substitution processes in metal clusters will rival that of our current knowledge for mononuclear systems (*170*).

ACKNOWLEDGMENTS

It is a pleasure to acknowledge the contributions to this research area of my past and present students and postdoctoral fellows, whose names are to be found among the references. I also wish to thank Professors J. D. Atwood, F. Basolo, T. L. Brown, M. Y. Darensbourg, G. R. Dobson, and E. L. Muetterties for supplying manuscripts or results prior to publication. The financial support of the National Science Foundation for the work carried out in this area in the author's laboratories is greatly appreciated. Finally, I am sincerely grateful to Mrs. Helen George, who has for the last ten years been a good friend and the person primarily responsible for my manuscripts reaching the editors' desks in legible form.

REFERENCES

1. C. A. Tolman, *Chem. Soc. Rev.* **1**, 337 (1972).
2. G. W. Parshall, *J. Mol. Catal.* **4**, 243 (1978).
3. G. Henrici-Olivé and S. Olivé, "Coordination and Catalysis," Verlag Chemie, New York, 1977.
4. F. Basolo, *Inorg. Chim. Acta* **50**, 65 (1981).
5. R. J. Angelici, *Organomet. Chem. Rev.* **3**, 173 (1968).
6. D. A. Brown, *Inorg. Chim. Acta, Rev.* **1**, 35 (1967).
7. H. Werner, *Angew. Chem., Int., Ed. Engl.* **7**, 930 (1968).
8. G. R. Dobson, *Acc. Chem. Res.* **9**, 300 (1978).

9. J. D. Atwood and T. L. Brown, *J. Am. Chem. Soc.* **98**, 3160 (1976).
10. C. H. Langford and H. B. Gray, "Ligand Substitution Processes." Benjamin, New York, 1965.
11. W. D. Covey and T. L. Brown, *Inorg. Chem.* **12**, 2820 (1973).
12. C. L. Hyde and D. J. Darensbourg, *Inorg. Chem.* **12**, 1286 (1973).
13. M. Y. Darensbourg and J. M. Hanckel, unpublished observations.
14. G. R. Dobson, unpublished observations.
15. G. R. Dobson, *Inorg. Chem.* **19**, 1413 (1980).
16. J. P. Day, F. Basolo, and R. G. Pearson, *J. Am. Chem. Soc.* **90**, 6927 (1968).
17. D. J. Darensbourg and H. L. Conder, *Inorg. Chem.* **13**, 374 (1974).
18. G. Cardaci, *Inorg. Chem.* **13**, 2974 (1974).
19. D. J. Darensbourg and M. J. Incorvia, *J. Organomet. Chem.* **171**, 89 (1979).
20. S. K. Malik and A. Poë, *Inorg. Chem.* **17**, 1484 (1978).
21. G. S. Hammond, *J. Am. Chem. Soc.* **77**, 334 (1955).
22. J. R. Graham and R. J. Angelici, *Inorg. Chem.* **6**, 2082 (1967).
23. J. A. Connor, J. P. Day, E. M. Jones, and G. K. McEwen, *J. Chem. Soc., Dalton Trans.* p. 347 (1973).
24. A. D. Allen and P. F. Barrett, *Can. J. Chem.* **46**, 1655 (1968).
25. J. A. Connor and G. A. Hudson, *J. Organomet. Chem.* **73**, 351 (1974).
26. H. Wawersik and F. Basolo, *Inorg. Chim. Acta* **3**, 113 (1969).
27. J. D. Atwood and T. L. Brown, *J. Am. Chem. Soc.* **97**, 3380 (1975).
28. R. G. Pearson, *J. Am. Chem. Soc.* **85**, 3533 (1963).
29. F. Zingales, F. Canziani, and F. Basolo, *J. Organomet. Chem.* **7**, 461 (1967).
30. M. Graziani, F. Zingales, and U. Belluco, *Inorg. Chem.* **6**, 1582 (1967).
31. G. R. Dobson and L. A. H. Smith, *Inorg. Chem.* **9**, 1001 (1970).
32. D. J. Darensbourg and R. L. Kump, *Inorg. Chem.* **17**, 2680 (1978).
33. D. J. Darensbourg and A. H. Graves, *Inorg. Chem.* **18**, 1257 (1979).
34. M. J. Wovkulich and J. D. Atwood, *J. Organomet. Chem.* **184**, 77 (1979).
35. M. J. Wovkulich, S. J. Feinberg, and J. D. Atwood, *Inorg. Chem.* **19**, 2608 (1980).
36. R. J. Dennenberg and D. J. Darensbourg, *Inorg. Chem.* **11**, 72 (1972).
37. F. A. Cotton, D. J. Darensbourg, and W. H. Ilsley, *Inorg. Chem.* **20**, 578 (1981).
38. C. A. Tolman, *Chem. Rev.* **77**, 313 (1977).
39. J. W. Kang, R. F. Childs, and P. M. Maitlis, *J. Am. Chem. Soc.* **92**, 720 (1970).
40. F. A. Cotton, D. J. Darensbourg, S. Klein, and B. W. S. Kolthammer, *Inorg. Chem.* **21**, 294 (1982).
41. D. J. Darensbourg and R. Sachs, unpublished observations.
42. F. A. Cotton, D. J. Darensbourg, and B. W. S. Kolthammer, *Inorg. Chem.* **20**, 4440 (1981).
43. J. von Pickardt, L. Rösch, and H. Schumann, *Z. Anorg. Allg. Chem.* **426**, 66 (1976).
44. H. C. Clark, *Isr. J. Chem.* **15**, 210 (1977).
45. F. A. Cotton and G. Wilkinson, "Advanced Inorganic Chemistry," 4th ed. Wiley (Interscience), New York, 1980.
46. M. Y. Darensbourg and J. M. Hanckel, *Organometallics* **1**, 82 (1982).
47. M. Y. Darensbourg and J. M. Hanckel, *J. Organomet. Chem.* **217**, C9 (1981).
48. M. Y. Darensbourg, D. J. Darensbourg, and H. L. C. Barros, *Inorg. Chem.* **17**, 297 (1978).
49. F. Zingales, A. Chiesa, and F. Basolo, *J. Am. Chem. Soc.* **88**, 2707 (1966).
50. A. Salzer, *J. Organomet. Chem.* **107**, 79 (1976).
51. A. Salzer, *J. Organomet. Chem.* **117**, 245 (1976).
52. G. Huttner and S. Lange, *Acta Crystallogr., Sect. B* **28**, 2049 (1972).

53. A. Bond, M. Bottrill, M. Green, and A. J. Welch, *J. Chem. Soc., Dalton Trans.* p. 2372 (1977).
54. M. Y. Darensbourg and E. L. Muetterties, *J. Am. Chem. Soc.* **100,** 7425 (1978).
55. H. Wawersik and F. Basolo, *J. Am. Chem. Soc.* **89,** 4626 (1967).
56. H. G. Schuster-Woldan and F. Basolo, *J. Am. Chem. Soc.* **88,** 1657 (1966).
57. D. F. Keeley and R. E. Johnson, *J. Inorg. Nucl. Chem.* **11,** 33 (1959).
58. F. Basolo and A. Wojcicki, *J. Am. Chem. Soc.* **83,** 520 (1961).
59. O. Crichton, A. J. Rest, and D. J. Taylor, *J. Chem. Soc., Dalton Trans.* p. 167 (1980).
60. M. J. Wax and R. G. Bergman, *J. Am. Chem. Soc.* **103,** 7028 (1981).
61. C. Reichardt, "Solvent Effects in Organic Chemistry." Verlag Chemie, New York, 1979.
62. C. P. Casey and W. D. Jones, *J. Am. Chem. Soc.* **102,** 6156 (1980).
63. C.-Y. Chang, C. E. Johnson, T. G. Richmond, Y.-T. Chen, W. C. Trogler, and F. Basolo, *Inorg. Chem.* **20,** 3167 (1981).
64. M. Y. Darensbourg, H. L. Conder, D. J. Darensbourg, and C. Hasday, *J. Am. Chem. Soc.* **95,** 5919 (1973).
65. E. E. Siefert and R. J. Angelici, *J. Organomet. Chem.* **8,** 374 (1967).
66. R. J. Angelici and J. R. Graham, *J. Am. Chem. Soc.* **88,** 3658 (1966).
67. J. Ewen and D. J. Darensbourg, *J. Am. Chem. Soc.* **97,** 6874 (1975).
68. F. R. Jensen and R. C. Kiskis, *J. Am. Chem. Soc.* **97,** 5820 (1975).
69. J. A. Ewen and D. J. Darensbourg, *J. Am. Chem. Soc.* **98,** 4317 (1976).
70. D. J. Darensbourg and T. L. Brown, *Inorg. Chem.* **17,** 1679 (1968).
71. D. J. Darensbourg and J. A. Ewen, *Inorg. Chem.* **20,** 4168 (1981).
72. C. M. Ingemanson and R. J. Angelici, *Inorg. Chem.* **7,** 2646 (1968).
73. H. Werner, *J. Organomet. Chem.* **94,** 285 (1975), and references therein.
74. H. Werner, W. Beck, and H. Engelmann, *Inorg. Chim. Acta* **3,** 331 (1969).
75. J. L. Atwood and D. J. Darensbourg, *Inorg. Chem.* **16,** 2314 (1977).
76. D. J. Darensbourg, *Inorg. Chem.* **18,** 2821 (1979).
77. F. A. Cotton, D. J. Darensbourg, A. Fang, B. W. S. Kolthammer, D. Reed, and J. L. Thompson, *Inorg. Chem.* **20,** 4090 (1981).
78. G. C. Lalor, *J. Chem. Soc. A* p. 1 (1966).
79. D. A. Sweigart, M. Gower, and L. A. P. Kane-Maguire, *J. Organomet. Chem.* **108,** C15 (1976).
80. G. R. John, L. A. P. Kane-Maguire, and D. A. Sweigart, *J. Organomet. Chem.* **120,** C47 (1976).
81. J. Halpern, *J. Chem. Educ.* **45,** 372 (1968).
82. D. J. Darensbourg, B. J. Baldwin, and J. A. Froelich, *J. Am. Chem. Soc.* **102,** 4688 (1980).
83. F. A. Cotton and C. S. Kraihanzel, *J. Am. Chem. Soc.* **84,** 4432 (1964).
84. R. N. Perutz and J. J. Turner, *Inorg. Chem.* **14,** 262 (1975).
85. D. J. Darensbourg, H. H. Nelson III, and M. A. Murphy, *J. Am. Chem. Soc.* **99,** 896 (1977).
86. D. J. Darensbourg, *J. Organomet. Chem.* **174,** C70 (1979).
87. D. J. Darensbourg and B. J. Baldwin, *J. Am. Chem. Soc.* **101,** 6447 (1979).
88. R. L. Kump and L. J. Todd, *J. Chem. Soc., Chem. Commun.* p. 292 (1980).
89. R. L. Kump and L. J. Todd, *J. Organomet. Chem.* **194,** C43 (1980).
90. R. L. Kump and L. J. Todd, *Inorg. Chem.* **20,** 3715 (1981).
91. D. J. Darensbourg and H. H. Nelson, III, *J. Am. Chem. Soc.* **96,** 6511 (1974).
92. J. Nasieiski, P. Kirsch, and L. Wilputte-Steinert, *J. Organomet. Chem.* **27,** C13 (1971).
93. G. Platbrood and L. Wilputte-Steinert, *J. Organomet. Chem.* **70,** 393 (1974).
94. G. Platbrood and L. Wilputte-Steinert, *J. Organomet. Chem.* **70,** 407 (1974).

95. M. Wrighton and M. A. Schroeder, *J. Am. Chem. Soc.* **95**, 5764 (1973).
96. M. A. Schroeder and M. S. Wrighton, *J. Organomet. Chem.* **74**, C29 (1974).
97. R. N. Perutz and J. J. Turner, *J. Am. Chem. Soc.* **97**, 4791, 4800 (1975).
98. J. D. Black and P. S. Braterman, *J. Am. Chem. Soc.* **97**, 2908 (1975).
99. J. K. Burdett, R. N. Perutz, M. Poliakoff, and J. J. Turner, *J. Chem. Soc., Chem. Commun.* p. 157 (1975).
100. P. R. Hoffman and K. G. Caulton, *J. Am. Chem. Soc.* **97**, 4221 (1975).
101. J. K. Burdett, *Inorg. Chem.* **14**, 375 (1975).
102. J. K. Burdett, M. A. Graham, R. N. Perutz, M. Poliakoff, A. J. Rest, J. J. Turner, and R. F. Turner, *J. Am. Chem. Soc.* **97**, 4805 (1975).
103. D. J. Darensbourg, G. R. Dobson, and A. Moradi-Araghi, *J. Organomet. Chem.* **116**, C17 (1976).
104. G. Boxhoorn, G. C. Schoemaker, D. J. Stufkens, A. Oskam, A. J. Rest, and D. J. Darensbourg, *Inorg. Chem.* **19**, 3455 (1980).
105. R. S. Berry, *J. Chem. Phys.* **32**, 933 (1960).
106. D. J. Darensbourg and A. Salzer, *J. Organomet. Chem.* **117**, C90 (1976).
107. D. J. Darensbourg and A. Salzer, *J. Am. Chem. Soc.* **100**, 4119 (1978).
108. D. J. Darensbourg, L. J. Todd, and J. P. Hickey, *J. Organomet. Chem.* **137**, C1 (1977).
109. J. M. Kelly, H. Hermann, and E. Koerner von Gustorf, *J. Chem. Soc., Chem. Commun.* p. 105 (1973).
110. J. M. Kelly, D. V. Bent, H. Hermann, D. Schulte-Frohlinde, and E. Koerner von Gustorf, *J. Organomet. Chem.* **69**, 259 (1974).
111. G. R. Dobson and J. C. Rousche, *J. Organomet. Chem.* **179**, C42 (1979).
112. S. W. Kirtley, M. A. Andrews, R. Bau, G. W. Grynkewich, T. J. Markis, D. L. Tipton, and B. R. Whittlesey, *J. Am. Chem. Soc.* **99**, 7154 (1977).
113. W. Majunke, D. Leibfritz, T. Mack, and H. tom Dieck, *Chem. Ber.* **108**, 3025 (1975).
114. D. J. Darensbourg, *Inorg. Chem.* **18**, 14 (1979).
115. F. A. Cotton, D. J. Darensbourg, S. Klein, and B. W. S. Kolthammer, *Inorg. Chem.* **21**, 2661 (1982).
116. D. J. Darensbourg and R. Sachs, unpublished observations.
117. A. R. Rossi and R. Hoffmann, *Inorg. Chem.* **14**, 365 (1975).
118. J. D. Atwood and T. L. Brown, *J. Am. Chem. Soc.* **98**, 3155 (1976).
119. M. A. Cohen and T. L. Brown, *Inorg. Chem.* **15**, 1417 (1976).
120. D. L. Lichtenberger and T. L. Brown, *J. Am. Chem. Soc.* **100**, 366 (1978).
121. P. A. Bellus and T. L. Brown, *J. Am. Chem. Soc.* **102**, 6020 (1980).
122. A. Wojcicki and F. Basolo, *J. Am. Chem. Soc.* **83**, 525 (1961).
123. W. Hieber and K. Wollmann, *Chem. Ber.* **95**, 1552 (1962).
124. B. F. G. Johnson, J. Lewis, J. R. Miller, B. H. Robinson, P. W. Robinson, and A. Wojcicki, *J. Chem. Soc. A* p. 522 (1968).
125. T. L. Brown, *Inorg. Chem.* **7**, 2673 (1968).
126. D. J. Darensbourg, M. Y. Darensbourg, and R. J. Dennenberg, *J. Am. Chem. Soc.* **93**, 2807 (1971).
127. D. J. Darensbourg, N. Walker, and M. Y. Darensbourg, *J. Am. Chem. Soc.* **102**, 1213 (1980).
128. D. J. Darensbourg, M. Y. Darensbourg, and N. Walker, *Inorg. Chem.* **20**, 1918 (1981).
129. F. A. Cotton, D. J. Darensbourg, and B. W. S. Kolthammer, *J. Am. Chem. Soc.* **103**, 398 (1981).
130. D. J. Darensbourg, M. B. Fischer, R. E. Schmidt, Jr., and B. J. Baldwin, *J. Am. Chem. Soc.* **103**, 1297 (1981).

131. D. J. Darensbourg, A. Rokicki, and M. Y. Darensbourg, *J. Am. Chem. Soc.* **103**, 3223 (1981).

132. D. J. Darensbourg and A. Rokicki, *ACS Symp. Ser.* **152**, 107 (1981).

133. D. J. Darensbourg and R. Kudaroski, *182nd Nat. Meet., Am. Chem. Soc., 1981* INOR No. 50 (1981).

134. F. A. Cotton, D. J. Darensbourg, B. W. S. Kolthammer, and R. Kudaroski, *Inorg. Chem.* **21**, 1656 (1982).

135. D. J. Darensbourg and A. Rokicki, *Organometallics* (in press).

136. T. L. Brown, *J. Mol. Catal.* **12**, 41 (1981), and references therein.

137. K.-Y. Hui and B. L. Shaw, *J. Organomet. Chem.* **124**, 262 (1977).

138. D. J. Darensbourg and J. A. Froelich, *J. Am. Chem. Soc.* **100**, 338 (1978).

139. T. L. Brown and P. A. Bellus, *Inorg. Chem.* **17**, 3726 (1978).

140. D. E. Morris and F. Basolo, *J. Am. Chem. Soc.* **90**, 2531 (1968).

141. D. J. Darensbourg and R. L. Kump, *J. Organomet. Chem.* **140**, C29 (1977).

142. D. J. Darensbourg and J. A. Froelich, *J. Am. Chem. Soc.* **99**, 5940 (1977).

143. D. J. Darensbourg, R. R. Burch, Jr., and M. Y. Darensbourg, *Inorg. Chem.* **17**, 2677 (1978).

144. B. D. Dombek and R. J. Angelici, *J. Am. Chem. Soc.* **98**, 4110 (1976).

145. M. Poliakoff, *Inorg. Chem.* **15**, 2892 (1976).

146. D. J. Darensbourg, R. Kudaroski, and W. Schenk, *Inorg. Chem.* **21**, 2488 (1982).

147. J. C. Bailar, Jr., *J. Inorg. Nucl. Chem.* **8**, 165 (1958).

148. P. Ray and N. K. Dutt, *J. Indian Chem. Soc.* **20**, 81 (1943).

149. C. S. Springer and R. E. Sievers, *Inorg. Chem.* **6**, 852 (1967).

150. N. Serpone and D. G. Bickley, *Prog. Inorg. Chem.* **17**, 391 (1972).

151. L. G. Vanquickenborne and K. Pierloot, *Inorg. Chem.* **20**, 3673 (1981).

152. R. Hoffmann, J. M. Howell, and A. R. Rossi, *J. Am. Chem. Soc.* **98**, 2484 (1976).

153. G. R. Dobson, K. J. Asali, J. L. Marshall, and C. R. McDaniel, Jr., *J. Am. Chem. Soc.* **99**, 8100 (1977).

154. G. R. Dobson and K. J. Asali, *J. Am. Chem. Soc.* **101**, 5433 (1979).

155. G. R. Dobson and K. J. Asali, *Inorg. Chem.* **20**, 3563 (1981).

156. A. M. Bond, D. J. Darensbourg, E. Mocellin, and B. J. Stewart, *J. Am. Chem. Soc.* **103**, 6827 (1981).

157. T. L. Brown, *Ann. N. Y. Acad. Sci.* **333**, 80 (1980).

158. D. R. Kidd and T. L. Brown, *J. Am. Chem. Soc.* **100**, 4095 (1978).

159. B. H. Byers and T. L. Brown, *J. Am. Chem. Soc.* **99**, 2527 (1977).

160. N. W. Hoffman and T. L. Brown, *Inorg. Chem.* **17**, 613 (1978).

161. M. Absi-Halabi and T. L. Brown, *J. Am. Chem. Soc.* **99**, 2982 (1977).

162. B. H. Byers and T. L. Brown, *J. Am. Chem. Soc.* **97**, 947 (1975).

163. S. B. McCullen, H. W. Walker, and T. L. Brown, *J. Am. Chem. Soc.* **104**, 4007 (1982).

164. A. Poë, *Inorg. Chem.* **20**, 4029, 4032 (1981), and references therein.

165. H. Wawersik and F. Basolo, *Inorg. Chim. Acta* **3**, 113 (1969).

166. D. Sonnenberger and J. D. Atwood, *J. Am. Chem. Soc.* **102**, 3484 (1980).

167. J. D. Atwood, *Inorg. Chem.* **20**, 4031 (1981).

168. S. P. Schmidt, W. C. Trogler, and F. Basolo, *Inorg. Chem.* **21**, 1698 (1982).

169. Some of the published reports dealing with this subject include K. J. Karel and J. R. Norton, *J. Am. Chem. Soc.* **96**, 6812 (1974); S. K. Malik and A. Poë, *Inorg. Chem.* **17**, 1484 (1978); G. Bor, U. K. Dietler, P. Pino, and A. Poë, *J. Organomet. Chem.* **154**, 301 (1978); D. J. Darensbourg and M. J. Incorvia, *ibid.* **171**, 89 (1979); J. P. Candlin and A. C. Shortland, *ibid.* **16**, 289 (1969); S. K. Malik and A. Poë, *Inorg. Chem.* **18**, 1241

(1979); D. J. Darensbourg and M. J. Incorvia, *ibid.* **19**, 2585 (1980); **20**, 1911 (1981); D. Sonnenberger and J. D. Atwood, *ibid.* p. 3243; G. F. Stuntz and J. R. Shapley, *J. Organomet. Chem.* **213**, 389 (1981); D. J. Darensbourg and B. J. Baldwin-Zuschke, *Inorg. Chem.* **20**, 3846 (1981); D. J. Darensbourg, B. S. Peterson, and R. E. Schmidt, Jr., *Organometallics* **1**, 306 (1982); D. Sonnenberger and J. D. Atwood, *J. Am. Chem. Soc.* **104**, 2113 (1982); D. J. Darensbourg and B. J. Baldwin-Zuschke, *ibid.* p. 3906.

170. E. L. Muetterties, R. R. Burch, Jr., and A. M. Stolzenberg, *Annu. Rev. Phys. Chem.* **33**, 89 (1982).

ADVANCES IN ORGANOMETALLIC CHEMISTRY, VOL. 21

1,4-Diaza-1,3-butadiene (α-Diimine) Ligands: Their Coordination Modes and the Reactivity of Their Metal Complexes

GERARD VAN KOTEN and KEES VRIEZE

Anorganisch Chemisch Laboratorium
J. H. van 't Hoff Instituut
University of Amsterdam
Amsterdam, The Netherlands

I

INTRODUCTION

Molecules containing the 1,4-diaza-1,3-butadiene skeleton have attracted much interest because of both their versatile coordination behavior and the interesting properties of their metal complexes. In particular, extensive chemistry has been carried out with 2,2'-bipyridine and phenanthroline, which are both known to coordinate to metal centers in the chelate bonding mode (1).

Relatively less well investigated, but increasingly of interest to various research groups, is the coordination chemistry of the most simple representative of this class of compounds, i.e., the 1,4-disubstituted 1,4-diaza-1,3-butadienes, $RN=CR'-CR''=NR$. These compounds are particularly fascinating since they have a flexible $N=C-C=N$ skeleton, they appear to have unusual electron donor and acceptor properties as compared to the above-mentioned bidentate nitrogen donors, and they can potentially act in a variety of coordination modes. The latter bonding modes involve not only the lone pairs of the N atoms but also the π-$C=N$ bonds.

Recently other aspects have also come to light that involve the chemical activation of R-DAB[1] and the subsequent stoichiometric and catalytic reactions in which the activated ligand plays a crucial role in the reaction processes.

In view of these novel developments it was considered worthwhile to review in depth the synthesis, structures, and properties of the various types of 1,4-diaza-1,3-butadiene complexes known at present. Furthermore, applications of these compounds in organic synthesis and catalysis will be discussed. Attention will be devoted to the possible relation(s) between the type of coordination and the type of reaction occurring. Our survey is restricted to complexes of the R-DAB ligand in which the R group is connected to N via a carbon atom and does not cover complexes of 2,2'-bipyridines or 2-pyridinecarbaldehydeimines.[2]

---

[1] Most 1,4-diaza-1,3-butadienes that are known have the general formula $RN=CR'CR''=NR$ and herein this will be abbreviated to R-DAB(R', R''). The important subgroup of this class is $RN=CHCH=NR$ [R-DAB(H,H)] but for economy of space if the R grouping is specifically stated then the form R substituent-DAB is used and this implies proton substitution at the α diimine carbon atoms, e.g., t-Bu-N=CHCH=N-t-Bu and HN=CHCH=NH become t-Bu-DAB and H-DAB, respectively. For the general case applying to all variously substituted 1,4-diaza-1,3-butadienes, including the rarely encountered $RN=CR'CR''=NR'''$ species, the abbreviation used is R-DAB.

[2] An account of the metal-1,4-diaza-1,3-butadiene research which has been carried out in the authors' laboratory is presented in Ref. 2.

II

THE 1,4-DIAZA-1,3-BUTADIENE (R-DAB) LIGAND

A. *Preparation*

1,4-Diaza-1,3-butadienes (R-DAB)[1] may be prepared by condensation reactions involving either glyoxals (3–7), α-ketoaldehydes (8–10), or α,β-diketones (8–11) with primary amines RNH_2. In the case of methyl-glyoxal it has been established (8) that the reaction with amine proceeds in two steps [see Eqs. (1) and (2)]. In the first step the α-imino ketone formed is only stable enough to be isolated when R is a bulky group such as t-Bu (8–10). This result suggests that this reaction occurs with high chemiospecificity (8) and this is probably due to the higher reactivity of the aldehyde group toward amines than the keto group.

$$O=C(Me)C(H)=O + RNH_2 \rightleftarrows O=C(Me)C(H)=NR + H_2O \tag{1}$$

$$R = t\text{-Bu or } EtMe_2C$$

$$O=C(Me)C(H)=NR + R'NH_2 \leftrightarrows R'N=C(Me)C(H)=NR + H_2O \tag{2}$$

$$R' = i\text{-Pr}$$

Further reaction of the α-amino ketone occurs only with less bulky amines [Eq. (2)] (8) resulting in formation of an asymmetric R,R'-DAB(Me,H) ligand.

Some of the R-DAB(R', R'') ligands are not very stable as free molecules (e.g., R = R' = R'' = Me) (11) and these must be synthesized in the coordination sphere of a metal. Examples will be discussed in Section III,D,1.

B. *Structural and Bonding Features*

It was concluded from NMR spectra (12), dipole moments (13), and IR spectra (14) that the R-DAB molecule exists in solution in the E (anti) configuration at both C=N double bonds, while the conformation of the central C—C bond is predominantly s-trans.[3] It was deduced that the N=C—C=N dihedral angle lies between 90 and 140° (13).

In the gas phase, according to electron diffraction analysis, t-BuN=CHCH=Nt-Bu has for the majority of the molecules a gauche conformation with respect to the central C—C bond with a torsion of about

---

[3] The terms s-trans and s-cis refer to torsional isomers around the central carbon–carbon bond of the 1,3-diene skeleton.

$65°$ from the s-cis[3] form. However, a small amount of the s-trans form is also calculated to be present (*15, 16*).

Recently we determined the structure of c-HexN=CHCH=Nc-Hex (c-Hex-DAB) in the solid state by X-ray crystallographic structure analysis in order to obtain insight into the C=N and C—C bond lengths and angles in the free molecule (see Table I and Section IV,B,1). The structure revealed a perfectly flat N=C—C=N skeleton in the *E-s-trans-E* conformation (*17*). The similarity between the central C—C distances in c-Hex-DAB and isostructural 1,3-butadienes (see Table I) is particularly striking and indicates that we are dealing with a pure $C(sp^2)$—$C(sp^2)$ bond. In Table I the bond lengths and angles in the N=C—C=N unit of a series of related 1,4-diaza-1,3-butadiene molecules are given.

NDDO (neglect of diatomic differential overlap) (*16*), CNDO/2 (*15*) and *ab initio* calculations (*18, 19*) on the conformational structures of 2,2'-bipyridine, H-DAB and Me-DAB show that the s-trans form is indeed expected to be the most stable one (*16*). For example, for 2,2'-bipyridine an energy difference of 26.8 kJ mol^{-1} was calculated between the less stable planar s-cis form and the more stable planar s-trans conformation. The destabilization of the s-cis form is caused by the interaction of the lone pairs and by the steric hindrance of the ortho hydrogen atoms. Values ranging between 20 and 28 kJ mol^{-1} were calculated to be necessary to overcome the rotational barrier to produce the cis arrangement that is present in the chelate form of many α-diimine–transition metal compounds.

Inspection of molecular models shows that substitution of methyl groups at the central C atom destabilizes the *E-s-trans-E* conformer in particular when the R substituent is triply branched at C^α. In this case the *E-s-cis-E* conformer becomes relatively much more stable. Notable exceptions are the so-called 1,4-diaza-1,3-butadien-2-ylmetal complexes in which one of the central C atoms is σ-bonded to a square planar trans-ClL$_2$Pd moiety, e.g., PdCl[C(=NC$_6$H$_4$(OMe-p)-C(Me)=NC$_6$H$_4$OMe-p](PPh$_3$)$_2$ (*20*) (see Table I). Since the Pd coordination plane is almost perpendicular to the planar *E-s-trans-E* skeleton it is internally recognized as a very small substituent. Accordingly, complexes are known containing the 1,4-diaza-1,3-butadien-2-yl ligand in both the *E-s-trans-E* and the *E-s-cis-E* conformation.

Also of interest are the relative donor–acceptor properties of the various organic molecules containing the N=C—C=N skeleton. NDDO calculations of the LUMO (lowest unoccupied metal orbital) energies indicated that the π-acceptor capacity increases in the order 2,2'-bipyridine < 2-pyridinecarbaldehyde-*N*-methylimine < R-DAB (*26*).

Finally, it should be mentioned that some UV-PES spectra in combination with UV spectra have been recorded (*11, 27*) for some R-DAB

TABLE I
BOND LENGTHS AND ANGLES IN THE N=C—C'=N UNIT OF FREE α-DIIMINES[a]

Compound	Reference	C—C' (Å)	C=N (Å)	N—C[b] (Å)	⟨N=C—C'⟩ (°)
c-Hex-DAB[c]	21	1.4571(23)	1.2576(22)	1.4561(23)	120.80(17)
2,2'-Bipyridine	22	1.50			
2,2'-Biquinoline[c,d]	23	1.492(3)	1.323(2)[e]		116.8(2)
8,8'-Biquinoline	24	1.495(2)			
4-MeOC$_6$H$_4$-DAB{trans-[Pd(PPh$_3$)$_2$Cl], Me}[c]	20	1.51(2)	1.26(2)/1.29(2)	1.44(3)/1.40	113.9(1.3)
t-Bu-DAB[f]	18	1.496(20)[g]	1.283(6)		117.3(4.0)
H$_2$C=C(H)C(H)=CH$_2$	25	1.48(1)[h]	1.341[i]		

[a] Single crystal X-ray diffraction analysis.
[b] Carbon of R group.
[c] s-trans conformation ($\theta \simeq 0°$).
[d] H(3)- - -N' 2.44(1) Å.
[e] Part of the aromatic ring system.
[f] Electron diffraction analysis of the molecule in the gas phase.
[g] $\theta \simeq 65°$.
[h] Central C—C bond length.
[i] C=C bond length.

ligands. The results are not very revealing, but comparisons indicate increasing polarity in the C=X bond on going from 1,3-butadienes to 1,4-diaza- and 1,4-dioxo-1,3-butadienes (R-DABs and glyoxals respectively) (*11*).

III

METAL–1,4-DIAZA-1,3-BUTADIENE COMPLEXES: SYNTHESIS, STRUCTURE, AND BONDING

A. *Introduction*

The reports of metal complexes with R-DAB ligands date back to 1953 when Krumholz (*28*) described the synthesis of some ferrous complexes, e.g., $[Fe(Me-DAB)_3]I_2$. The unusual stability and characteristic color being ascribed to the presence of π bonding between the metal and the nitrogen atoms. Since then numerous examples of metal–R-DAB complexes have been synthesized and their bonding studied by spectroscopic and theoretical methods.

A consistent structural feature of these complexes appeared to be the chelate bonding of the R-DAB ligand. Since the free R-DAB molecule exists in the *E-s-trans-E* conformation (see Section II,B) this implies that upon coordination to the metal center rotation around the central C—C bond must have taken place to give the *E-s-cis-E* conformation present. In fact it is very surprising that it was not before 1978 that the first examples were found of the other possible interactions with metal centers by Frühauf, Vrieze, and van Koten (*29, 30*). The earlier reports by Kliegman and co-workers concerning the behavior of R-DAB molecules toward perchloric acid had already pointed to the possible existence of other coordination modes (*12, 31*). They found that in addition to the monobasic behavior of most R-DAB molecules those with bulky R groups (e.g., o-tolyl) appeared to be dibasic in nature. It was proposed that the monobasic behavior is due to formation of a five-membered, highly stabilized, planar ring system. This is only possible if the N=C—C=N skeleton can assume the *E-s-cis-E* conformation according to Eq. (3).

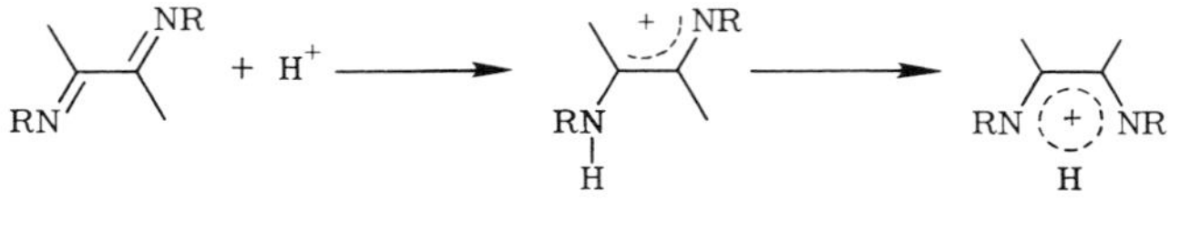

$$\tag{3}$$

By contrast, if the R groups are bulky the N atoms are assumed to be blocked from interaction with the proton in the E-s-cis-E conformation (4). In these cases the R-DAB ligand is diprotonated in the E-s-trans-E conformation.

It is exactly this influence of the steric nature of the R substituents of the R-DAB ligand on the stability of the conformation of the $N=C-C=N$ skeleton that affects the type of metal interaction found (32). Other factors are the nature of the metal center itself and the coordinated ligands. The various bonding modes found for the 1,4-diaza-1,3-butadiene ligand are shown in Fig. 1. The planar E-s-trans-E conformation will be particularly suited either for coordination to one metal center via the lone pair of one N atom (σ-N monodentate) or for a bridging coordination mode between metal centers via the lone pairs on each of the N atoms (σ-N,σ-N' bridging). Furthermore, conformations ranging from gauche to planar E-s-cis-E also allow the involvement of one π C=N bond in addition to the two lone pairs on the N atoms resulting in a bridging coordination mode (σ-N,μ^2-N',η^2-CN'). Finally the planar E-s-cis-E conformer can be either chelate bonded to one metal center (σ,σ-N,N' chelate) or chelate bonded to one metal atom and η^2-bonded via both π C=N bonds to a second metal center thus attaining a bridge bonding mode (σ-N,σ-N',η^2-CN,η^2-CN'). It is obvious that pure η^2-C=N,η^2-C=N' bonded R-DAB ligands in the E-s-cis-E conformation are only possible when both σ lone pairs are first involved in bonding to either one or two metal centers (31, 33).

In the following sections these various coordination modes will be successively treated thus revealing the fascinating versatile coordination behavior of this ligand. This versatility becomes particularly evident when one compares the coordination modes of the R-DAB ligand with the single σ,σ-N,N' chelate coordination mode observed for related 2,2'-bipyridine. The R-DAB ligand has the unique property that it enables the metal center to adjust its electron density by changing its point of attachment to the $N=C-C=N$ system (2); the latter donating either 2, 4, 6, or 8 electrons.

B. *Monodentate (σ-N;2e) 1,4-Diaza-1,3-butadienes*

So far stable complexes containing monodentate R-DAB have only been reported for the square planar d^8-metals Pd^{II}, Pt^{II}, and Rh^I (29, 32, 34, 35). Compounds of the type *trans*-PdX$_2$(R-DAB)$_2$ have been obtained from the reaction of PdX$_2$(PhCN)$_2$ with t-Bu-DAB or EtMe$_2$C-DAB (29). The NMR spectra of these compounds were consistent with the trans structure shown in Fig. 2. The R-DAB ligand is coordinated via the N lone pair in

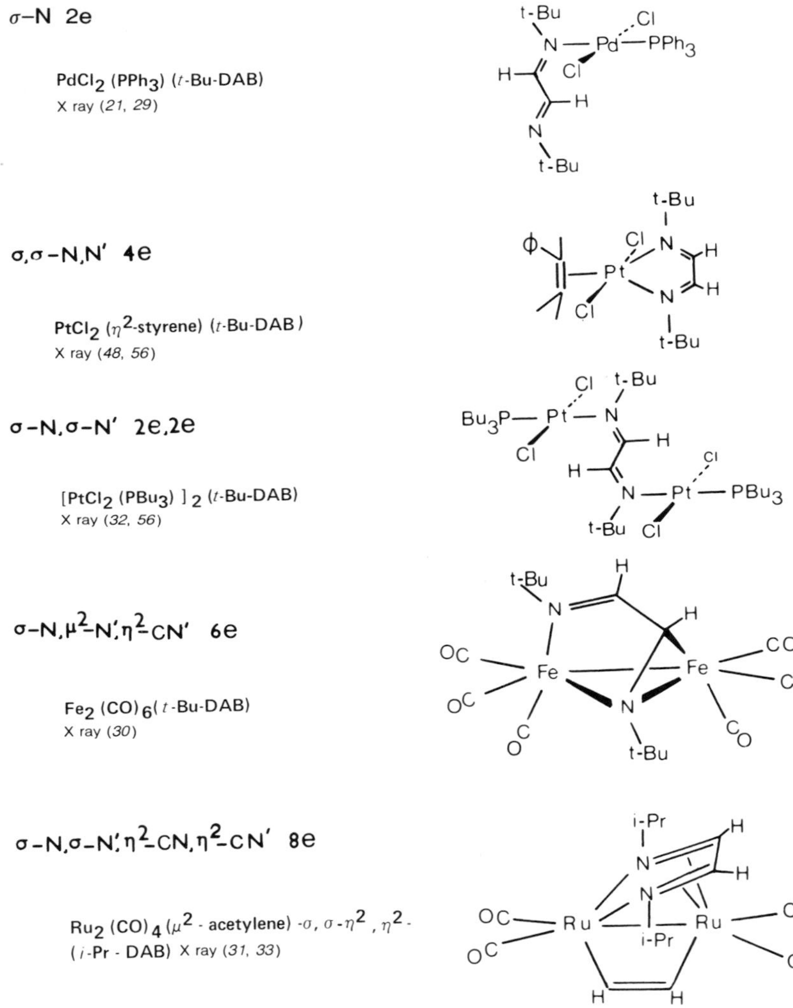

FIG. 1.   Examples of metal–R-DAB complexes, illustrating the various coordination possibilities.

the *s*-trans conformation with the coordinated C=N site in the *E* conformation and as a result the N=C—C=N skeleton is almost planar. This conclusion is based on the anomalously low field shift of the H^β proton upon coordination. The magnetic anisotropy of the planar complex would necessarily result in a deshielding of H^β located close to the metal above the coordination plane in this conformation.

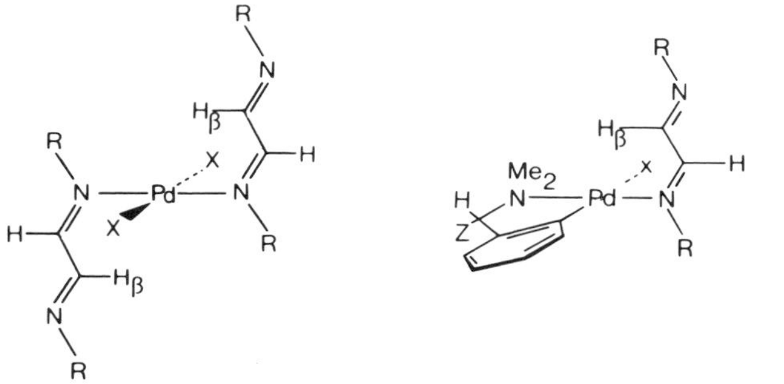

(a)   (b)

FIG. 2. Proposed structure (a) of *trans*-PdX$_2$(R-DAB)$_2$ (R = *t*-Bu, EtMe$_2$C; X = Cl, Br) (*29*), and (b) of *trans*-*N*-PdX(C$_6$H$_4$CH(Z)NMe$_2$)(R-DAB) (R = *t*-Bu, EtMe$_2$C; X = Cl, Br; Z = H, (*S*)-Me) (*36*).

That the planar *E-s*-trans-*E* conformation is indeed preferred for monodentate bonded R-DAB (as well as for σ-N,σ-N' bridging R-DAB, see Section III,C) can be concluded from the structure in the solid state of *trans*-PdCl$_2$(PPh$_3$)(*t*-Bu-DAB) (*21*) (see Fig. 1 and Table II). The *t*-Bu-DAB ligand is coordinated via one N atom while the second imino-N atom is free. H$^\beta$ resides above the Pd^{II} coordination plane at a calculated distance of 2.6 Å. One important aspect of this structure is the fact that the N=C—C=N skeleton is somewhat bent toward the Pd coordination plane in order to minimize contact of the *t*-Bu group with the cis ligands.

Complexes similar to *trans*-PdCl$_2$(PPh$_3$)(*t*-Bu-DAB) have been prepared via the bridge splitting reactions shown in Eq. (4) which proceed via formation of complexes with 2:1 M:R-DAB molar ratios (*29*).

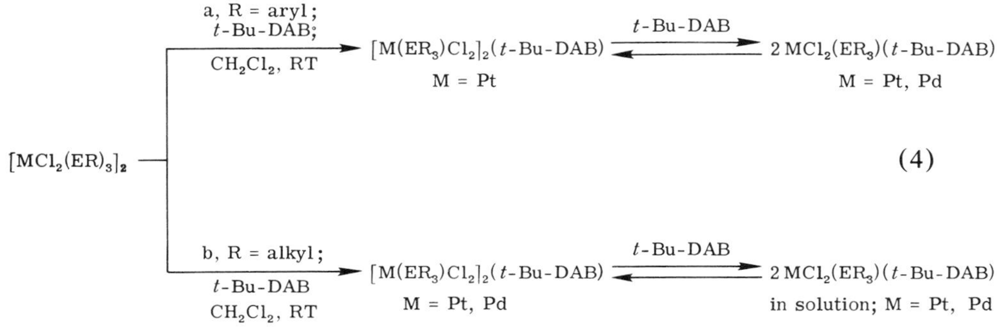

$$[MCl_2(ER_3)]_2$$

a, R = aryl;
t-Bu-DAB;
CH$_2$Cl$_2$, RT   →   [M(ER$_3$)Cl$_2$]$_2$(*t*-Bu-DAB) $\underset{}{\overset{t\text{-Bu-DAB}}{\rightleftarrows}}$ 2 MCl$_2$(ER$_3$)(*t*-Bu-DAB)
                              M = Pt   M = Pt, Pd

                                                                                 (4)

b, R = alkyl;
t-Bu-DAB
CH$_2$Cl$_2$, RT   →   [M(ER$_3$)Cl$_2$]$_2$(*t*-Bu-DAB) $\underset{}{\overset{t\text{-Bu-DAB}}{\rightleftarrows}}$ 2 MCl$_2$(ER$_3$)(*t*-Bu-DAB)
                              M = Pt, Pd   in solution; M = Pt, Pd

Both the 2:1 and 1:1 Pt–R-DAB complexes with a trans triarylphosphine or -arsine ligand are stable and isolable but in the case of Pd this

TABLE

STRUCTURAL FEATURES OF R-DAB–METAL

Bonding mode; number of electrons donated. Compound	Distances (Å) and angles (°)		
	M—N	N—M—N	C=N
A. Monodentate (σ-N; 2e)			
trans-$PdCl_2(PBu_3)$(*t*-Bu-DAB)	2.130(6)		1.264(10)[b] 1.239(10)[c]
B. Bridging (σ—N,σ—N′; 2e + 2e)			
[*trans*-$PtCl_2(PBu_3)]_2$(*t*-Bu-DAB)	2.214(10)		1.27(3)
C. Chelating (σ,σ—N,N′; 4e)			
$Mo(CO)_4$(R-DAB) R = *i*-Pr	2.263 2.276	N.R.	1.277 1.283
R = 2,6-(*i*-$Pr)_2C_6H_3$-DAB	2.238 2.222	N.R.	1.275 1.288
$MoCl(\eta^3\text{-}C_4H_7)(CO)_2$(*c*-Hex-DAB)	2.237(4)	72.78(15)	1.283(7)
$WBr(\eta^3\text{-}C_3H_5)(CO)_2$(*c*-Hex-DAB)	2.219(10)	72.38(34)	1.303(16)
$MnBr(CO)_3$(*c*-Hex-DAB)	2.057(14) 2.050(15)	78.05(55)	1.294(27) 1.274(30)
$ReCl(CO)_3$(*i*-Pr-DAB)	2.258(18) 2.232(19)	72.72(73)	1.345(36) 1.264(39)
Mn(*t*-Bu-$DAB)_2$	2.06	80.5	1.32
Ru(*p*-$MeOC_6H_4$-$DAB)_3$	2.06(3) mean 2.04–2.10	74.4–78.7	1.34(5) mean 1.28–1.38
$RuCl_2$(*i*-Pr-$DAB)_2$	2.000(6)–2.051(6)	78.4(3) and 77.8(3)	1.291 mean
Rh(*i*-Pr_2CH-DAB)$(CO)_2$-$RhCl_2(CO)_2$	2.118(5) 2.109(5)	78.8(2)	1.282(8) 1.304(8)
$Fe(CO)_3$(2,6-*i*-$Pr_2C_6H_3$-DAB)	1.927(3)	80.1(1)	1.329(5)
$Fe(NO)_2$(*t*-Bu-DAB)	2.03	79.8	1.26
Fe(CO)(*i*-Pr-DAB)(2,3-$Me_2C_4H_4$)	1.930(1)	81.0(1)	1.311(2)
Ni(*c*-Hex-$DAB)_2$	1.924 mean	83.0 mean	1.321 mean
Ni(xylyl-$DAB)_2$	1.928 mean	83.1 mean	1.342 mean
$Ni(CO)_2[Me_2N$-DAB(Me, Me)]	1.97(3) 1.99(5)	81.0	1.22(6) 1.22(5)
NiBr(metalated *i*-Pr_2CH-DAB)	1.820(13)[d] 1.995(14)[e]	82.0	1.294(24)

$C-C'^a$	$C-C'-N$	$N-C^\alpha$	Section	Reference
1.485(9)	124.17(67)	1.494(8)	III,B (see Fig. 1)	*21, 29*
1.48(2)	118.2(13)	1.52(2)	III,C (see Fig. 1)	*32*
1.443	N.R.	N.R.	III,D,2,b	*37*
1.467	N.R.	N.R.	III,D,2,b	*37*
1.448(7)	118.05(49)	1.494(6)	III,D,2,b	*38*
1.466(17)	116.28(106)	1.506(14)	III,D,2,b	*39*
1.490(22)	112.12(1.86) 118.43(1.80)	1.453(17) 1.473(22)	III,D,2,c	*40*
1.378(45)	116.95(2.62)	1.508(37) 1.462(32)	III,D,2,c	*41*
1.38	N.R.	N.R.	III,D,2,c	*42*
1.37(5) mean 1.33–1.45	N.R.	N.R.	III,D,2,d	*43*
1.40 mean	117.7 mean	N.R.	III,D,2,d	*44*
1.466(8)	118.1(6)	1.507(7) 1.491(7)	III,D,2,e	*45*
1.390(5)	114.2(3)	1.441(4)	III,D,2,d	*110*
1.45	N.R.	N.R.	III,D,2,d	*42*
1.405(3)	N.R.	1.407(2)	III,D,2,d	*111b*
1.400 mean	N.R.	1.479 mean	III,D,2,f	*9*
1.374 mean	116.06 mean	1.421 mean	III,D,2,f	*10*
1.54(4)	110.4 119.3		III,D,2,f	*46*
1.454(24)	111.4	1.504(20) 1.483(22)	III,D,2,f	*47*

(Continued)

TABLE II

Bonding mode; number of electrons donated. Compound	Distances (Å) and angles (°)		
	M—N	N—M—N	C=N
Ni$_2$(μ-Br)$_2$(i-PrCH-DAB)$_2$	1.93(3) mean	83.1(8)	1.29(3) mean
PtCl$_2$(η^2-styrene)(t-BuDAB)	2.20(3) 2.31(3)	7.47(10)	1.28(4)
PtCl$_2$(η^2-ethylene)-[Me(H)N-DAB]	2.221(10)	72.0(4)	1.289(15)

D. Bridging (σ-N,μ^2-N',η^2-CN'; 6e)

Compound	M—M	M^1—N^1	M^1—N^2	M^2—N^2
Fe$_2$(CO)$_6$(c-Hex-DAB)	2.597(1)	1.991(3)	1.972(3)	1.930(3)
Ru$_2$(CO)$_4$(i-Pr-DAB)$_2$	3.308(1)^g	2.16(1)	2.11(1)	2.14(1)
MnCo(CO)$_5$(μ-CO)(t-Bu-DAB)	2.639(3)	2.094(9)	2.048(9)	1.891(9)

E. Bridging (σ,σ-N,N',η^2-CN,η^2-CN'; 8e)

Compound	M—M	M^1—N^1 / M^1—N^2	M^2—N^1 / M^2—N^2	M^2—C^1 / M^2—C^2
Ru$_2$(CO)$_4$(μ-C$_2$H$_2$)(i-Pr-DAB)	2.936(1)	2.117(6) 2.111(6)	2.226(7) 2.225(6)	2.226(7) 2.226(7)
Ru$_4$(CO)$_8$(i-Pr-DAB)$_2$	i	2.07(1) 2.20(1)	2.17(2) 2.18(2)	2.26(2) 2.24(2)
Mn$_2$(CO)$_6$[Me-DAB(Me,Me)]	2.615(1)	1.997(3) 1.995(3)	2.111(3) 2.108(3)	2.147(4) 2.137(4)

[a] Central C atoms (imino-carbon atoms)
[b] Coordinated imino-N atom.
[c] Free imino-N atom.
[d] Trans to Br.
[e] Trans to C.

is only so for the 1 : 1 complexes. By contrast, the use of trialkylphosphines or -arsines yielded the 2 : 1 platinum or palladium complexes as stable solids [Eq. (4)] and 1 : 1 complexes were formed in solution. The intermolecular exchange processes between the dinuclear and mononuclear species and free R-DAB are slow on the NMR time scale (29). However, ^{1}H, ^{13}C, ^{15}N, ^{31}P, and ^{195}Pt NMR studies of ^{15}N labeled (35) and nonlabeled (29) com-

(*Continued*)

C—C'^{a}	C—C'—N	N—C^a	Section	Reference
1.38(3) mean	117(2) mean	1.48(3) mean	III, D, 2, f	47, 147
1.51(5)		1.46(4)	III,D,2,f	48
1.50(2)	116(1)	N.R.	III,D,2,f	49

M²—C²	C²—N²	N¹=C¹	C¹—C²—N²	C¹—C²	θ^f	Section	Reference
2.069(3)	1.397(4)	1.280(5)	N.R.	1.435(5)	12.2°	III,E (See Fig. 1)	30
2.14(1)	1.43(1)	1.30(1)	115.9(6)	1.45(1)	5.0	III,E	50
2.065(11)	1.358(16)	1.260(16)	115.2(11)	1.405(15)	11.0	III,E	51

C¹—N¹ C²—N²	C¹—C²		Section	Reference
1.395(10)	1.396(11)	≃0°^{h}	III,F (See Fig. 1)	31
1.451(9)				
1.39(2)	1.42(3)		III,F	31
1.43(2)	1.40(3)			52
1.392(5)	1.407(5)		III,F	53
1.388(4)				

^{f} Dihedral angle between N¹=C¹ and N²—C².
^{g} Ru- - -Ru distance.
^{h} Ru¹N¹N² plane makes dihedral angle of 14° with the N¹C¹C²N² plane.
^{i} Ru³(in metallocycle)—Ru¹(between metallocycles) 2.838(2); Ru²(in metallocycle)—Ru¹ 2.848(2); Ru²- - -Ru³ 2.994(2); Ru⁴(nonbridged)—Ru² 2.838(2); Ru⁴—Ru³ 2.846(2) Å.

pounds showed that only at low temperature (slow exchange limit) is the R-DAB ligand monodentate bonded and rigid with a ground state conformation deduced to be similar to that in *trans*-PdCl$_2$(PPh$_3$)(*t*-Bu-DAB) (Fig. 1). At room temperature the metal is rapidly changing its point of attachment by the process shown in Fig. 3.

At −55°C the spectrum belonging to isomers A and A' is observed (char-

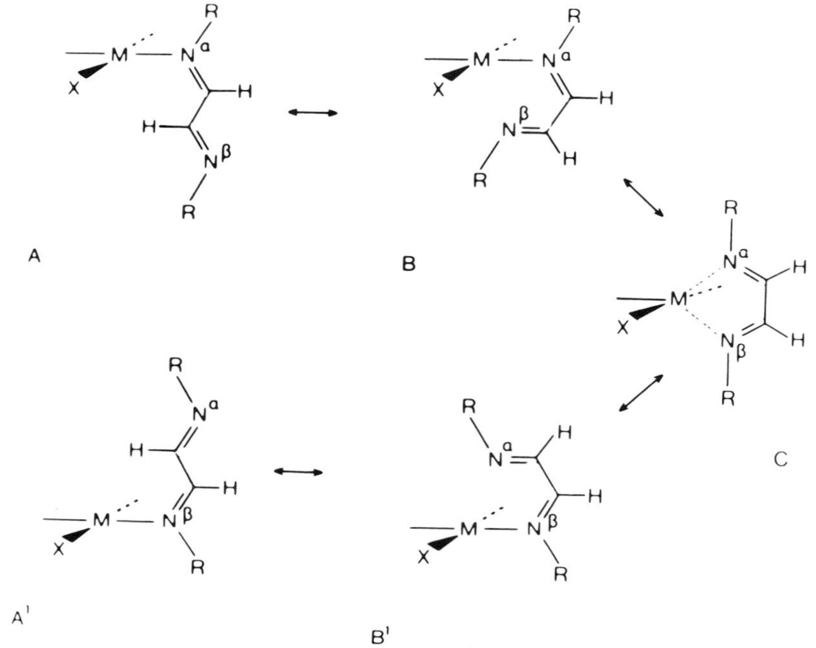

FIG. 3. Proposed mechanism for the fluxional behavior of *trans*-$MX_2(PR_3')$(R-DAB) (*29*) and *trans*-N-MX(C_6H_4CH(Z)NMe_2)(R-DAB) (*36*) (M = Pt or Pd; X = Cl, Br, or I; R' = Ph, Bu; R = *t*-Bu or $EtMe_2C$; Z = H or (*S*)-Me) complexes in solution (changing the point of attachment of the metal to the R-DAB ligand via σ-N ⇌ σ,σ-N,N' rearrangement).

acterized by the low field shift of H^β). At room temperature a situation is reached in which the process A ⇌ A' via B is rapid on the NMR time scale. This process involves *E* to *Z* inversion at the free N site and rotation around the central C—C bond in order to bring the lone pair into the coordination sphere of the metal. In the intermediate or transition state B (cf. Ref. *35*) the central metal is rehybridized from square planar to a trigonal bipyramidal configuration.

Replacement of one Cl and one phosphine ligand in the [$PdCl_2$-(ER_3')]$_n$(R-DAB) (E = P,As) complexes by a carbon–nitrogen donor ligand leading to the compounds shown in Fig. 2b even further destabilizes the Pd—N (imine) interaction. For these compounds, which can only be studied in solution, an intramolecular process similar to that outlined for the 1 : 1 complexes in Fig. 3 has been established (*36*).

The decrease of the M—N bond strength going from the Pd–R-DAB complexes PdX_2(R-DAB)$_2$, [$PdX_2(PR_3')$]$_2$(R-DAB), $PdX_2(PR_3')$(R-DAB) to $PdX(C_6H_4CH_2NMe_2$-2)(R-DAB) (R = *t*-Bu, $EtMe_2C$; X = Cl, Br, or

I; R' = Bu) can be explained by an increase in electron density caused by the ligands trans to the imine-N atom thereby reducing the possibility of σ donation from this N atom to the metal (36).

There is evidence for intermediates containing σ-monodentate bonded R-DAB molecules, e.g., the 1 : 1 Et_3Al–R-DAB complexes which are stable only at temperatures below −10°C (for R = t-Bu δH^α is 7.65 and δH^β 8.90 ppm again pointing to an E-s-trans-E conformation for the monodentate bonded ligand) (54). The course of these reactions at room temperature is discussed in Section VI,B.

Finally, the reaction of $M(CO)_5THF$ (M = Cr,Mo) with Ph-DAB at −60°C afforded $M(CO)_5$(Ph-DAB) in which, according to IR and NMR spectra, the Ph-DAB ligand is σ-N (2e) bonded to the bulky $M(CO)_5$ group. This complex is converted above −20°C to $M(CO)_4$(Ph-DAB) (55).

C. Bridging (σ-N,σ-N'; 2e + 2e) 1,4-Diaza-1,3-butadienes

It has already been pointed out that the σ-N monodentate and the σ-N,σ-N' bridging bonding modes of the R-DAB ligand are very much related because both have the E-s-trans-E conformation of the N=C—C=N skeleton as a common structural feature. As for the σ-N monodentate R-DAB complexes the structure of one example of a bridge bonding mode, i.e., of stable $[PtCl_2(PBu_3)]_2$(t-BuDAB) [Eq. (4)] has been established by X-ray structure determination (32, 56) (see Fig. 1). Indeed this structure contains a planar ClPtN=C—C=NPtCl skeleton while H^β resides at a calculated Pt- - -H^β distance of 2.6 Å which again is within the distance of 3.2 Å expected for van der Waals contacts. These structural features are retained in solution as can be concluded from, for example, ^{15}N and ^{195}Pt NMR data (35) and the characteristic low field shift of H^β in the ^1H NMR spectrum (32).

Likewise the 1 : 1 complex containing the cyclometalated Pd unit (see Fig. 2b) can be converted to a dinuclear complex $\overset{\displaystyle\frown}{Pd(C_6H_4CH-}$ $(Z)NMe_2$-2)X$]_2$(R-DAB) (Z = H or (S)-Me; X = Cl, Br, I; R = t-Bu, $EtMe_2C$) that can be isolated when R = t-Bu. This compound has a similar structure to $[PtCl_2(PBu_3)]_2$(t-Bu-DAB) i.e., an E-s-trans-E PdN= C—C=NPd skeleton with the N ligands in each Pd coordination plane in trans position (36).

Other complexes with bridging R-DAB ligands, again comprising metals from the d^8 series, have been derived from the bridge splitting reactions [cf. Eqs. (4)] of $[MCl(\eta^3$-allyl)]$_n$ (M = Pd, n = 2; M = Pt, n = 4) with various R-DAB ligands (57). In the presence of $NaClO_4$ exclusively complexes $[M(\eta^3$-allyl)(R-DAB)]ClO_4 (M = Pd; R = C_6H_4OMe-p; M = Pt; R = Ph,R' = Me) were isolated containing σ,σ-N,N' chelate bonded R-

DAB [see Eq. (5)]. However, in the absence of strong anions complexes with $[PdCl(\eta^3\text{-allyl})]_2(\text{R-DAB})$ stoichiometry were observed pointing to the presence of $\sigma\text{-N},\sigma\text{-N}'$ bridging R-DAB ligands. These complexes were only stable for the t-Bu-DAB ligand and then only when in apolar solvents. Replacement of one H for Me destabilized the dinuclear species and produced (in methanol) the ionic complex $\{Pd(\eta^3\text{-allyl})[\text{R-DAB(H,Me)}]\}$ $[PdCl_2(\eta^3\text{-allyl})]$ ($R = C_6H_4OMe\text{-}p$) with a $\sigma,\sigma\text{-N,N}'$ chelate bonded R-DAB (H,Me) ligand in the cation. This is in line with our earlier suggestion (see Section III,A) that the presence of methyl groups at the central C atoms stabilized the E-s-cis-E conformer relative to its s-trans isomer. On the other hand it must be recalled that for these complexes the bridging bonding mode can also be assumed because the Pd center has a square planar coordination geometry.

$$\text{R-DAB} + [PdCl(\eta^3\text{-Meall})]_2 \rightleftharpoons [PdCl(\eta^3\text{-Meall})]_2(\text{R-DAB})$$

$$[Pd(\eta^3\text{-Meall})(\text{R}-\text{DAB})][PdCl_2(\eta^3\text{-Meall})]$$

$$\tag{5}$$

With regard to the above it is not surprising that the mononuclear Pd^0 compound $Pd(\eta^2\text{-olefin})(\text{R-DAB})$ can be converted to a dinuclear Pd^0 compound with a $\sigma\text{-N},\sigma\text{-N}'$ bridge bonding mode by addition of a coordinating molecule [see Eq. (6)]. The dinuclear compound (containing three coordinate Pd^0) is stable because the metal has the required planar geometry. Equation (6) furthermore shows that oxidative addition of $[Pd(\eta^2\text{-olefin})(t\text{-Bu-DAB})]$ with methylallyl chloride produces $[PdCl(\eta^3\text{-Meall})]_2(t\text{-Bu-DAB})$ (58).

$$2\ Pd(\eta^2\text{-olefin})(t\text{-Bu-DAB}) + \text{olefin} \longrightarrow [Pd(\eta^2\text{-olefin})]_2(t\text{-Bu-DAB})$$

$$\text{MeallCl} \downarrow - \text{olefin and } t\text{-Bu-DAB} \tag{6}$$

$$\{[PdCl(\eta^3\text{-Meall})]_2(t\text{-Bu-DAB}\}\qquad\qquad \text{olefin} = \text{dmf(emf)}$$

Bridging R-DAB ligands have also been observed in Rh^I chemistry (34, 59–61). Bridge splitting reactions of $Rh(CO)_2(\mu\text{-Cl})_2Rh(CO)_2$ with R-DAB afforded complexes with $\{[RhCl(CO)_2]_2(\text{R-DAB})\}$ ($R = t$-Bu, $EtMe_2C$) stoichiometry. The actual species and ratios present in solution is dependent on the branching at C^α and C^β (34). It appeared that due to the weaker Rh–R-DAB bonding there exists in solution an equilibrium mixture of the dinuclear species $RhCl(CO)_2[\mu\text{-}(\sigma\text{-N},\sigma\text{-N}')\text{-R-DAB}]\text{-}RhCl(CO)_2$ and the ionic species $[Rh(CO)_2(\text{R-DAB})][RhCl_2(CO)_2]$.

If the ligand is t-Bu-DAB the dinuclear complex is the major species. Intermolecular exchange between these dinuclear and ionic Rh species is fast on the NMR time scale at room temperature (34) (Section III,D,2,e).

Bridging R-DAB has also been reported in the reaction of the 1,4-diaza-1,3-butadien-2-ylpalladium compound with $[RhCl(CO)_2]_2$ shown in Eq. (7) (60).

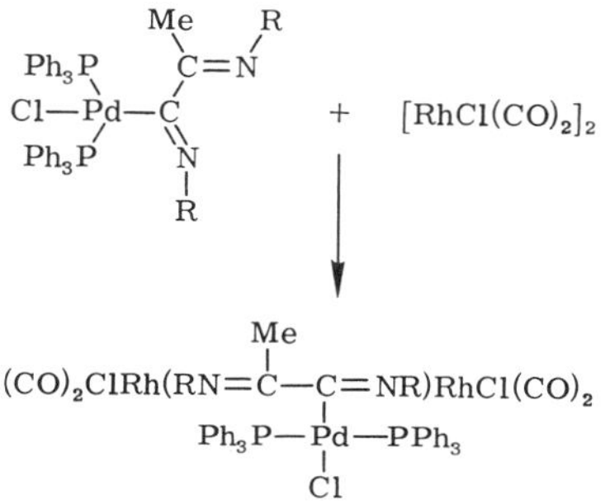

$$
\begin{array}{c}
\overset{\displaystyle Me \qquad R}{\underset{\displaystyle}{}} \\
\overset{Ph_3P}{\underset{Ph_3P}{}} \diagdown \overset{C=N}{\diagup} \\
Cl-Pd-C \\
\diagup \qquad \diagdown \\
N \\
| \\
R
\end{array}
\qquad + \qquad [RhCl(CO)_2]_2
$$

$$
\downarrow
$$
(7)

$$
\begin{array}{c}
Me \\
| \\
(CO)_2ClRh(RN=C-C=NR)RhCl(CO)_2 \\
Ph_3P-Pd-PPh_3 \\
| \\
Cl
\end{array}
$$

$$R = C_6H_4OMe\text{-}p$$

D. Chelate Bonded (σ,σ-N,N';4e) 1,4-Diaza-1,3-butadienes

The chemistry and structural aspects of metal complexes containing chelated bonded R-DAB have been well explored in contrast to the other bonding modes that have been realized only recently. Accordingly a vast amount of information is available that will be covered in two sections. First, the commonly applied synthetic routes will be discussed. Second, the specific synthetic and structural features will be treated with the complexes arranged according to the group to which the metal belongs. In this latter section the main results and general conclusions emerging from the sometimes detailed investigations of the bonding features of these complexes by MO calculations, resonance Raman spectroscopy (RR), and NMR and ESR spectroscopy will also be put forward.

1. General Synthetic Methods

Most complexes have been prepared by mixing a metal salt with the R-DAB ligand in the required molar ratio. Examples are the syntheses of $MCl_2(p\text{-}MeOC_6H_4\text{-}DAB)$ complexes of Zn, Cd, and Hg (62).

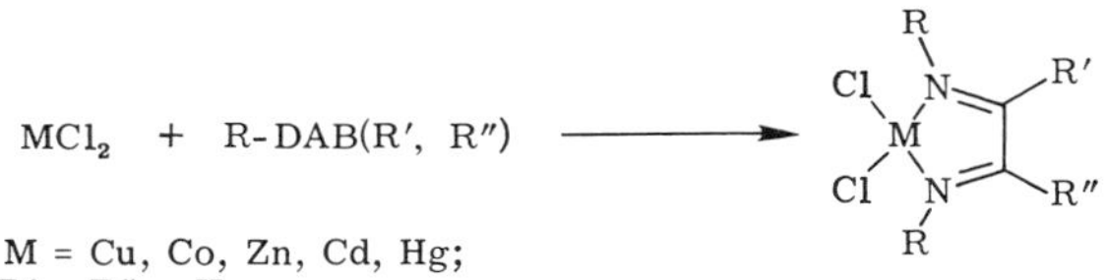

$$
MCl_2 \; + \; R\text{-}DAB(R', R'') \; \longrightarrow \;
\begin{array}{c}
R \\
\diagdown \\
Cl \quad N \diagdown_{R'} \\
\diagdown M \diagup \\
Cl \quad N \diagup \\
\diagup \qquad \diagdown_{R''} \\
R
\end{array}
$$
(8)

M = Cu, Co, Zn, Cd, Hg;
R' = R'' = H;
R' = H, R'' = $trans$-$PdCl(PPh_3)_2$

Likewise reaction of a 1,4-diaza-1,3-butadien-2-ylmetal complex with $CuCl_2$, $CoCl_2$, and $ZnCl_2$ gives the corresponding 1:1 complexes. This route is summarized in Eq. (8) (*60, 63–67*). Diarylzinc–R-DAB complexes have been similarly prepared (*68*) (see Section III,D,2,g).

For the synthesis of metal carbonyl complexes various routes have been reported. Methods that are not of general applicability (being unique for a given compound) are described in Section III,D,2.

a. Thermal Reaction of the Metal Hexacarbonyl Complexes with R-DAB. Reaction of $Mo(CO)_6$ with Ph-DAB(Me,Me) afforded at 80°C via a slow substitution of CO the corresponding $Mo(CO)_4[Ph-DAB(Me,Me)]$ complexes (*69, 70*).

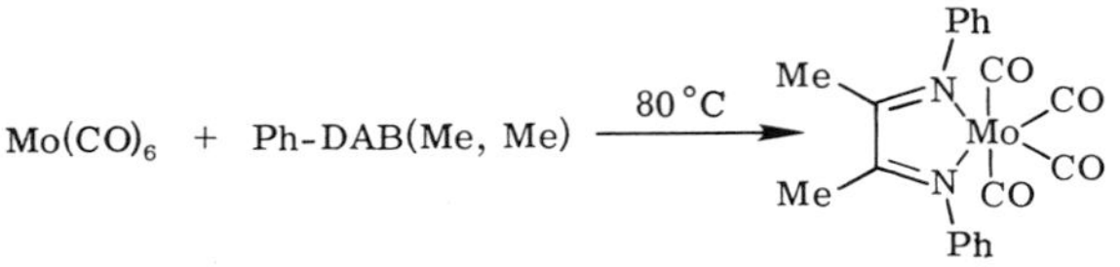

$$Mo(CO)_6 + \text{Ph-DAB(Me, Me)} \xrightarrow{80\,°C} \qquad (9)$$

b. Ligand Substitution of Metal Carbonyl Derivatives. Since reaction (9) at elevated temperatures often leads to side reactions a better procedure involves the use of metal carbonyl derivatives containing at least one ligand that is weakly bonded. Two such reactions are shown in Eqs. (10) (*71*) and (11) (*55*).

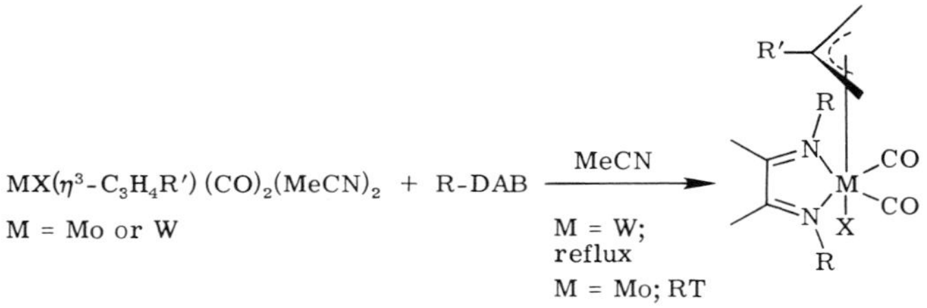

$$\text{MX}(\eta^3\text{-C}_3\text{H}_4\text{R}')(CO)_2(MeCN)_2 + \text{R-DAB} \xrightarrow{MeCN} \qquad (10)$$
$$M = \text{Mo or W} \qquad\qquad\qquad\quad M = W;\ \text{reflux}$$
$$M = \text{Mo; RT}$$

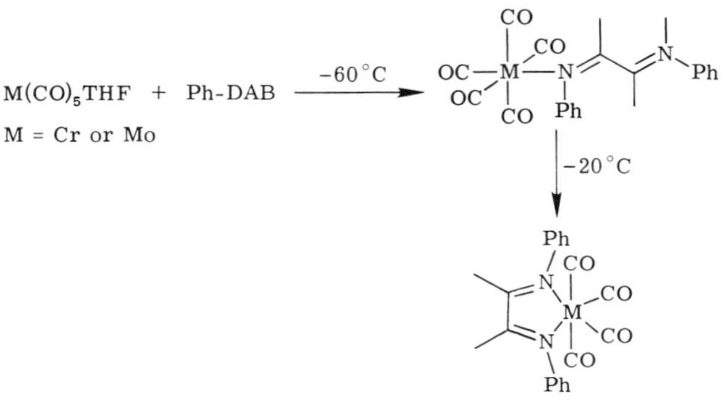

$$M(CO)_5THF + \text{Ph-DAB} \xrightarrow{-60\,°C} \qquad\qquad\qquad (11)$$
$$M = \text{Cr or Mo} \qquad\qquad\qquad\qquad\qquad \downarrow -20\,°C$$

It has been shown that the conversion of the σ-N to a σ,σ-N,N' bonded Ph-DAB ligand in Eq. (11) is a fast reaction.

   c. *Substitution in the Ligand Sphere of a Metal.*   Some R-DAB ligands are either not known as free molecules or have only a very limited stability. For these ligands the synthesis of complexes by an *in situ* preparation is a well-known approach and are illustrated in Eqs. (12) (*11, 69*) and (13) (*72*).

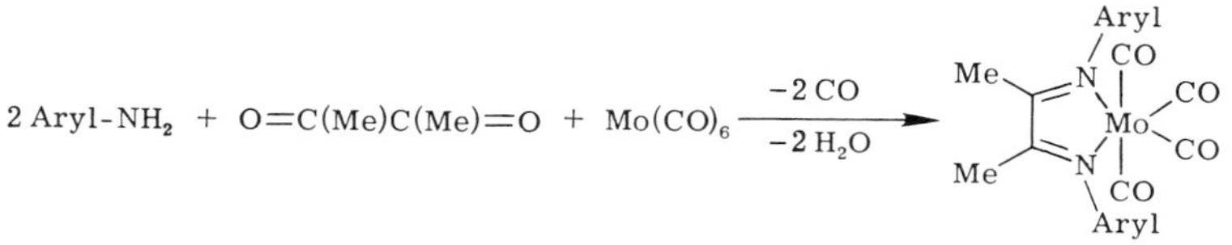

$$2 \text{ Aryl-NH}_2 + O{=}C(Me)C(Me){=}O + Mo(CO)_6 \xrightarrow[-2\,H_2O]{-2\,CO} \quad (12)$$

$$+ \; MCl_2 + p\text{-RC}_6H_4NH_2 \xrightarrow[\text{reflux}]{\text{glacial AcOH}} \quad (13)$$

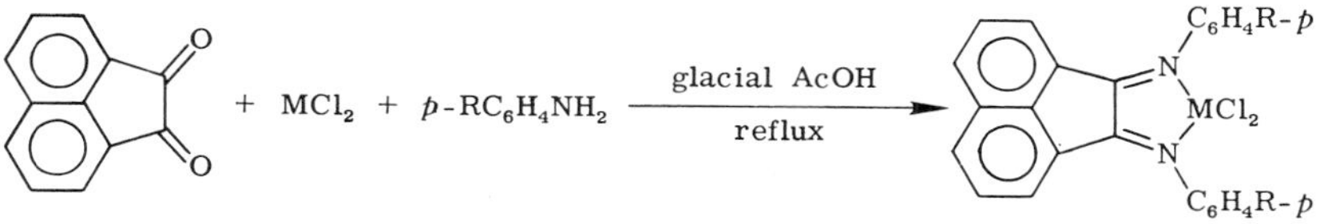

R = e.g., Me, NO$_2$
M = Zn, Co, Ni, or Cu

   Another interesting reaction involves the exchange of an amino for a benzyl group as illustrated in Eq. (14) (*69*).

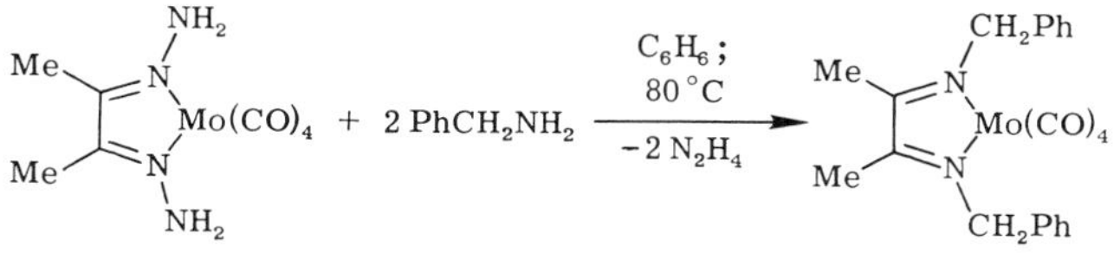

$$\cdots\, Mo(CO)_4 + 2\,PhCH_2NH_2 \xrightarrow[-2\,N_2H_4]{\substack{C_6H_6;\\80\,°C}} \cdots\, Mo(CO)_4 \quad (14)$$

   d. *Via Reduction of Metal Halides in the Presence of R-DAB.*   A large number of zerovalent metal 1,4-diaza-1,3-butadiene complexes are accessible via reactions of R-DAB with zerovalent metal–ligand complexes, or via reduction of metal–ligand complexes with Grignard or aluminum compounds in the presence of R-DAB. Examples are shown in Eqs. (15)–(18).

$$Ni(COD)_2 + 2R\text{-DAB} \rightarrow Ni(R\text{-DAB})_2 + 2COD \; (9, 10, 73, 74) \quad (15)$$

$$Pt(COD)_2 + R\text{-DAB} \rightarrow Pt(COD)(R\text{-DAB}) + COD \; (75) \quad (16)$$

$$Pd(DBA)_2 + R\text{-DAB} + \text{olefin} \rightarrow Pd(R\text{-DAB})(\eta^2\text{-olefin}) \; (58) \quad (17)$$

$$Cr(acac)_3 + 2R\text{-}DAB \xrightarrow[\text{THF}]{3e^-} Cr(R\text{-}DAB)_2 \; (73) \tag{18}$$

2. Structural and Bonding Aspects

a. Groups IIIA–VA. Only limited information is available concerning the complex formation of R-DAB ligands with the early transition metals. Some work has been done directed to the synthesis of $TiCl_4(R\text{-}DAB)$ (R = e.g., *i*-Pr, *t*-Bu, *c*-Hex, $C_6H_4OMe\text{-}p$) complexes. These are insoluble in apolar solvents and decompose upon attempted recrystallization from polar solvents such as Me_2SO and DMF (*76*). σ,σ-N,N′ chelate bonding for the R-DAB in these complexes has been proposed but IR data could not unambiguously preclude the fact that these complexes could be oligomers rather than monomers.

Unsuccessful attempts were undertaken to reduce these $TiCl_4(R\text{-}DAB)$ complexes to $Ti^0(R\text{-}DAB)_2$. In contrast, blue-green colored $V(i\text{-}Pr\text{-}DAB)_3$, which appeared to be relatively stable, was obtained from the reaction of VCl_3 with *i*-Pr-DAB and sodium in THF (*76*).

b. Group VIA. Complexes of the d^6 metals Cr, Mo, and W containing exclusively R-DAB ligands have been reported for Cr (*74, 76*). The number of R-DAB ligands bonded to Cr is dependent on the type of substituents present, e.g., tetracoordinate $Cr(R\text{-}DAB)_2$ was obtained for R = $i\text{-}Pr_2CH$ and *t*-Bu, and hexacoordinate $Cr(R\text{-}DAB)_3$ for R = *i*-Pr. Analogous complexes of zerovalent Mo have not been isolated although some evidence for the synthesis of $Mo(i\text{-}Pr_2CH\text{-}DAB)_2(MeCN)_2$ containing cis-positioned MeCN ligands was obtained (*76*).

Stable R-DAB–metal carbonyl complexes were obtained starting from hexacarbonyls [cf. Eq. (9)] and complexes with $M(CO)_4(R\text{-}DAB)$ stoichiometry have been reported for all three metals (Cr: *55, 77–80*; Mo: *55, 69, 77–91*; W: *55, 77–80, 88*). These compounds have the symmetry properties of the C_{2v} point group, they are strongly colored and the bidentate nitrogen donor ligand shows strong π-bonding interaction with the group VIA metal (*vide infra*). The molecular geometry of $Mo(CO)_4(R\text{-}DAB)$ (R = *i*-Pr and 2,6-$(i\text{-}Pr)_2C_6H_3$) (*37*) has been established by X-ray structural determinations (see Table II).

Thermally stable $Mo(CO)_4(R\text{-}DAB)$ undergoes single CO substitution with tertiary phosphines in boiling benzene leading to $Mo(CO)_3$-$(PR'_3)(R\text{-}DAB)$ (R′ = Ph,Bu (*70, 78, 81, 84, 86–88, 92, 93*) R = alkyl or aryl).

In the case of thermally less stable complexes the route shown in Eq. (19) can be followed (*94*).

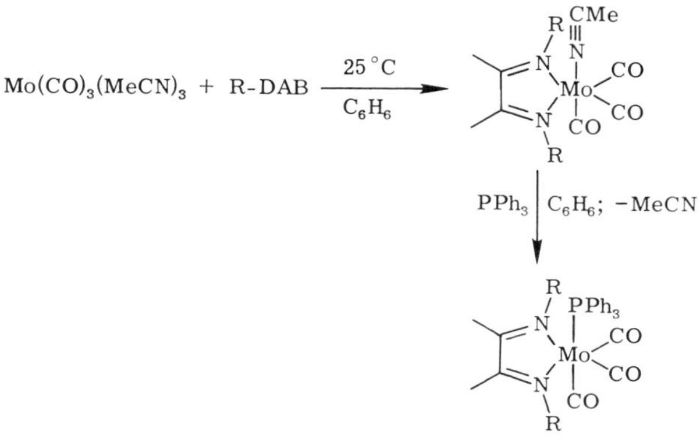

$$Mo(CO)_3(MeCN)_3 + R\text{-}DAB \xrightarrow[C_6H_6]{25\,^\circ C}$$

$$(19)$$

$$PPh_3 \Big| C_6H_6; -MeCN$$

The interesting point of this synthesis is that exclusively the cis product is formed. In Eq. (20) the synthesis of a bisphosphine complex is shown that likewise occurs with high stereospecificity (95).

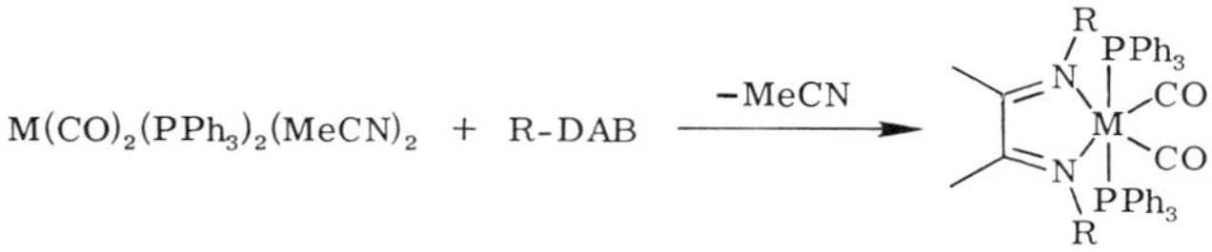

$$M(CO)_2(PPh_3)_2(MeCN)_2 + R\text{-}DAB \xrightarrow{-MeCN} \qquad (20)$$

Other $Mo(CO)_2(PR'_3)_2(R\text{-}DAB)$ complexes where $R' = Bu$ (81, 93), Ph (85), and Et (87) have been reported.

The Cr and Mo complexes $M(CO)_4(R\text{-}DAB)$, in which $R = i\text{-}Pr$, have a sufficiently high photoreactivity to form the monosubstituted $M(CO)_3[P(OMe)_3](i\text{-}Pr\text{-}DAB)$ complexes in the presence of $P(OMe)_3$ upon irradiation within the MLCT (metal-to-ligand charge transfer) band (80). In contrast no photosubstitution is observed for the W compound when R is p-tolyl.

$MX(R'\text{-}All)(CO)_2(R\text{-}DAB)$ complexes of the d^6 metal–R-DAB complexes are obtained by treating the acetonitrile complexes $MX(\eta^3\text{-}C_3H_4R')(CO)_2(MeCN)_2$ (M = Mo, W; $R' = H$ or Me and $X = Cl$ or Br) with R-DAB (R = alkyl or aryl) according to Eq. (10) (71). The oxidative addition reaction of $fac\text{-}M(CO)_3L_2L'$ (M = Mo, W) with $R'C_3H_4X$, which was shown to be successful when L_2 is phenanthroline or 2,2'-bipyridine (96), does not proceed when L_2 is R-DAB (71).

The structures of $MoCl(\eta^3\text{-}C_3H_4Me)(CO)_2(c\text{-}Hex\text{-}DAB)$ (38) and of $WBr(\eta^3\text{-}C_3H_5)(CO)_2(c\text{-}Hex\text{-}DAB)$ (39) have been solved by X-ray structure determinations (see Table II) and these are schematically shown in Eq. (10). The observation that the cis positioned Cl atom can be exchanged for Br, I, or SCN whereas no reaction was observed with, for example, $HgCl_2$ was ascribed to steric hindrance imposed by the N substituents (71).

$MoX(\eta^3-C_3H_5)(CO)_2(c-Hex-DAB)$ can be converted to cationic complexes with Ag^I or Tl^IBF_4 in the presence of a suitable ligand such as pyridine (71). An interesting aspect of these complexes is their structural similarity to the corresponding 2,2'-bipyridine and phenanthroline complexes but the reactivity of the R-DAB complexes is far less.

A stable trinuclear molybdenum–mercury compound was obtained via the reaction shown in Eq. (21) (97).

$$2K^+[Mo(CO)_4(R-DAB)] + HgCl_2 \longrightarrow 2KCl + 2CO + [Mo(CO)_3(R-DAB)]_2Hg$$

$$R = t\text{-Bu}, i\text{-Pr}$$

(21)

The bonding of the $M(CO)_4(R-DAB)$ (M = Cr, Mo, W) complexes has been extensively investigated. These studies were induced in particular by the intense colors they exhibit in solution ranging from purple to orange. Indeed all compounds have an absorption in the visible region with an ϵ value of 7.000–17.500 liters mol^{-1} cm in cyclohexane (69). This absorption originated from a metal-to-ligand charge transfer (MLCT) transition, i.e., electron transfer from filled metal d orbitals into the empty π^* orbital of the ligand (69, 84, 98). This transfer is possible because the planarity of the five-membered metallocycle allows overlap of metal d and π^* orbitals.

Combined data from resonance Raman spectra, magnetic circular dichroism measurements, and UV spectra showed that the CT band comprised four separate electronic transitions (55, 78, 99). The relevant part of a tentative MO scheme for the five allowed CT transitions is shown in Fig. 4 together with their polarization characteristics. Apart from the five symmetry allowed transitions there is one symmetry forbidden ($a_2 \leftarrow a_1$) while ($b_2 \leftarrow a_1$) is overlap forbidden (55).

The maximum of the CT band of the $Mo(CO)_4(R-DAB)$ complexes shifts with changing polarity of the solvent (69, 83–86,[4] 91, 100, 101), so-called solvatochromism, which is due to the fact that the ground state molecule is polar. This dependence is illustrated by Table III and has been used to quantify the polarity of a series of solvents (86).

Strong solvatochromism only occurs for electronic transitions in which electron transfer takes place along the dipole moment vector that for $Mo(CO)_4(R-DAB)$ coincides with the z-vector shown in Fig. 4. This was indicated by the observation that the resonance Raman effect is more

---

[4] In Ref. 85 the term negative is erroneously used when positive solvatochromism is meant. The latter refers to a shift of the maximum to shorter wavelength on going from apolar to polar solvent.

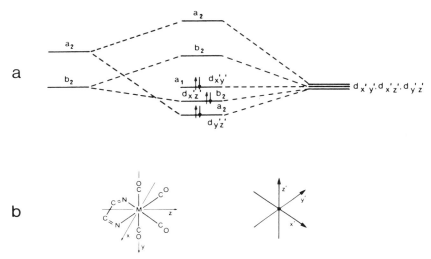

FIG. 4. (a) Part of a tentative MO scheme relevant to the CT transitions between the metal and the R-DAB ligand in M(CO)$_4$(R-DAB) complexes together with the metal-to-R-DAB CT transitions and their polarization direction (55). (b) Applied coordinate transformation for obtaining symmetry adapted orbitals.

pronounced in the z-polarized transition b$_2$ ← b$_2$ for the more polar solvents, i.e., in polar solvents the electrons are more localized on the metal and this involves a shift of negative charge along the dipole moment vector. This conclusion was also supported by the results of ^{1}H and ^{13}C NMR studies involving the use of ^{15}N-enriched Mo(CO)$_4$(Me-DAB) (55).

Since the solvatochromism effect depends on the degree of stabilization of the ground state and destabilization of the excited state the R substituents are of much importance because they are responsible for the π-accepting properties of the R-DAB ligand. It has been found that a decrease in π-accepting ability is paralleled by an enhanced solvatochromic effect (84) (see Table IV).

Extensive resonance Raman studies of Mo(CO)$_4$(t-Bu-DAB) (55) showed that the CT transitions are not purely metal-to-R-DAB in character, as has previously been assumed (69, 100), but that orbitals of the cis carbonyl groups appear to be mixed in the first excited states of the complex (55). This is rationalized by Fig. 5 which shows the π* R-DAB orbital overlap with the cis-CO π* orbitals (cf. solvatochromic effect of cis and trans ν COs). On the basis of this orbital scheme the selective photosubstitution of the cis CO for phosphine ligands in M(CO)$_4$(R-DAB) complexes could also be explained (vide supra). The decreasing reactivity

TABLE III

Solvent Dependence (Solvatochromy) of the CT Absorption and ν(CO)
(cm^{-1}) of Mo(CO)$_4$(t-Bu-DAB)[a]

Solvent	CT absorption[b]	ν(CO)^{cis}		ν(CO)^{trans}	
		A$_1$	B$_1$	A$_1$	B$_1$
Dimethyl sulfoxide	21,120	1878	1828	2023	1908
Dimethylformamide	20,960	1880	1832	2024	1908
Acetone	20,400	1887	1836	2024	1912
Methanol	19,880	1887	1838	2024	1913
Dioxane	19,230	1892	1842	2022	sh
Chloroform	18,930	1890	1838	2924	1915
Tetrachloromethane	18,020	1910	1855	2024	1915
Cyclohexane	17,745	1914	1864	2024	1923

[a] Data taken from Ref. 85.
[b] ν_m (cm^{-1}).

going from Cr to W is due to a decrease of delocalization of the MLCT excited state over the cis COs as the central metal atom becomes larger (extent of overlap decreases) (80, 102).

Similar studies have been carried out for M(CO)$_{4-n}$L$_n$(R-DAB) complexes (92). One of the more general conclusions is that, when comparing the complexes with n = 0, 1, and 2 with L = PPh$_3$, the effect of the solvent on the maximum of the CT band increases, thus pointing to decreasing π-accepting properties of the ligand systems (81, 87, 94) (cf. Table IV). Accordingly, there is a question as to whether the central metal atom in M(CO)$_2$L$_2$(R-DAB) complexes is oxidized to some extent (92, 94, 95).

TABLE IV

Relationship between Solvatochromic Effect and
π-Accepting Ability of the R-DAB
Ligand in Mo(CO)$_4$(R-DAB)[a]

R	ν_m(DMF)[b] (cm^{-1})	ν_m(C$_6$H$_6$)[b] (cm^{-1})	$\Delta\nu$ (cm^{-1})
c-Hex	21,275	18,553	2722
i-Pr	20,325	18,622	1703
c-Pr	20,100	18,553	1547
Me	20,000	18,762	1238
Ph	17,606	16,570	1036
C$_6$H$_4$Me-o	18,215	17,391	824

[a] Data taken from Ref. 84.
[b] Maximum of the CT absorption.

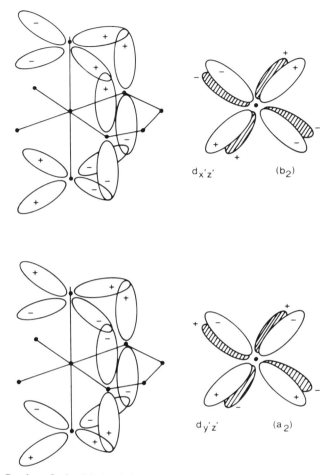

$d_{x'z'}$ (b$_2$)

$d_{y'z'}$ (a$_2$)

FIG. 5. Overlap of π^* orbitals of the cis CO groups and the π^* orbitals of the R-DAB ligand. The metal d-orbital combinations which give π back bonding with these π^* orbitals are shown on the right (55).

IR, UV, and ^{1}H, ^{13}C and ^{31}P NMR studies of a series of Mo(CO)$_{4-n}$(PR$_3'$)$_n$L$_2$ (n = 0, 1; R$'$ = OMe, Ph, Et, t-Bu, c-Hex, and n = 2; R$'$ = Et, L = i-Pr-DAB, i-PrPyca and 2,2$'$-bipyridine) showed that the π-back bonding between Mo and L is strongest for i-Pr-DAB (87). This conclusion concerning the better π-accepting properties of the R-DAB ligands was also obtained from comparison of these neutral complexes with their paramagnetic monoanions (81) (see Section V).

Finally, attention has been paid to the conformational stability of $Mo(CO)_4(R\text{-}DAB)$. Using ^{13}C NMR spectroscopy (79) cis–trans exchange of the CO ligands was found while absorption and resonance Raman spectra (83) indicated a distortion of the C_{2v} conformation at 77 K. CO exchange takes place via a transition state with pseudo C_{4v} symmetry (see Section IV,B,3 for the influence of R on the activation barrier for this process).

The interesting trinuclear molybdenum–mercury compound shown in Eq. (21), which has been used for the C—C coupling reaction of two R-DAB ligands (Section VI,C) has been proposed to have the structure shown below; a linear Mo—Hg—Mo arrangement with the R-DAB ligands $\sigma,\sigma\text{-}N,N'$ chelate bonded to Mo (97) [cf. structure of $Mn(CO)_5M'(CO)_3(R\text{-}DAB)$, Section III,D,2,c].

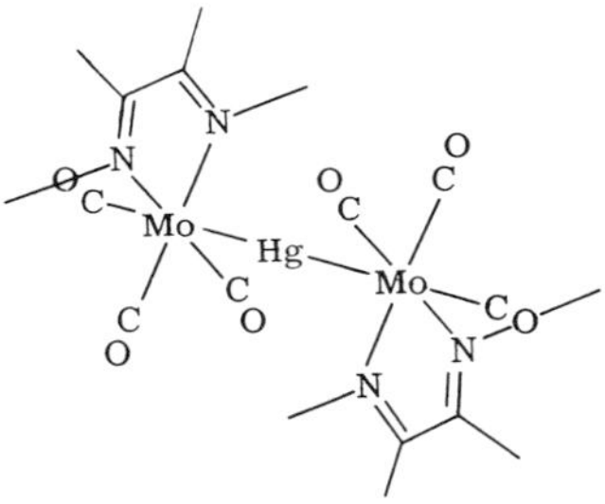

c. Group VIIA.   The reaction of (t-Bu-DAB)Na with manganese acetylacetonate affords in excellent yields the 15-electron $Mn(t\text{-}Bu\text{-}DAB)_2$ complexes (42).

Carbonyl complexes of d^7 metals with R-DAB are formed via the ligand substitution reaction shown in Eq. (22).

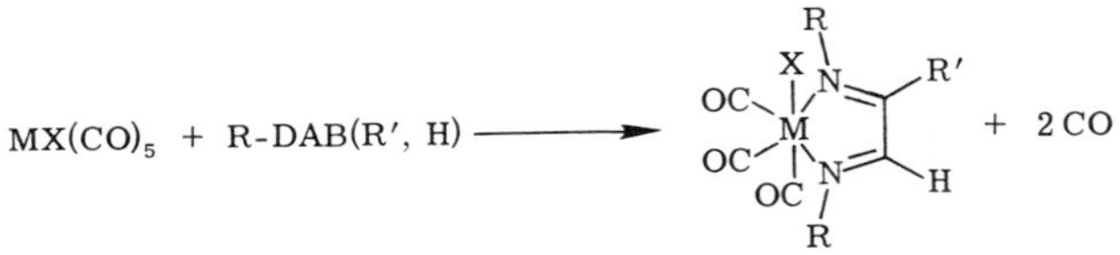

$$MX(CO)_5 \ + \ R\text{-}DAB(R', H) \longrightarrow \qquad + \ 2\,CO \tag{22}$$

(M = Re, X = Cl, R = i-Pr, c-Hex, p-tolyl;
M = Mn, X = Br, R = i-Pr, t-Bu (R' = H or Me), Ph, p-ClC$_6$H$_4$, p-MeOC$_6$H$_4$)

The complexes of Mn are formed more easily than those of Re. Whereas complex formation occurs readily at 30°C with alkyl-DAB ligands in ether the synthesis of aryl-DAB complexes requires prolonged heating in toluene–heptane (77). The exchange of chloride for iodide in the Mn and Re complexes was possible via reaction with KI (77). Similarly other X groups, e.g., CN^- and $MeCO_2^-$ can be exchanged (77).

MnX(CO)$_3$(t-Bu-DAB) reacts with AgBF$_4$ under a CO atmosphere (1.5 atm) to give the cationic complex [Mn(CO)$_4$(t-Bu-DAB)]BF$_4$ (77) which is isoelectronic with the corresponding d^6 metal complexes (see Section III,D,2,b).

Dinuclear complexes Mn(CO)$_5$M'(CO)$_3$(R-DAB) (M' = Mn, Re) are formed according to Eq. (23). In view of the high yields, it is assumed that these complexes are formed via a nucleophilic mechanism (103). The formation of Mn$_2$(CO)$_{10}$ can be explained by a redistribution reaction involving Mn(CO)$_5^-$ and Mn(CO)$_5$M'(CO)$_3$(R-DAB).

$$Na[Mn(CO)_5] \; + \; M'X(CO)_3(R\text{-}DAB[R', H]) \longrightarrow \; \text{—Mn——M—} \tag{23}$$

(M' = Re, R = i-Pr, c-Hex, p-tolyl; M' = Mn, R = i-Pr,
R' = H or Me, p-tolyl, C$_6$H$_4$OMe-p)

Mn(t-Bu-DAB)$_2$ has a tetrahedral structure containing two σ,σ-N,N' chelate bonded t-Bu-DAB ligands (see Table II) (42). The dihedral angle (θ) between the two almost planar MnN=C—C=N chelate rings being 90°.

MnBr(CO)$_3$(c-Hex-DAB) (40) and ReCl(CO)$_3$(i-Pr-DAB) (41) are isostructural as regards the direct metal–coordination sphere and are also isostructural with MnBr(CO)$_3$L complexes in which L is 2,2'-bipyridine or 1,10-phenanthroline (77). The fac configuration of the X and N ligands in these complexes is schematically shown in Eq. (22).

^{1}H and ^{13}C NMR spectroscopic data pointed to a structure for the Mn(CO)$_5$M'(CO)$_3$(R-DAB) complexes consisting of a Mn—M' bond with the R-DAB ligand σ,σ-N,N' bonded to M' and the chelate plane perpendicular to the Mn—M' axis [see structure in Eq. (23)]. This structure is closely related to the homodinuclear M$_2$(CO)$_8$L (M = Mn, Re; L = 1,10-phenanthroline, 2,2'-biquinoline) complexes (103). It is assumed that steric factors are more important for the stability of the MM'(CO)$_8$(R-DAB) complexes than electronic factors on the basis of the much greater stability of the complex with R = i-Pr over the R = t-Bu analog (103).

The fact that in the MM'(CO)$_8$(R-DAB) complexes intramolecular attack of one of the C=N bonds on the Mn center does not occur contrasts with instability of the intermediate MnCo(CO)$_7$(R-DAB) formed in the reaction of Co(CO)$_4^-$ with MnBr(CO)$_3$(R-DAB). In this Mn—Co dinuclear species intramolecular attack occurs converting the initially σ,σ-N,N' chelate bonded R-DAB ligand to a σ-N,μ^2-N',η^2-CN' bridge bonded ligand (see Section IV,A,3). Like the d^6 metal complexes the MX(CO)$_3$(R-DAB) and Mn(CO)$_5$M'(CO)$_3$(R-DAB) complexes are all highly colored in solution and have complex electronic absorption spectra in the visible and near UV (*103*). The $MX(CO)_3(R\text{-}DAB)$ complexes show positive solvatochromism (*77*) which indicates that there is a strong resultant dipole moment from the R-DAB ligand to the metal and that the electron transfer during the CT transition is antiparallel to it. Extensive resonance Raman investigations of a series of $ReCl(CO)_3(p\text{-}Tol\text{-}DAB)$ complexes showed that upon excitation within both MLCT and intraligand transitions resonance enhancement of a cis-carbonyl stretching mode also occurs (*104*).

d. Group VIII; d^8 Metals Fe, Ru and Os. Of the d^8 metals only for Fe has a series of zerovalent R-DAB complexes been reported and the synthetic route is outlined in Eq. (24) (*105*).

$$FeCl_2 + R\text{-}DAB \longrightarrow FeCl_2(R\text{-}DAB) \xrightarrow[-2\ NaCl]{\substack{2\ Na; \\ R\text{-}DAB}} \quad \text{(24)}$$

$[R = t\text{-}Bu,\ c\text{-}Hex,\ i\text{-}Pr,\ (i\text{-}Pr)_2CH,\ C_6H_4Me\text{-}o, C_6H_3Me_2\text{-}o, p\ ]$

$Fe(R\text{-}DAB)_2$ complexes are very soluble in apolar solvents and take up CO reversibly to give $Fe(CO)(R\text{-}DAB)_2$ which is stable only in solution in a CO atmosphere (*105, 106*). Excess CO irreversibly generates $Fe(CO)_3(R\text{-}DAB)$ and free R-DAB (*105*). Likewise, the reaction with $(CN)_2$ giving rise to formation of $Fe^{II}(CN)_2(R\text{-}DAB)_2$ is irreversible (R = small alkyl grouping) (*107*). No reaction is observed for R = *t*-Bu.

Much effort has been put into the study of $Fe_2(CO)_9$ and $Fe(CO)_5$ with R-DAB (*30, 105, 106, 108, 109*). According to a reinvestigation (*110*) the thermal reaction of $Fe_2(CO)_9$ with *t*-Bu-DAB forms $Fe(CO)_3(t\text{-}Bu\text{-}DAB)$, $t\text{-}Bu\text{-}\overline{NC(H)}{=}C(H)N(t\text{-}Bu)\overline{C}{=}O$ (2-imidazolinone) and $Fe(CO)_5$ in equimolar amounts (Fig. 22 in Section VI,A,1). $Fe(CO)_3(R\text{-}DAB)$ can also be prepared via the photochemical reaction of $Fe(CO)_5$ in the presence of the R-DAB ligand (*30*). In this reaction dinuclear products $Fe_2(CO)_6(R\text{-}DAB)$, which can be isolated, are also formed (see Section III,E).

$Fe(NO)_2(R\text{-}DAB)$ (R = *i*-Pr or Ph) has been reported without any comment concerning its preparation (*107*).

Photolysis of $Fe(CO)_3(R\text{-}DAB)$ in the presence of 1,3-dienes affords complexes of the type $Fe(CO)(\eta^4\text{-}1,3\text{-}diene)(R\text{-}DAB)$ liberating two moles of CO. This conversion occurs in two steps of which the first is thermally reversible. Only for 1,3-dienes with weak π-acceptor properties could such complexes be obtained (*111*).

$$Fe(CO)_3[R\text{-}DAB(Me, Me)] + 1,3\text{-diene} \underset{\substack{\Delta \\ +CO}}{\overset{\substack{h\nu \\ -CO}}{\rightleftharpoons}} Fe(CO)_2(\eta^2\text{-diene})[R\text{-}DAB(Me, Me)]$$

$$\downarrow\uparrow h\nu; -CO$$

(25)

R = i-Pr, c-Hex, C_6H_4OMe-p

Mononuclear Ru and Os derivatives of the type $M(CO)_3(R\text{-}DAB)$ have been reported only for Ru with special sterically demanding R groups (*112*) (see Scheme 6 in Section IV,A,2).

A zerovalent $Ru(p\text{-}MeOC_6H_4\text{-}DAB)_3$ has been reported as the product from the reaction of p-$MeOC_6H_4$-DAB with either $RuH_2(PPh_3)_4$ or $RuH(C_6H_4PPh_2)(PPh_3)_2(C_2H_4)$ in toluene at 80°C (*43*). So far analogous Os compounds are not known.

One of the most thoroughly investigated classes of compounds are those of general formula $[Fe^{II}(R\text{-}DAB)_3]X_2$ (*28, 89, 90, 107, 113–131*). The complexes with R = Me are generally made by the condensation reaction of methylamine with glyoxal in the presence of the Fe^{II} salt [cf. Eqs. (12) and (13)].

Ru^{II} complexes were obtained from reactions of $RuHCl(PPh_3)_3$ with i-Pr-DAB affording $RuHCl(PPh_3)_2(i\text{-}Pr\text{-}DAB)$ (*132*). The latter compound reacts in MeOH at 50°C to give $RuH_2(PPh_3)_2(i\text{-}Pr\text{-}DAB)$ which could also be prepared via the reaction of $RuH_4(PPh_3)_3$ with i-Pr-DAB. In Scheme 1 the other products that have been obtained are summarized.

$[Ru^{II}(R\text{-}DAB)_3]X_2$ is produced quantitatively and isolated readily as the Cl^-, BPh_4^-, or PF_6^- salt by oxidation of $Ru(R\text{-}DAB)_3$ with molecular oxygen, I_2, HCl, or even excess R-DAB (*43*).

Reduction of $RuCl_3 \cdot 3H_2O$ by metallic zinc in the presence of i-Pr-DAB afforded $RuCl_2(i\text{-}Pr\text{-}DAB)_2$ (*131*). The corresponding low-spin complex $Fe(SCN)_2(c\text{-}Hex\text{-}DAB)_2$ was obtained from the reaction of $FeCl_2$ with c-Hex-DAB in the presence of KSCN (*131*).

$Fe(R\text{-}DAB)_2$ are 16-electron tetrahedral species that are paramagnetic (R = t-Bu; 2.98 BM in benzene) and very sensitive to oxygen (*105*). $Ru(R\text{-}DAB)_3$ has been formulated on the basis of an X-ray structure determination (see Table II) as a formally 20-electron octahedral complex. This

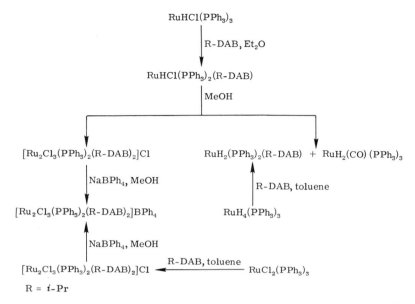

SCHEME 1. Various related reactions of $RuHCl(PPh_3)_3$, $RuCl_2(PPh_3)_3$, and $RuH_4(PPh_3)_3$ (*132, 133*).

complex, contrary to expectations, is diamagnetic (*43*), the analogous $Fe(Bipy)_3$ complex being paramagnetic (*134*). It is proposed that the ruthenium compound has a temperature-dependent spin-free spin-paired equilibrium with a small population of the paramagnetic state (*43*).

For $Fe(CO)_3(R-DAB)$ it was assumed that, depending on the substituents, the R-DAB ligand was bonded either in the σ,σ-N,N' chelate mode or in a σ-N,η^2-C=N' fashion (*106, 109*). However, later it was shown that all mononuclear $M(CO)_3(R-DAB)$ complexes, as was earlier suggested (*108*), contain σ,σ-N,N' chelate bonded R-DAB ligands (*110*). A recent X-ray structure determination of $Fe(CO)_3(2,6-i-Pr_2C_6H_3-DAB)$ (see Table II) revealed that the Fe center is square pyramidally surrounded. The basal plane contains two CO and the σ,σ-N,N' chelate bonded R-DAB ligand while the third CO ligand resides in apical position. An interesting feature is the position of the aryl group perpendicular to the chelate plane (*110b*).

The $Fe(CO)(\eta^4$-1,3-diene$)(R-DAB)$ complexes have, based on ^1H and ^{13}C NMR, IR, and UV-visible spectra, square-pyramidal structures with basal–basal coordination of the R-DAB and 1,3-diene ligands [see structure in Eq. (25)] (*111*). This has been established by the structure in the solid of $Fe(CO)(i-Pr-DAB)(2,3-Me_2C_4H_4)$ (*111b*).

Structural information concerning $Fe(NO)_2(R-DAB)$ and $Fe(CN)_2(R-DAB)$ (107) is not available.

The octahedral Fe^{II} complexes $[Fe(R-DAB)_3]X_2$ were some of the first R-DAB–metal complexes to be studied. In particular Krumholz and co-workers established the stability of the chelate ring in these spin-paired complexes (125, 135). The IR spectra, normal coordinate analysis (136), resonance Raman spectra, and excitation profiles (114) have all been subjects of interest.

For the Ru complexes shown in Scheme 1 the R-DAB ligands are proposed to be σ,σ-N,N' chelate bonded. Recently a η^2,η^2-bonded R-DAB ligand in the E-s-cis-E conformation was considered as a possibility in trans-$Ru(PPh_3)_2(H)_2(p$-OMeC$_6$H$_4$-DAB) (137). This structure is unlikely in view of the unfavorable interaction of the lone pairs of the N atoms (see Section III,A).

Finally, the structure in the solid state of $RuCl_2(i$-Pr-DAB)$_2$ (see Table II) shows a cis arrangement of the two σ,σ-N,N' chelate bonded i-Pr-DAB ligands in agreement with that deduced from ^{1}H NMR experiments (131).

e. Group VIII; d^9 Metals Co, Rh and Ir. To date R-DAB complexes of Ir have not been reported and the number of complexes containing Co and Rh is still very limited. Calculations have been carried out on cobalt complexes CoL_3^{3+} (L = 1,10-phenanthroline, 2,2'-bipyridine, Me-DAB) show that aliphatic diimines, which have the better π-acceptor ability in comparison to 2,2'-bipy, cause a higher net charge on the metal and a stronger stabilization of the combining π orbitals (138, 139). This conclusion is in agreement with results of earlier spectroscopic studies (cf. Section III,D,2,b).

Up to now the known isolated complexes of Co are homo- or hetero-dinuclear species of which the former contain σ,σ-N,N' chelate bonded R-DAB ligands (51, 140). The heterodinuclear complexes, which have 6-electron R-DAB ligands bridging the metal pair, are discussed in Section III,E.

Complexes of the type $Co_2(CO)_6(R-DAB)$ have been obtained from the reaction of $Co_2(CO)_8$ with R-DAB. Reaction of $Co_2(CO)_6(p$-Tol-DAB) with p-Tol-DAB in ether resulted in further substitution of CO while heating the complex in hexane caused dimerization (140) [Eq. (26)].

$$
\begin{array}{ccc}
& p\text{-Tol-DAB} & \\
Co_2(CO)_8 & \xrightarrow{} & Co_2(CO)_6(p\text{-Tol-DAB}) \\
& \diagdown & \diagup \quad \diagdown \\
p\text{-Tol-DAB} \diagup \begin{array}{c}\Delta; \\ n\text{-hexane}\end{array} & & \quad p\text{-Tol-DAB} \\
\tfrac{1}{2}\,Co_4(CO)_8(p\text{-Tol-DAB})_2 & & Co_2(CO)_4(p\text{-Tol-DAB})_2
\end{array}
\qquad (26)
$$

^{1}H and ^{13}C NMR spectroscopy of $Co_2(CO)_6(R\text{-}DAB)$ revealed the characteristic chemical shifts of σ,σ-N,N' chelate bonded R-DAB ligands (cf. Section IV,B,2) while the resonance pattern, furthermore, showed isochronous signals for the two R—N=CH halves of the R-DAB ligand in agreement with the schematic structure shown below. A series of $Co_2(CO)_6L$ complexes (in which L is a bidentate N donor ligand) (*141*) as well as $Fe_2(CO)_7(2,2'\text{-bipy})$ (*142*) are known to have this structure. The nature of the 2e–2c Co—Co bond has been studied by UV-visible and resonance Raman spectroscopy [observation of band at 145 cm^{-1} assigned to $\nu(Co-Co)$] (*51*). In contrast to the pronounced solvatochromic shifts observed for d^6 and d^7 metal carbonyl R-DAB complexes such shifts were not found for $Co_2(CO)_6(R\text{-}DAB)$ (*51*).

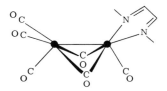

The striking difference of the coordination mode of the R-DAB ligand in the homo- and heterodinuclear complexes (4e σ,σ-N,N' versus 6e σ-N,μ^2-N',η^2-C=N'), which can be explained on the basis of the different geometries in the coordination polyhedra, is discussed in Section IV,A,3.

Finally, entirely different heterodinuclear Co–Pd complexes have been derived from the reaction of the (1,4-diaza-3-methyl-1,3-butadien-2-yl)palladium(II) complexes with $CoCl_2$ [see Eq. (8)]. In these complexes the Pd-substituted R-DAB ligands are also σ,σ-N,N' bonded (*63, 143*).

The R-DAB-to-Rh interaction has been found to be vary between σ-N monodentate, σ-N,σ-N' bridging and σ,σ-N,N' chelate coordination modes depending on the nature of R (*29, 34, 59, 61*). The former two have been discussed in Section III,B and C.

Complexes with σ,σ-N,N' bonded R-DAB ligands are encountered when the R group is doubly or triply branched at C^α in combination with substitution at the imino-C atoms [R' is Me or $PdCl(Ph_2PCH_2CH_2PPh_2)$ [see Eq. (27)].

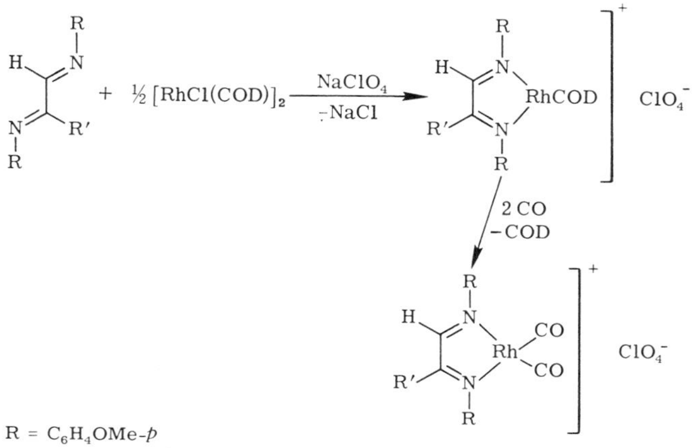

$$R = C_6H_4OMe\text{-}p$$
$$R' = Me \text{ or } PdCl(Ph_2PCH_2CH_2PPh_2)$$

The X-ray structure of $[Rh(CO)_2(i\text{-}Pr_2CH\text{-}DAB)][RhCl_2(CO)_2]$ (see Table II) contains a stable Rh-cationic species with a σ,σ-N,N' chelate bonded $i\text{-}Pr_2CH\text{-}DAB$ ligand (45) in agreement with earlier expectations (34, 59, 61).

Violet $RhCl(CO)(\eta^2\text{-}C_2H_4)(R\text{-}DAB)$ (R = $t\text{-}Bu$ and $EtMe_2C$) have been isolated according to the bridge splitting reactions [see Eq. (28)] or by a ligand substitution reaction (see Scheme 2) (61). Extensive ^1H and ^{13}C NMR spectroscopic studies revealed a structure for this complex type analogous to that found for $PtCl_2(\eta^2\text{-}olefin)(R\text{-}DAB)$ (Section III,D,2,f): i.e., a trigonal bipyramidal array with the σ,σ-N,N' chelate bonded R-DAB and the olefin residing in the equatorial plane (61).

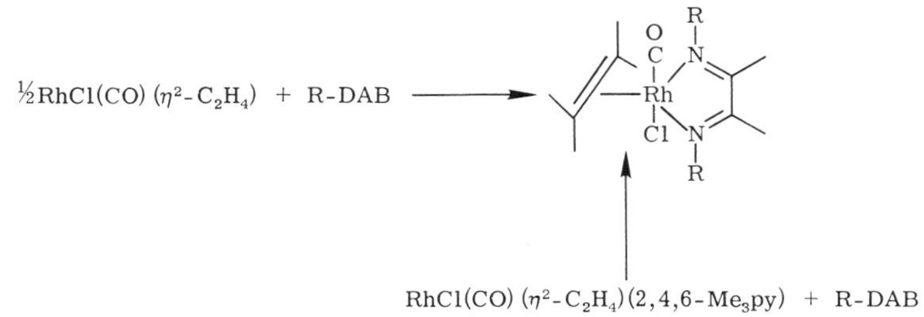

$$\tfrac{1}{2}RhCl(CO)\,(\eta^2\text{-}C_2H_4) + R\text{-}DAB \longrightarrow \qquad (28)$$

$$RhCl(CO)\,(\eta^2\text{-}C_2H_4)(2,4,6\text{-}Me_3py) + R\text{-}DAB$$

$RhCl(CO)(\eta^2\text{-}C_2H_4)(R\text{-}DAB)$ readily loses ethylene thus providing four-coordinate $RhCl(CO)(R\text{-}DAB)$ (R = $t\text{-}Bu$, $EtMe_2C$) in which the R-DAB is σ,σ-N,N' bonded (see Scheme 2). The dissymmetry in the square plane was clearly reflected by the anisochronous ^1H and ^{13}C resonances of the two $R{-}N{=}CH$ halves (61).

Addition of another equivalent of R-DAB to $Rh_2Cl_2(CO)_4(R\text{-}DAB)$ or the bridge splitting reaction of R-DAB with $[RhCl(CO)]_2$ in hexane resulted in formation of $RhCl(CO)_2(R\text{-}DAB)$ (R = t-Bu, $EtMe_2C$) which, however, cannot be isolated (61). In solution a trigonal bipyramidal structure is proposed containing σ,σ-N,N' bonded R-DAB but the geometry of this complex could not be conclusively established.

Spin saturation experiments on mixtures of $RhCl(CO)(\eta^2\text{-}C_2H_4)(R\text{-}DAB)$ with R-DAB (involving ^1H resonances of free and coordinated R-DAB) showed that intermolecular exchange of R-DAB occurs (61). The $RhCl(CO)_2(R\text{-}DAB)$ species are less stable in solution than in the solid state. Exchange reactions for these five coordinate Rh-R-DAB species are shown in Scheme 2.

f. Group VIII; d^{10} Metals Ni, Pd, and Pt. In contrast to the relatively few metal d^9 species, the complex chemistry of d^{10} metals with R-DAB is rather developed. The synthesis of zerovalent Ni complexes has been extensively investigated. The most important synthetic routes are given in Scheme 3. The synthesis starting from $Ni(CO)_4$ occurs at elevated temperatures and cannot be used for thermally unstable R-DAB ligands (144). Furthermore, these reactions sometimes stop at the stage of $Ni(CO)_2(R\text{-}DAB)$ as has been shown for R = $i\text{-}Pr_2CH$ (9, 100). $Ni(COD)_2$ is a better

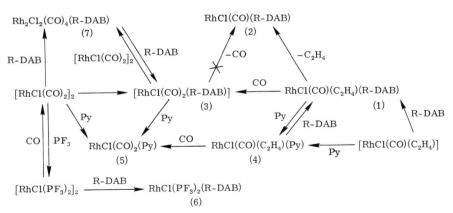

SCHEME 2. Reactions of rhodium(I)-R-DAB [R = t-Bu (1a–3a, 6a, 7a) or $EtMe_2C$ (1b–3b, 6b, and 7b)] complexes in $CHCl_3$ or hexane (61).

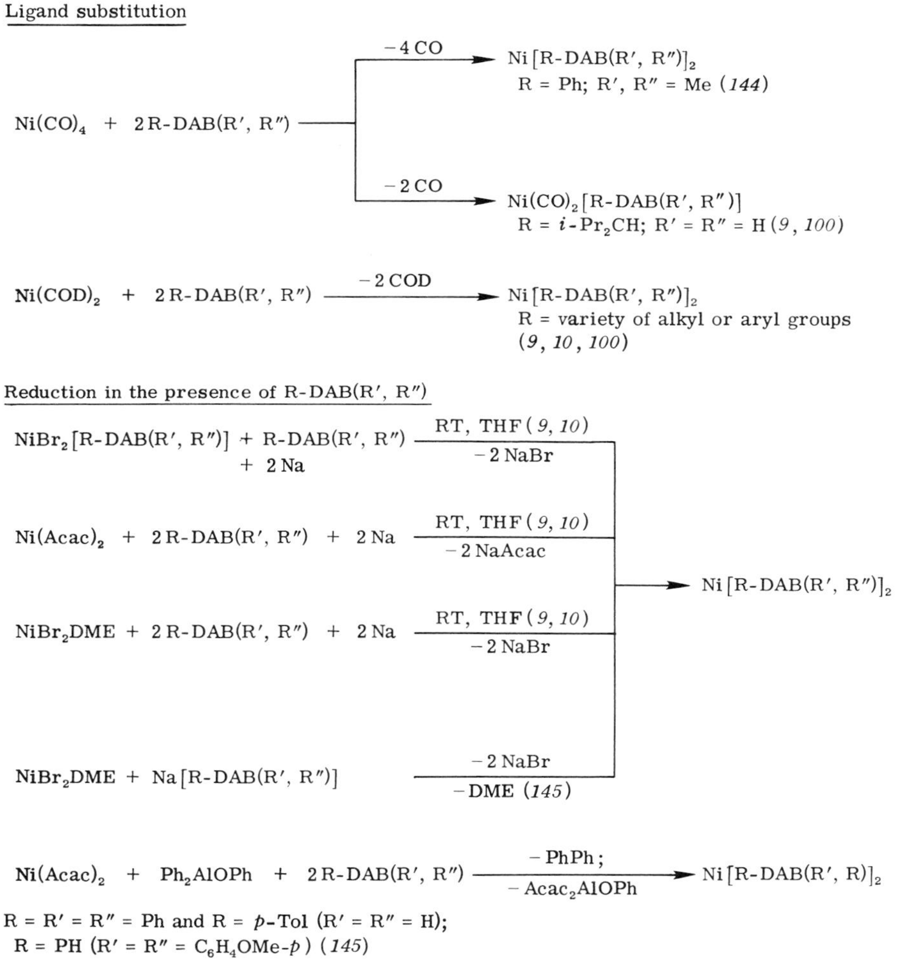

Ligand substitution

Ni(CO)$_4$ + 2 R-DAB(R', R'') ⟶

    $\xrightarrow{-4\,CO}$ Ni[R-DAB(R', R'')]$_2$
               R = Ph; R', R'' = Me (*144*)

    $\xrightarrow{-2\,CO}$ Ni(CO)$_2$[R-DAB(R', R'')]
               R = *i*-Pr$_2$CH; R' = R'' = H (*9*, *100*)

Ni(COD)$_2$ + 2 R-DAB(R', R'') $\xrightarrow{-2\,COD}$ Ni[R-DAB(R', R'')]$_2$
                                               R = variety of alkyl or aryl groups
                                               (*9*, *10*, *100*)

Reduction in the presence of R-DAB(R', R'')

NiBr$_2$[R-DAB(R', R'')] + R-DAB(R', R'') $\xrightarrow[-2\,NaBr]{RT,\ THF\ (9,\ 10)}$
                + 2 Na

Ni(Acac)$_2$ + 2 R-DAB(R', R'') + 2 Na $\xrightarrow[-2\,NaAcac]{RT,\ THF\ (9,\ 10)}$

                                       ⟶ Ni[R-DAB(R', R'')]$_2$

NiBr$_2$DME + 2 R-DAB(R', R'') + 2 Na $\xrightarrow[-2\,NaBr]{RT,\ THF\ (9,\ 10)}$

NiBr$_2$DME + Na[R-DAB(R', R'')] $\xrightarrow[-DME\ (145)]{-2\,NaBr}$

Ni(Acac)$_2$ + Ph$_2$AlOPh + 2 R-DAB(R', R'') $\xrightarrow[-Acac_2AlOPh]{-PhPh;}$ Ni[R-DAB(R', R)]$_2$

R = R' = R'' = Ph and R = *p*-Tol (R' = R'' = H);
R = PH (R' = R'' = C$_6$H$_4$OMe-*p*) (*145*)

SCHEME 3. Reactions leading to Ni^0-R-DAB complexes.

starting material although also in this case with sterically hindered groups (e.g., R = *i*-Pr$_2$CH) substitution beyond the Ni(COD)(R-DAB) stage is not possible (*9*). The preferred method appears to be one of the reductive routes shown in Scheme 3 (*9*, *47*, *70*, *145*, *146*). Following such a route the interesting tetra(organyl)cyclobutadienenickel complex shown in Eq. (29) has been synthesized.

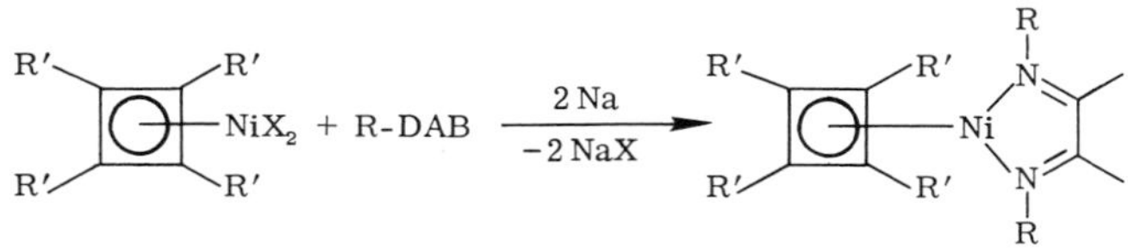

$$R = c\text{-Hex}, i\text{-Pr}; \quad R' = Ph, Me; \quad X = Cl, Br \ (146)$$

A combined ligand substitution–reductive elimination reaction of $NiEt_2(2,2'\text{-bipy})$ complexes in the presence of Ph-DAB(Ph,Ph) resulted in formation of Ni[Ph-DAB(Ph,Ph)]$_2$ (145). The complex Ni[Ph-DAB-(Ph,Ph)](PPh$_3$)$_2$ can be isolated from the 1:1 reaction of R-DAB with Ni(PPh$_3$)$_2$(C$_2$H$_4$). However, this complex readily undergoes further substitution of the PPh$_3$ ligands (70, 73).

NiX$_2$(R-DAB) complexes that are obtained from the 1:1 reaction of NiX$_2$ with R-DAB (47) are good starting materials for zerovalent Ni-R-DAB species (see Scheme 3).

Reaction of NiBr$_2$(i-Pr$_2$CH-DAB) with o-tolylmagnesium bromide results (probably via an o-tolylnickel intermediate from which toluene is eliminated) in formation of a new Ni—C bond (Eq. 30) (47, 147).

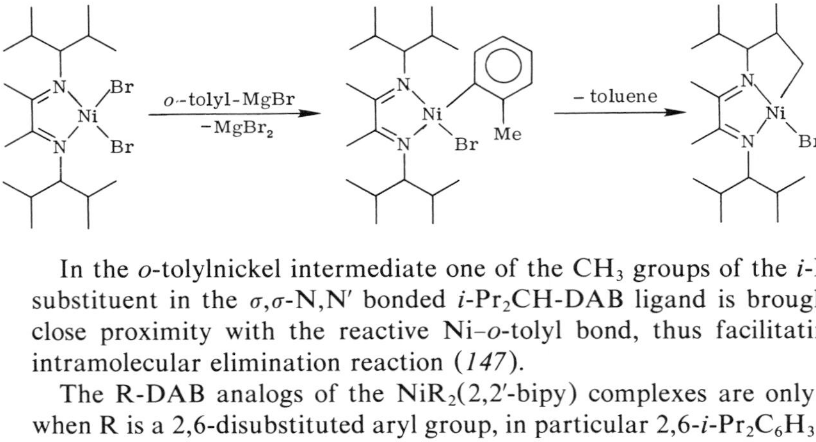

In the o-tolylnickel intermediate one of the CH$_3$ groups of the i-Pr$_2$CH substituent in the σ,σ-N,N' bonded i-Pr$_2$CH-DAB ligand is brought into close proximity with the reactive Ni–o-tolyl bond, thus facilitating the intramolecular elimination reaction (147).

The R-DAB analogs of the NiR$_2$(2,2'-bipy) complexes are only stable when R is a 2,6-disubstituted aryl group, in particular 2,6-i-Pr$_2$C$_6$H$_3$ (148), and can be prepared via routes shown in Eq. (31).

$$NiBr_2(R\text{-DAB}) + 2\,R'MgBr \xrightarrow[-80\,°C]{Et_2O} NiR_2'(R\text{-DAB}) + 2\,MgXBr$$

$$\Big\uparrow Et_2O \Big| -80\,°C \qquad (31)$$

$$Ni(Acac)_2 + R\text{-DAB} + 2\,R'MgBr$$

X = Br, Acac

The lower stability of the NiMe$_2$(R-DAB) complexes compared to the 2,2'-bipy analogs is ascribed to a destabilization of the Ni—C bond brought about by a distortion from the square planar geometry as a result of steric interaction of the Me groups and the bulky R substituents (*148*). However, electronic factors may also contribute to a destabilization of the NiR$_2$(R-DAB) complexes.

Selective reduction of NiBr$_2$(*i*-Pr$_2$CH-DAB) with sodium in ether affords the Ni^I dimer [NiBr(*i*-Pr$_2$CH-DAB)]$_2$ (*47*). Three further routes to this product are available as shown in Scheme 4.

X-Ray structure determinations of Ni(R-DAB)$_2$ [R = *c*-Hex (*9*) and 2,6-Me$_2$C$_6$H$_3$ (*10*)] (see Table II) have been carried out and revealed structures containing σ,σ-N,N' bonded R-DAB ligands. However, the structures differ with respect to the angle between the two chelate rings. For the compound with R = *c*-Hex this angle is 88.3° whereas a value of 44.5° is found for R = 2,6-Me$_2$C$_6$H$_3$. This difference has been explained on the basis of the tendency of Ni to favor a planar diamagnetic (i.e., a formally Ni^{II}) arrangement with 1,4-heterodienes which leads for the sterically smaller 2,6-Me$_2$C$_6$H$_3$ groupings to an intermediate square planar–tetrahedral arrangement. In this respect it is interesting that isoelectronic Ni[(3,5-Me$_2$C$_6$H$_3$)$_2$N$_4$]$_2$ has a corresponding angle θ of 90°. Since the tetraazadiene ligand is an even better π-electron acceptor than the diazabutadiene ligand (*75*, *148a*) this suggests that other factors may be operative. Both structures are shown in Fig. 6.

An attempt has been made to classify the Ni(R-DAB)$_2$ complexes with various R groups according to the formal oxidation state on the Ni center by ESCA spectroscopy (*149*). However, the Ni (2*p* 3/2) data do not appear to provide reliable information, in contrast to the Pt (4*f* 7/2) data for

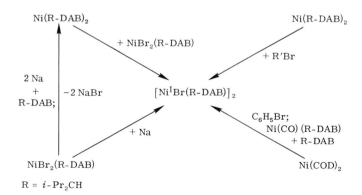

SCHEME 4. Routes to [Ni^{I}Br(R-DAB)]$_2$ (*47*).

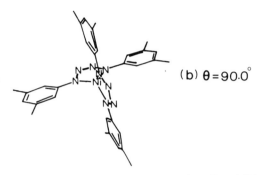

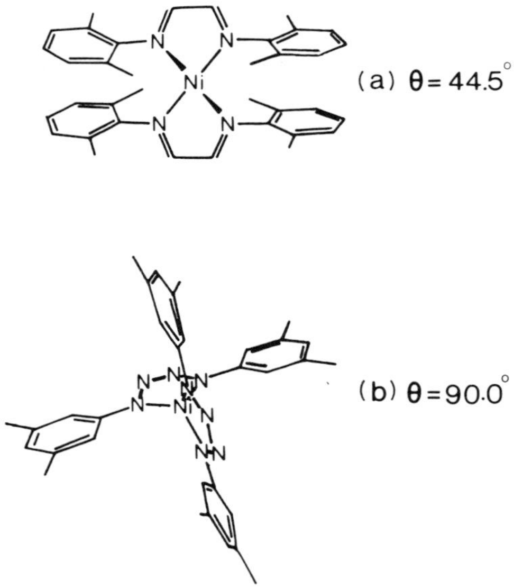

(a) θ = 44.5°

(b) θ = 90.0°

FIG. 6. Structure (a) of isoelectronic nickel-R-DAB (*9, 10*) and (b) nickeltetraazadiene (*148a*) complexes; θ is the angle between the chelate planes.

which such classification in ranges for Pt⁰, Ptᴵᴵ, and Ptᴵⱽ seems meaningful (*150*).

The structure of [NiBr(*i*-Pr₂CH-DAB)]₂ has been established by an X-ray diffraction study (see Table II and Scheme 4).

The structures of Ni(CO)₂(R-DAB) complexes can be deduced from the crystal structure of a corresponding diacetyl(bis-dimethylhydrazon)nickel complex which is shown in Fig. 7 (*46*). Apart from the tetrahedral geometry around Ni and the σ,σ-N,N′ bonded α-diimine ligand this structure shows the influence of the methyl substituents on the imino-C atoms on the ro-

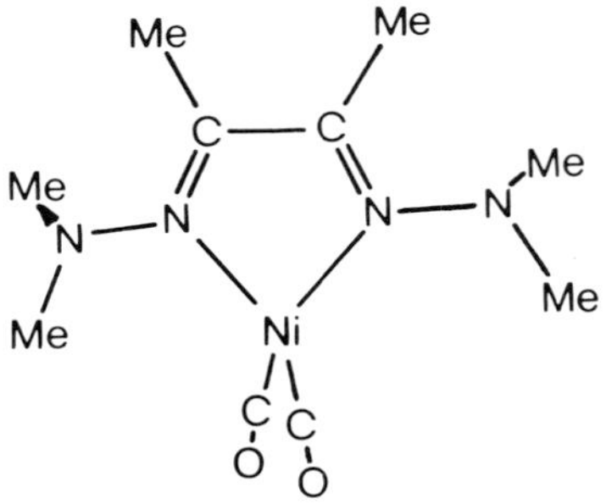

FIG. 7.   Molecular geometry of Ni(CO)₂[Me₂N-DAB(Me,Me)]₂ (*46*).

tamer conformation of the Me_2N groups (both methyl groups of the Me_2N grouping being directed away).

Zerovalent complexes of palladium containing σ,σ-N,N' bonded R-DAB have been obtained by the simultaneous reaction of activated olefins and a R-DAB ligand with $Pd_3(TTAA)_3$ [cf. (58)] as the preferred source of Pd^0 [Eq. (32)].

$$\tfrac{1}{3}Pd_3(TTAA)_3 + \text{olefin} + t\text{-Bu-DAB} \xrightarrow[-TTAA]{} Pd(\eta^2\text{-olefin})(t\text{-Bu-DAB})$$

(32a)

olefin = e.g., Dmf

Stable combinations of the η^2-olefin and R-DAB ligands are formed with strong σ-donating R groups and a strong π-electron density accepting olefin (58).

$Pt(COD)_2$ reacts with R-DAB to give tetrahedral Pt(COD)-(R-DAB) (151) complexes that are strikingly similar to the $Pt(COD)(R_2N_4)$ (R_2N_4 = arylN=N—N=Naryl) complexes for which ESCA measurements indicate a Pt^{II} formal oxidation state (150). However, ^{13}C NMR measurements reveal $^1J(Pt-C)$ values comparable with those of Pt(COD)(R-DAB) and other known zerovalent platinum cyclooctadiene complexes (150, 151).

Divalent Pd- and Pt-R-DAB complexes with σ,σ-N,N' bonded R-DAB ligands have been reported for the (1,4-diaza-3-methyl-1,3-butadien-2-yl)palladium(II) ligands as shown below (60).

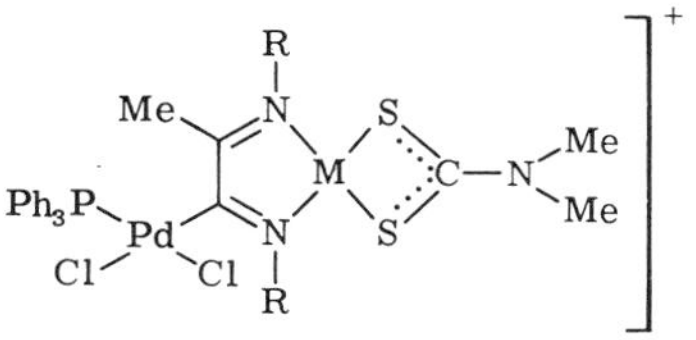

R = C_6H_4OMe-p; M = Pt, Pd

Cis-MCl_2(R-DAB) can exclusively be obtained for M = Pd starting from trans-$MCl_2(PhCN)_2$ with R-DAB (R = t-Bu, $EtMe_2C$) (29). The corresponding Pt compounds are accessible via olefin dissociation from the five-coordinate $PtCl_2(\eta^2$-olefin)(R-DAB) complexes (29) that can be obtained according to Eq. (32) (48).

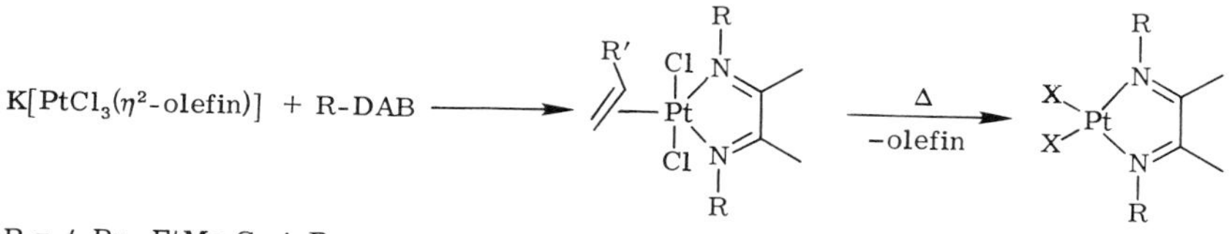

$K[PtCl_3(\eta^2\text{-olefin})] + \text{R-DAB} \longrightarrow$ (32b)

R = t-Bu, $EtMe_2C$, i-Pr

^{1}H and ^{13}C NMR measurements confirmed the five-coordinate structure for these compounds which is schematically shown in Eq. (32b). When styrene is the η^2-coordinated olefin, the ground state structure in the solid state (see Table II) established that the C=C skeleton is in the trigonal plane (48, 56). This in-plane conformation is a reflection of the stabilizing metal back bonding capability of the olefin on the σ,σ-N,N' chelate bonding mode. Complexes containing a phosphine instead of an η^2-olefin undergo σ-N $\rightleftarrows$ σ-N' fluxional processes (see Fig. 3) in which the five-coordinate situation is an intermediate or transition state (30, 35).

Stabilization of Pt–olefin and Pt–alkyne bonds due to effective back bonding from the metal to empty π^* orbitals of the unsaturated hydrocarbon in electron-rich five-coordinate complexes have been demonstrated in a number of cases: e.g., PtCl$_2$(η^2-C$_2$H$_4$)[Ph(Me)N-DAB(Me,Me)] (49, 152) PtMeCl(η^2-RC≡CR)(2,2'-bipy) (153), PtCl$_2$(η^2-C$_2$H$_4$)[(RN-(H)CH$_2$)$_2$] (48, 153a) and PtCl$_2$(η^2-C$_2$H$_4$)(2-RN=C(H)py) (154).

The structural aspects and reactivity of the five-coordinate PtX$_2$(η^2-olefin)(R-DAB) and analogous RhX(CO)(η^2-C$_2$H$_4$)(R-DAB) complexes have been studied in detail since they represent model compounds for the five-coordinate intermediates (activated complexes) found in ligand displacement reactions from four-coordinate complexes: i.e., trans-MX$_2$A(η^2-olefin) + B $\rightleftarrows$ trans-MX$_2$B(η^2-olefin) + A (48, 155). Moreover, the barrier to rotation of the olefin around the olefin–Pt axis in PtX$_2$(η^2-olefin)(R-DAB) was studied (48) and found to be similar to that observed in square planar Pt–olefin complexes. This has been attributed to the planarity of the five-membered chelate ring, which lowers the barrier for bending of the Cl atoms toward the R-DAB ligand when the olefin passes through an upright conformation (see Fig. 8). It is interesting that the above-mentioned Pt^0(COD)(R-DAB) and Pt^0(η^2-olefin)(R-DAB) complexes, which lack the two axial ligands, have the same stereochemistry in the trigonal plane as has been found for the five-coordinate Pt^{II} and Rh^I complexes.

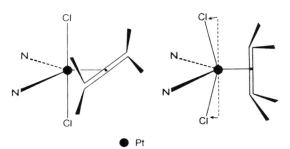

● Pt

FIG. 8. Bending back of the Cl atoms in the TBP structure of PtCl$_2$(η^2-olefin)(R-DAB) when the olefin passes the upright conformation during the rotation process (48).

In $PtCl_2(\eta^2$-olefin$)(6$-R'-py-2-CH=NR$)$ the N-coordination sites of the bidentate ligand are dissimilar (*154*). Owing to the resulting asymmetry in the equatorial plane of the trigonal bipyramidal structure of these complexes it was possible to use ^{13}C NMR spectroscopy to establish the ground state structure with an in-plane C=C olefinic moiety. Furthermore, it was shown that changing the 6-H for a 6-Me group introduced a considerable increase of the steric constraints in the equatorial plane. This can explain the increased stability of the 6-Me complexes with respect to Pt–olefin bond dissociation: a process that would afford *cis*-$PtCl_2(6$-R'-py-2-CH=NR$)$. In the latter square planar complex the 6-R' substituent comes much closer to one of the cis-Cl groups (*154*).

Finally, the above observations were confirmed by studying the ^{15}N labeled compounds $PtCl_2(\eta^2$-styrene$)(t$-Bu-DAB-$^{15}N_2)$ (*35*).

An important aspect of these five-coordinate $PtX_2(\eta^2$-olefin$)(R$-DAB$)$ complexes is that the axial halogen atoms as well as the equatorial η^2-olefin (e.g., C_2H_4) and σ,σ-N,N' chelate bonded R-DAB ligands can be replaced with retention of the trigonal bipyramidal structure (see Scheme 5) (*155, 156*). Halogen exchange is initiated by formation of an ionic intermediate $[PtX(\eta^2$-$C_2H_4)(R$-DAB$)]X$. The reversible exchange of the equatorial ligands with olefins, R-DAB, or 1,2-diaminoethane ligands is proposed to proceed via five-coordinate intermediates containing a σ-N monodentate bonded R-DAB or 1,2-diaminoethane ligand (*155, 156*) (see Fig. 9).

It was argued on the basis of stereochemical and electronic grounds that the R-DAB and 1,2-diaminoethane ligands are better suited to this reversible σ-N $\rightleftarrows$ σ,σ-N,N' bonding than py-2-CH=NR or 2,2'-bipyridine (*156*).

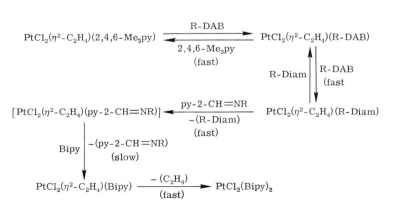

SCHEME 5. Ligand substitution in five-coordinate platinum(II) complexes [R-Diam is $RN(H)CH_2CH_2N(H)R$] (*156*).

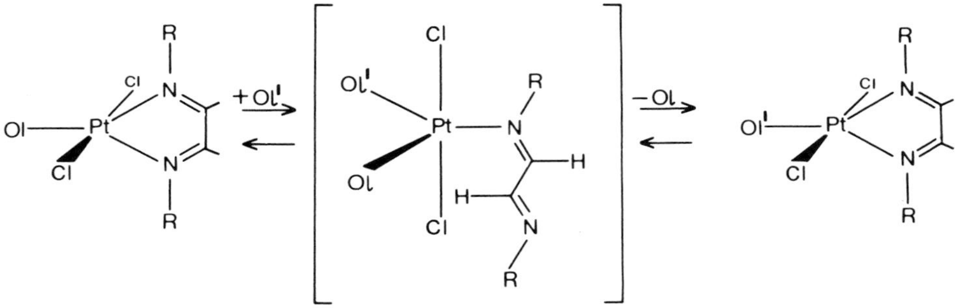

FIG. 9. Mechanism proposed for olefin exchange in PtCl$_2$(η^2-olefin)L$_2$ complexes. Similar intermediates are proposed for the bidentate N-donor ligand exchange: replace in the figure ol' by L$_2'$ for the incoming ligand and ol by R-DAB as leaving group (156).

g. *Groups IB and IIB.* Reactions of copper(I) halides with R-DAB(R',R″) (R = t-Bu, c-Hex, C$_6$H$_4$Y-p with Y = H, OH, NMe$_2$, or Me; R' = R″ = H or Me) afford highly colored monomeric (in nitrobenzene) CuX[R-DAB(R',R″)] complexes. It has been put forward that the trigonal coordination at copper is reflected in spectroscopic measurements (14). A UV–visible spectroscopic study using various solvents showed that for these complexes the typical solvatochromic behavior and high intensity of the long-wavelength absorption (MLCT band) found for Mo^0 (see Section III,D,2,b) is absent. This is explained by extensive participation of metal d and ligand π^* orbitals in the ground state, which removes the typical CT character in the case of Cu^I (14).

Later the reaction of copper(I) perchlorate with t-Bu-DAB in acetonitrile was studied (157). It was shown that in this solvent the reaction proceeds in two steps involving first formation of Cu(t-Bu-DAB)$^+$ (in which the Cu$^+$ center is probably further coordinated by two MeCN ligands) and in the second a four-coordinate species Cu(t-Bu-DAB)$_2^+$. On the basis of a comparison of the optical spectra, Zelewsky et al. suggest that the above-mentioned CuX(R-DAB) complexes can be better formulated as being [CuX$_2$][Cu(R-DAB)$_2$] species (157).

The CuX(R-DAB) complexes have been used in reactions directed to formation of pure K(R-DAB) (see Section V).

Reaction of tetranuclear Cu$_4$(C$_6$H$_4$Me-p)$_4$ (158) with t-Bu-DAB afforded a stable red-colored complex with Cu$_3$(C$_6$H$_4$Me-p)$_3$(t-Bu-DAB) stoichiometry, the structure of which is under study (159). In contrast, reaction of Cu$_2$Li$_2$(C$_6$H$_4$Me-p)$_4$(Et$_2$O)$_2$ (160) with t-Bu-DAB forms at low temperature a Cu$_2$Li$_2$(C$_6$H$_4$Me-p)$_4$(t-Bu-DAB)$_2$ complex that decomposes at room temperature into a variety of products (159).

So far R-DAB complexes of Ag^I and Au^I or Au^{III} have not been reported. That α-diimine and also R-DAB complexes for these metals are feasible is indicated by the stable tetrahedral cations formed with 6-R'-py-2-CH=NR ligands [see Eq. (33)] (161).

$$AgO_3SCF_3 \; + \; 2 \qquad\qquad\qquad\longrightarrow \qquad\qquad\qquad O_3SCF_3^- \qquad (33)$$

R' = Me, H; R = t-Bu, (S)-C(H)MePh

The stereochemistry of these complexes in methanol solutions has been studied by ^{109}Ag INEPT NMR spectroscopy (161a) and ^{15}N NMR of ^{15}N labeled complexes (161, 161a).

A series of MX_2(R-DAB) complexes where M is Zn, Hg, and Cd have been prepared [see Eq. (8)] containing σ,σ-N,N' chelate bonded R-DAB (R = Ph and the central C—C bond is part of a camphor skeleton (162) or R = c-Hex, p-MeOC$_6$H$_4$, t-Bu (62).

The radical complexes [M(R-DAB)]X (M = Zn, Mg, Ca) will be discussed in Section V.

Organozinc complexes show a remarkable variance in stability. Whereas Zn(C$_6$H$_4$Me-p)$_2$(t-Bu-DAB) is stable up to 130°C, ZnEt$_2$(t-Bu-DAB) is stable only below −50°C and above this temperature a selective ethyl transfer from Zn to N occurs [see Section V and VI,B (e.g., Fig. 23)] (68).

E. Complexes Containing Bridging σ-N,μ²-N',η²-C=N' (6e) Bonded 1,4-Diaza-1,3-butadiene Ligands

There are relatively few examples of compounds containing the R-DAB ligand bonded in a σ-N,μ^2-N',η^2-C=N' (6e) fashion. Frühauf et al. (30) reported the first examples of such a bonding mode for a number of compounds Fe$_2$(CO)$_6$(R-DAB). The structure of Fe$_2$(CO)$_6$(t-Bu-DAB) shows that two electrons donated by one σ-N bonded N=C group [1.260(5) Å] and four electrons via the other N=C group [1.397(4) Å] that bridges the two Fe atoms. Formally the last N=C group donates two electrons through the bridging N atom and two electrons by η^2-bonding to one Fe center (30) (see Fig. 10 and Table II). It has now been shown on the basis of crystallographic and/or spectroscopic evidence that this 6e-donor mode of R-DAB, which is then always in the E-s-cis-E conformation, exists for Fe$_2$(CO)$_6$(R-DAB) [R = t-Bu (30, 50, 112, 163), c-Hex (30, 163), i-Pr

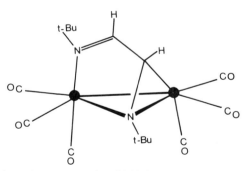

FIG. 10. Schematic structure of M₂(CO)₆(t-Bu-DAB) (30). (●) Fe, Ru, Os.

(112, 163)], Ru₂(CO)₆(R-DAB) [R = t-Bu (50, 112, 163), c-Hex (50, 163), i-Pr (50, 112, 163)], and Os₂(CO)₆(R-DAB) [R = t-Bu, i-Pr (112)].

An analogous structure has been reported for Fe₂(CO)₆[PhCH-(Me)NC(H)C(OEt)=O] (164) (Fig. 11). The N-α-methylbenzyl imi-noacetate donates two electrons via the O-atom [C=O = 1.326(8) Å] and four electrons via the C=N moiety [1.417(7) Å]. The Fe′—C, Fe′—N, and C—C bond lengths are 2.055(6), 1.927(5), and 1.435(10) Å, respectively. These data together with the Fe—Fe bond lengths of 2.551(1) Å are very similar indeed to those of Fe₂(CO)₆(c-Hex-DAB) (Table II) (30). Mixed complexes MM′(CO)₆-(R-DAB) have been obtained for M = Fe and M′ = Ru [prepared by the reaction of Fe(CO)₃(t-Bu-DAB) with Ru₃(CO)₁₂ in a three to one molar ratio (165)] and for M = Mn, Re and M′ = Co with R = t-Bu, i-Pr, and c-Hex (51). Binuclear ruthenium complexes Ru₂(CO)₄(R-DAB)₂ containing two σ-N,μ²-N′,η²-C=N′ (6e) bonded R-

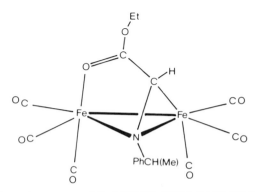

FIG. 11. Schematic structure of Fe₂(CO)₆[PhCH(Me)NCHC(OEt)=O] (164).

DAB groups [R = *i*-Pr, *c*-Hex, *p*-Tol (*50, 163*)] bridging a nonbonded Ru$_2$ pair have been reported. The σ-N,μ^2-N',η^2-C=N' bonding mode has also been realized for asymmetric R-DAB (H,Me) groups e.g., in Fe$_2$(CO)$_6$(R-DAB)(H,Me) (*50, 163*), Ru$_2$(CO)$_6$(R-DAB)(H,Me) (*50, 163*) and MnCo(CO)$_6$(R-DAB)(H,Me) (*51, 163*).

A common feature of all these compounds is that the C=N group, which is η^2-bonded to the metal atom, carries a H atom on the imino-C atom. It has not been possible as yet, except for one compound, to bind R-DAB ligands containing two Me groups on the imino-C atoms in a 6e-bonding mode. The single exception is MnCo(CO)$_6$(*c*-Pr-DAB)(Me,Me) (*51*). The methyl group clearly destabilizes η^2-C=N bonding as is also indicated by the fact that this 6e R-DAB compound is the only fluxional one. At ambient temperatures the two (Me)C=N groups exchange their points of attachment intramolecularly (Fig. 12). This movement, which probably proceeds via an intermediate containing the *c*-Pr-DAB(Me,Me) group σ,σ-N,N' (4e) chelated to Mn, suggests that metal–η^2-C=N bonding is here less strong than for the η^2-HC=N bonded groups.

Typical of the σ-N,μ^2-N',η^2-C=N' (6e) bonding mode is the bond lengthening of the η^2-C=N bonded imine group to about 1.40 Å, this is appreciably longer than the σ-N coordinated N=C group (about 1.28 Å). In Table II various relevant details are given for Fe$_2$(CO)$_6$(*t*-Bu-DAB) (*30*). MnCo(CO)$_6$(*t*-Bu-DAB) (*51*) and Ru$_2$(CO)$_4$(*i*-Pr-DAB)$_2$ (*50*). Furthermore, the ^{1}H and ^{13}C NMR chemical shifts of the η^2-bonded N=C lie at much higher fields than the signals of the σ-N bonded N=C group (Table V). These crystallographic and NMR aspects, returned to later in Section IV.B, clearly indicate intensive π-back donation into the π^*-N=C levels and negative charge density increase on the η^2-N=C(H) bonded groups.

An outline of the preparations of the various types of 6e R-DAB complexes now completes this Section with the mechanistic aspects being discussed in detail in Section IV, A.

In the first preparation such a complex was reported by Frühauf *et al.* (*30*) and involved the reaction of Fe(CO)$_3$(R-DAB) with Fe(CO)$_5$ or Fe$_2$(CO)$_9$ (*101*) (Eq. 34):

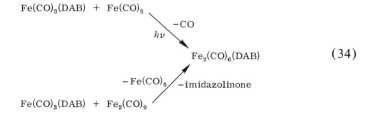

$$\text{(34)}$$

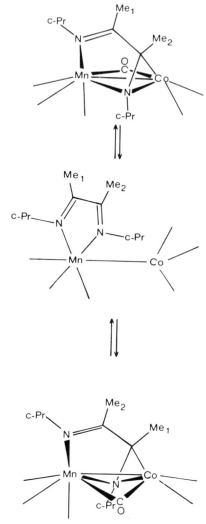

FIG. 12. Proposed route showing the fluxional behavior of MnCo(CO)$_6$[c-Pr-DAB-(Me,Me)] (51).

The mixed complex containing an Fe—Ru bond was prepared similarly (165) [Eq. (35)]:

$$Fe(CO)_3(DAB) + \frac{1}{3}Ru_3(CO)_{12} \rightarrow FeRu(CO)_6(DAB) + CO \qquad (35)$$

The preparation of Ru$_2$(CO)$_6$(R-DAB) (R = i-Pr, t-Bu, c-Hex) involves

TABLE V

RELEVANT ^{1}H AND ^{13}C NMR PARAMETERS OF (σ-N,μ^2-N',η^2-C=N'; 6e) BONDED R-DAB GROUPS IN COMPLEXES [M$_2$(CO)$_6$(R-DAB)] AND [Ru$_2$(CO)$_4$(R-DAB)$_2$]^{a,b}

Complex	Amino ^{1}H ($J \simeq 2$ Hz)	Imino ^{1}H	Amino ^{13}C	Imino ^{13}C
Fe$_2$(CO)$_6$(t-Bu-DAB)	3.33	7.63	60.1	175.1
Ru$_2$(CO)$_6$(t-Bu-DAB)	3.41	7.79	56.3	173.5
Os$_2$(CO)$_6$(t-Bu-DAB)	4.06	8.12	49.3	176.1
MnCo(CO)$_6$(t-Bu-DAB)	4.76	7.89	74.4	171.3
ReCo(CO)$_6$(t-Bu-DAB)	5.48	8.06	75.5	172.4
Fe$_2$(CO)$_6$(i-Pr-DAB)	3.30	7.52	58.9	173.3
Ru$_2$(CO)$_6$(i-Pr-DAB)	3.24	7.70	—	—
Os$_2$(CO)$_6$(i-Pr-DAB)	3.98	8.17	—	—
MnCo(CO)$_6$(i-Pr-DAB)	4.85	7.79	75.8	170.5
Ru$_2$(CO)$_4$(i-Pr-DAB)$_2$	4.08	8.23, 8.25	63.8	172.1
Fe$_2$(CO)$_6$(c-Hex-DAB)	—	—	63.7	173.6
Ru$_2$(CO)$_6$(c-Hex-DAB)	3.27	7.74	61.5	173.5
MnCo(CO)$_6$(c-Hex-DAB)	4.84	7.80	76.6	170.7
ReCo(CO)$_6$(c-Hex-DAB)	5.54	7.97	78.6	173.6
Ru$_2$(CO)$_4$(c-Hex-DAB)$_2$	4.31	8.43	—	—
MnCo(CO)$_6$(c-Pr-DAB)	4.88	7.65	79.6	172.1
MnCo(CO)$_6$[c-Pr-DAB(Me, Me)]	—	—	96.7	186.5

^{a} From Refs. *50, 51, 112,* and *163.*

^{b} Ppm from TMS (δ) in CDCl$_3$.

the treatment of Ru$_3$(CO)$_{12}$ with R-DAB and a similar reaction (*92*) was used for Os$_2$(CO)$_6$(R-DAB) (*50, 112*) [Eq. (36)].

$$\frac{2}{3}M_3(CO)_{12} + R\text{-DAB} \xrightarrow{\text{M = Ru, Os}} M_2(CO)_6(DAB) + 2CO \qquad (36)$$

Finally, the formation of MnCo(CO)$_6$(R-DAB) proceeds via two steps. The first step involves a nucleophilic substitution of Br$^-$ in MnBr(CO)$_3$(R-DAB) by Co(CO)$_4^-$ to form the Mn—Co bonded MnCo(CO)$_7$(R-DAB), which contains an R-DAB σ,σ-N,N' (4e) bonded to Mn. Subsequently there occurs an intramolecular substitution of one cobalt CO group by one HC=N group which then becomes η^2-C=N bonded to Co (*51*) [Eqs. (37a) and (37b)].

$$\text{MnBr(CO)}_3(\text{R-DAB}) + \text{Co(CO)}_4^- \rightarrow \text{MnCo(CO)}_7(\text{R-DAB}) + \text{Br}^- \qquad (37a)$$

$$\text{MnCo(CO)}_7(\text{R-DAB}) \rightarrow \text{MnCo(CO)}_6(\text{R-DAB}) + \text{CO} \qquad (37b)$$

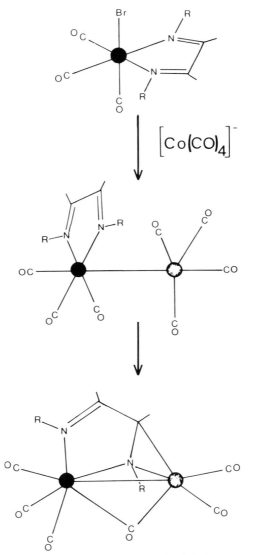

FIG. 13.   Reaction to give MnCo(CO)₆(R-DAB) (87).

This two step reaction (87) is shown in Fig. 13. Of interest is that in the final product one CO group is semibridging, as indicated by the structure determination of MnCo(CO)₆(t-Bu-DAB) and suggested by the ν(CO) (1880–2000 cm⁻¹) (Table VI), being in the range between terminal and bridging CO groups.

TABLE VI

$\nu(CO)$ Stretching Frequencies of $M_2(CO)_6(R\text{-}DAB)^{a,b}$

Complex	$\nu(CO)$ (cm^{-1})
$Fe_2(CO)_6(t\text{-}Bu\text{-}DAB)$	2053, 2003, 1981, 1969, 1945
$Ru_2(CO)_6(t\text{-}Bu\text{-}DAB)$	2069, 2030, 1994, 1983, 1961
$Os_2(CO)_6(t\text{-}Bu\text{-}DAB)$	2067, 2026, 1987, 1971, 1953
$Fe_2(CO)_6(i\text{-}Pr\text{-}DAB)$	2057, 2005, 1987, 1975, 1945
$Ru_2(CO)_6(i\text{-}Pr\text{-}DAB)$	2067, 2025, 1999, 1988, 1961
$Os_2(CO)_6(i\text{-}Pr\text{-}DAB)$	2069, 2028, 1990, 1974, 1955
$MnCo(CO)_6(t\text{-}Bu\text{-}DAB)$	2047, 2006, 1986, 1935, 1894^c
$ReCo(CO)_6(t\text{-}Bu\text{-}DAB)$	2049, 2013, 1989, 1937, 1932, 1876^c
$MnCo(CO)_6(i\text{-}Pr\text{-}DAB)$	2050, 2009, 1989, 1943, 1937, 1898^c

^{a} From Refs. *51* and *112*.
^{b} In *n*-pentane.
^{c} Semibridging CO.

F. Complexes Containing Bridging σ-N,σ-N',η^2-C=N,η^2-C=N' (8e) Bonded 1,4-Diaza-1,3-butadiene Ligands

Very recently examples of complexes in which the R-DAB ligand functions as a σ-N,σ-N',η^2-C=N,η^2-C=N' (8e) donor ligand have come to light. Firm crystallographic evidence was provided by the structures of $Ru_4(CO)_8(i\text{-}Pr\text{-}DAB)_2$ (*31, 52*) (Fig. 14a) and of $Ru_2(CO)_4(\mu^2\text{-}HC_2H)(i\text{-}Pr\text{-}DAB)$ (*31, 33, 166*) (Fig. 14b) and of $Mn_2(CO)_6[Me\text{-}DAB(Me,Me)]$ (*53*) (Fig. 15). The relevant ^1H NMR shifts are shown in Table VII. The

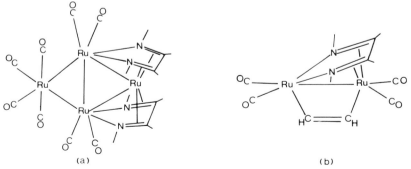

FIG. 14. Schematic structures (a) of $Ru_4(CO)_8(i\text{-}Pr\text{-}DAB)_2$ (*52*) and (b) of $Ru_2(CO)_4(\mu^2\text{-}HC_2H)$ (i-Pr-DAB) (*166*).

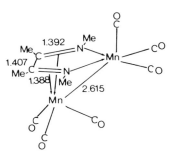

FIG. 15. Schematic structure of $Mn_2(CO)_6[Me\text{-}DAB(Me,Me)]$ (53).

overall reactions, which will be more extensively discussed in Section IV,A, are shown in Eqs. (38) and (39).

$$2Ru_2(CO)_6(R\text{-}DAB) \xrightarrow{N_2} Ru_4(CO)_8(R\text{-}DAB)_2 + 4CO \qquad R = i\text{-}Pr, c\text{-}Hex \qquad (38)$$

$$Ru_2(CO)_6(R\text{-}DAB) \xrightarrow{HC_2H} Ru_2(CO)_4(\mu^2\text{-}HC_2H)(R\text{-}DAB) + 2CO \qquad R = i\text{-}Pr, c\text{-}Hex \qquad (39)$$

To prepare the manganese compound $Mn_2(CO)_6[Me\text{-}DAB(Me,Me)]$ an unusual reaction was used involving the treatment of $Mn(CO)_4(CNMe)^-$ with MeI and the resultant yield was low (7%) (53). The diimine is formed on the metal atom(s) by C—C coupling of the isonitrile units and by addition of methyl groups (53).

TABLE VII

RELEVANT ^{1}H NMR PARAMETERS OF $(\sigma\text{-}N,\sigma\text{-}N',\eta^2\text{-}C=N,\eta^2\text{-}C=N';\ 8e)$
BONDED R-DAB LIGANDS[a,b]

Complex	δ^1H		
	$\eta^2\text{-}HC=N$	$\mu^2\text{-}HC_2H$	
$Ru_4(CO)_8(i\text{-}Pr\text{-}DAB)_2$	6.56		
$Ru_4(CO)_8(c\text{-}Hex\text{-}DAB)_2$	6.49		
$Ru_4(CO)_8(i\text{-}Bu\text{-}DAB)_2$	6.21		
$Ru_4(CO)_8(n\text{-}Pent\text{-}DAB)_2$	5.99, 6.20		
$Ru_3(CO)_8(i\text{-}Bu\text{-}DAB)$	5.86		
$Ru_3(CO)_8(n\text{-}Pent\text{-}DAB)$	5.89		
$Ru_2(CO)_4(\mu^2\text{-}HC_2H)(i\text{-}Bu\text{-}DAB)$	6.16	7.50	8.22
$Ru_2(CO)_4(\mu^2\text{-}HC_2H)(c\text{-}Hex\text{-}DAB)$	6.22	7.49	8.28

[a] From Refs. 33 and 52.
[b] In toluene-d_8 at RT.

The structural features of all three compounds clearly involve the presence of 8e-bonded R-DAB ligands (Table II). The N=C bond lengths (between 1.39 and 1.45 Å) indicate an appreciable bond lengthening of the C=N bonds in analogy to the η^2-C=N bonded imine groups of σ-N,μ^2-N',η^2-C=N' linked R-DAB groups. Meanwhile there is shortening of the central C—C bond of the R-DAB group (1.40–1.42 Å). While the diimine is symmetrically bonded in $Mn_2(CO)_6[Me$-DAB(Me,Me)], there appears to be some asymmetry in the two Ru complexes, since the N=C bond lengths differ (Table II). Interesting is the bonding in $Ru_4(CO)_8(i$-Pr-DAB)$_2$ in which one of the Ru atoms does not possess CO groups, but is exclusively bonded to the π system of the two five-membered $\overline{Ru-N=C-C=N}$ chelate rings resembling therefore a $M(C_5H_5)_2$ complex (52). The complex has a butterfly structure, which makes the two imino halves of each $\overline{M-N=C-C=N}$ skeleton nonequivalent. This was indeed confirmed by ^1H NMR at low temperatures of the imino ^1H signals. At higher temperatures these signals coalesce due to a flying movement of this butterfly (52).

In $Ru_2(CO)_4(\mu^2$-HC$_2$H)(i-Pr-DAB) one of the Ru atoms is coordinated to the π system of one $\overline{Ru-N=C-C=N}$ ring (Fig. 14b) (52). A similar situation is encountered in the manganese compound (53) that is isoelectronic and isostructural with $Fe_2(CO)_6[C(R)=C(R)-C(R)=C(R)]$ (167–170).

In the iron complex one of the Fe atoms is linked to the π system of the C_4 part of the $\overline{Fe-C=C-C=C}$ ring. Furthermore, the iron compound contains a semibridging carbonyl group in order to relieve the charge separation in the complex caused by the Fe → Fe donor bond. Although such a donor bond is not strictly necessary the manganese compound (53) indicated structural features not in disagreement with the presence of a semibridging CO group.

Recently a very interesting structure was found for $Ru_3(CO)_9(c$-Hex-DAB) isolated as a probable intermediate in the preparation of $Ru_2(CO)_6(c$-Hex-DAB) (17).

The structural features are shown in Fig. 16a. On the basis of simple electron counting one would expect the R-DAB ligand to be bonded as a σ-N,μ^2-N',η^2-C=N' (6e) bonded group in the way shown in Fig. 16b with two CO groups on the Ru atom with the μ^2-N bonded imine group and three CO groups on the Ru atom bonded to the η^2-C=N unit. However, this is not the case; instead one observes that, (Fig. 16a) although the diimine is clearly asymmetrically bonded, the ligand appears to be bonded approximately as an σ-N,σ-N',η^2-C=N,η^2-C=N' (8e) bonded ligand bridging a Ru—Ru pair with a Ru—Ru distance of 3.02 Å and C=N bond lengths of 1.33 and 1.45 Å. The bond length of 1.45 Å indicates extensive

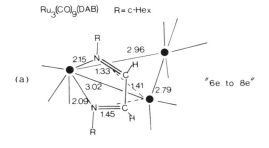

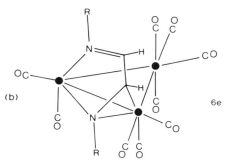

FIG. 16. Structural features of $Ru_3(CO)_9(c\text{-Hex-DAB})$, (a) as found by X-ray crystallography (17) and (b) as expected on the basis of electron counting.

back donation into the π^* C=N orbital. The C=N bond length of 1.33 Å is close to that of the η^2-C=N bonded imine group in $MnCo(CO)_6(t\text{-}$ Bu-DAB) and the σ-N bonded C=N groups found in $Ni(R\text{-DAB})_2$ (R = c-Hex, Xyl) and $Mn(t\text{-Bu-DAB})_2$ i.e., for electron-rich metal complexes containing σ,σ-N,N' chelated R-DAB groups (Table II). This length of 1.33 Å is therefore consistent with π-back bonding to the C=N group, though this is, however, not so pronounced as for the other C=N bond. Other features are the metal–metal bond lengths, which vary from 3.02 Å for the bridged Ru–Ru pair to 2.96 Å and 2.79 Å for the other Ru–Ru pairs. Note for comparison that the Ru–Ru bond length in $Ru_3(CO)_{12}$ is 2.854(4) Å (171).

According to the rules of Wade (172) addition of two electrons to a triangular cluster with 48 electrons (i.e., containing the 6e R-DAB ligand) should lead to the rupture of one of the three metal–metal bonds. For $Ru_3(CO)_9(R\text{-DAB})$ it is possible to describe an intermediate type of bonding situation with the R-DAB ligand between a 6e and 8e donor type of binding. As a result the LUMO situation on the Ru_3 triangle is sufficiently populated to cause the bond lengthening of two Ru_2 pairs as actually observed.

A point that is not clear is why the R-DAB ligand is not symmetrically bonded in an 8e fashion with relatively short $C=N$ bonds, instead of here with one very long one and one relatively short one. Further theoretical investigations are currently under way to help answer this question.

A final point to make here is that Adams (53) has clearly shown, albeit by a completely different reaction, that Me-DAB(Me,Me) can bond as an σ-N,σ-N',η^2-C$=$N,η^2-C$=$N' (8e) donor ligand. On the other hand it is almost impossible to prepare 6e R-DAB complexes with the η^2-N$=$C bonded imino group having a methyl group on the C atom (*vide infra*). In our view it might therefore be possible to use R-DAB(Me,Me) ligands to prepare metal complexes containing 8e-bonded diimines.

IV

STRUCTURAL ASPECTS

A. *The Influence of Steric and Electronic Factors on the Type of Products Formed*

It is obvious that factors such as the substituents R and R', the metal atom, and the other ligands bonded to the metal atom will influence the type of coordination of the R-DAB(R',R) ligands. In the following we shall discuss these aspects for various examples.

1. *The Relative Importance of σ-N (2e), σ-N,σ-N' (4e) and σ,σ-N,N' (4e) Bonding*

The first case in hand is the observation that stable five-coordinate complexes $PtCl_2(\eta^2$-olefin)(R-DAB) containing σ,σ-N,N' chelated R-DAB can be prepared in which the N$=$C$-$C$=$N plane is coplanar with the Pt-η^2-olefin unit (35, 48, 154–156). However, if the olefin is replaced by PPh_3, for example, a complex is formed of the composition *trans*-$PtCl_2(PPh_3)$(R-DAB) (29) in which the R-DAB ligand is σ-N (2e) monodentate bonded. This latter complex is fluxional by a process that involves an intramolecular change of the point of attachment from σ-N to σ-N' and vice versa via a five-coordinate intermediate with a chelate σ,σ-N,N' R-DAB ligand (see Fig. 3).

The rationalization is that the greater electron-accepting properties of the olefin allows more extensive back bonding from the relevant Pt orbitals into the π^* orbital of the olefin than is possible for the phosphine. The better π-back-bonding ability of the olefin compensates then for the increased charge density if one goes from 2e- to 4e-donation by the R-DAB ligand.

A case in which several species are in equilibrium is presented by the five-coordinate complex $RhCl(CO)_2(R\text{-}DAB)$ that in solution is in equilibrium with free R-DAB, four-coordinate ionic $[Rh(CO)_2(R\text{-}DAB)]$-$[RhCl_2(CO)_2]$, and with the dinuclear $[RhCl(CO)_2]_2(R\text{-}DAB)$ complex, which contains four-coordinate Rh^I and a bridging R-DAB ligand in the E-s-trans-E conformation (61). In this system the influence of R was studied extensively. It was found that the five-coordinate compound with a σ,σ-N,N' chelated R-DAB ligand is most stable for R containing a triply branched C^α (e.g., t-Bu). The stability most probably arises through an increased kinetic stability. A similar situation was found for the five-coordinate compounds $PtCl_2(\eta^2\text{-olefin})(R\text{-}DAB)$ for which the kinetic stability also increased on going from double to triple branching at C^α and from single to double branching at C^β (48).

Further research in the case of the Rh complexes showed that four-coordinate complexes with an ionic structure are favored if the imino-C atoms carry substituents and furthermore when the R groups on N are doubly or singly branched at C^α. When C^α is triply branched (t-Bu or $EtMe_2C$) dinuclear complexes of four-coordinate Rh^I with an σ-N,σ-N' bridging R-DAB ligand appear to dominate (34). Models show that while the t-Bu group interacts strongly with the cis-CO groups of $Rh(CO)_2(t\text{-}Bu\text{-}DAB)^+$ for the dimer one of the methyl groups interacts only slightly with the Rh^I atom (Fig. 17).

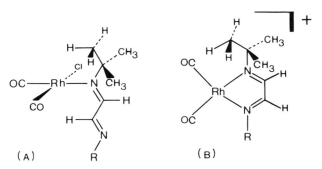

FIG. 17.   Steric interactions of the t-Bu substituents in the t-Bu-DAB ligand with the metal and the co-ligands; (A) σ-N monodentate t-Bu-DAB and (B) σ,σ-N,N' chelate t-Bu-DAB (61).

2. The Relative Importance of σ,σ-N,N' (4e) versus σ-N,μ²-N',η²-C=N' (6e) and σ-N,σ-N',η²-C=N,η²-C=N' (8e) Bonding

In Section III,E we have already discussed the reactions leading to $MnCo(CO)_6(R\text{-}DAB)$ and have made the observation that the introduction of methyl groups appears to destabilize the 6e type of bonding relative to the σ,σ-N,N' (4e) bonding type (51). Only for the very special R substituent c-Pr $MnCo(CO)_6[c\text{-}Pr\text{-}DAB(Me,Me)]$ was obtained, which at present cannot be properly explained. One expects that the methyl groups decrease the π-acceptor capacity of the heteroolefinic bond in analogy to similar observations for olefins. If so, it is then not immediately clear why the 8e-donor type of bonding can be stabilized as found in the case of $Mn_2(CO)_6[Me\text{-}DAB(Me,Me)]$ (53). In the last case it is, however, clear that we have a small methyl group, which is an example of an unbranched substituent. Also, there is only π back bonding from one Mn atom to the π system of the $Mn\overline{-N=C-C=N}$ ring, while in the 6e case of $Ru_2(CO)_6(R\text{-}DAB)$ the Ru atom has only π back bonding to about half of the five-membered $Ru\overline{-N=C-C=N}$ ring.

The influence of the R group is further strikingly demonstrated by the reactions of $Ru_3(CO)_{12}$ with R-DAB (52). A very tentative reaction scheme of the various reaction routes is shown in Scheme 6.

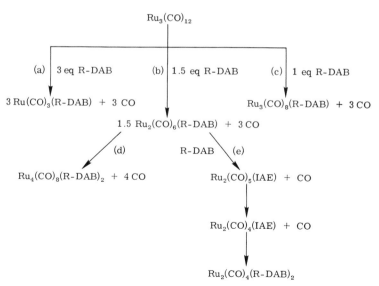

SCHEME 6.   Reactions of $Ru_3(CO)_{12}$ with R-DAB, (a) for R = Mes, Xyl, (b) for R = i-Pr, c-Hex, (c) for R = i-Bu, n-Pent, (d) for R = i-Pr, c-Hex (traces for R = i-Bu, n-Pent), and (e) for R = i-Pr, c-Hex [N.B. $Ru_2(CO)_4(R\text{-}DAB)_2$ for R = Ar is formed directly from $Ru_3(CO)_{12}$] (52).

The type of product is strongly dependent on the bulkiness of R and in particular on the branching at C^α and at C^β (*52*). If we assume that in the reaction of $Ru_3(CO)_{12}$ with R-DAB one of the first products will contain a σ,σ-N,N′ chelated R-DAB group, we may then consider several possibilities. When the R group is doubly branched at C^α and at C^β (e.g., R = 2,4,6-Mes; 2,6-Xyl; i-Pr$_2$CH) the N=C π bonds are blocked on both sides of the $\overline{M-N=C-C=N}$ skeleton, in which the R-DAB ligand is in the *E-s-cis-E* conformation. Therefore, it is impossible for another metal atom to reach the π system of the R-DAB ligand and as a consequence $Ru(CO)_3(R\text{-}DAB)$ is the final product if sufficient R-DAB is present (route a of Scheme 6). On the other hand, for R = n-Pent or i-Bu, which are singly branched at C^α and doubly branched at C^β, the π system at one side of the $\overline{M-N=C-C=N}$ ring is open and as a result the R-DAB ligand may use more coordination sites which then leads to the preferential formation of $Ru_3(CO)_8(R\text{-}DAB)$ (route c of Scheme 6) in which the diimine is bonded as an 8e σ-N,σ-N′,η^2-C=N,η^2-C=N′ ligand. Indeed, the R-DAB ligand in $Ru_3(CO)_8(R\text{-}DAB)$ appears to act as a symmetrically bonded ligand, since the imino H-substituents are equivalent and absorb at 5.86 and 5.89 ppm for R = i-Bu and n-Pent, respectively (*52*). These chemical shifts lie in the range observed for 8e-bonded R-DAB ligands (see Table VII). However, there is evidence that there are two isomers at lower temperatures of which the ratio is temperature dependent. The proposed equilibrium is shown in Fig. 18.

When R is doubly branched at C^α and singly branched at C^β (e.g., i-Pr, c-Hex, c-Pr, Ar) it was found that $Ru_2(CO)_6(R\text{-}DAB)$ (6e bonding mode) is the favored product (route b of Scheme 6). There is still sufficient room for the 8e-bonding mode, as demonstrated by the virtually quantitative conversion of $Ru_2(CO)_6(R\text{-}DAB)$ to $Ru_4(CO)_8(R\text{-}DAB)_2$ upon heating (*33, 52*), (see Section III,F and Table II for details). Finally, for R = t-Bu, i.e., triply branched at C^α there is only room for 6e bonding as indicated by the exclusive formation of thermally stable $Ru_2(CO)_6(R\text{-}DAB)$ (*52*).

After having discussed the various products as a function of the steric form of R it is of interest to consider in more detail the possible mechanisms

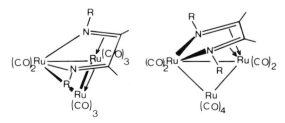

FIG. 18.   The two proposed isomers of $Ru_3(CO)_8(R\text{-}DAB)$ (*52*).

occurring in the various routes. Such a discussion is of course highly speculative, since it is very difficult to study the several chemical steps.

The first step, i.e., the reaction of $Ru_3(CO)_{12}$ with R-DAB has been studied kinetically by means of HPLC measurements (173). It was shown that this reaction with i-Pr-DAB is first order in metal carbonyl and first order in R-DAB ligand, while the overall reaction is of order two. The reaction with t-Bu-DAB is also of order two, but is much slower than for i-Pr-DAB. The rate determining step is the formation of an unstable intermediate, which might be $Ru_3(CO)_{11}$(R-DAB) or more probably $Ru_3(CO)_{10}$(R-DAB). Further kinetic evidence for the following steps could not be obtained (173). Of great interest is therefore the formation of $Ru_3(CO)_9$(c-Hex-DAB) during the preparation of $Ru_2(CO)_6$(c-Hex-DAB)], the structure of which has been discussed in Section III,F. Another look at the Ru_3 compound (Fig. 16a), in which the metal cluster has already been partially ruptured, shows in fact the coordination of a $Ru(CO)_3$(c-Hex-DAB) moiety to a $Ru_2(CO)_6$ entity. It can be easily seen that movement of one CO from one metal atom to the other and loss of $Ru(CO)_3$ would rationalize the formation of $Ru_2(CO)_6$(R-DAB) (route b). On the other hand, loss of one CO from $Ru_3(CO)_9$(R-DAB) would just give stable $Ru_3(CO)_8$(R-DAB) (route c) (Fig. 18). On present evidence, it seems unlikely that $Ru_2(CO)_6$(R-DAB) is formed from $Ru_3(CO)_8$(R-DAB), since these two species are clearly formed in competing reaction pathways (52).

A completely different possibility is that $Ru(CO)_3$(R-DAB) is always formed first and, for suitable R, this unstable monomer then reacts with $Ru_3(CO)_{12}$ to give $Ru_2(CO)_6$(R-DAB) and $Ru_3(CO)_9$(R-DAB) and thereafter $Ru_3(CO)_8$(R-DAB). Again it should be noted that the species ultimately formed depend on the branching at C^α and at C^β as discussed previously.

In the case of $Os_3(CO)_{12}$ the reactions are much slower than for Ru and $Os_2(CO)_6$(R-DAB) could be unambiguously characterized (112). In addition, $Os_3(CO)_9$(R-DAB) was isolated, and on the basis of its molecular formula was suggested to possess a 6e-donor diimine ligand (112), i.e., as shown in Fig. 16b. However, a structure analogous to $Ru_3(CO)_9$(c-Hex-DAB) (Fig. 16a) might also be possible.

Also it is worthwhile to note again that $Fe_2(CO)_6$(R-DAB) is prepared from $Fe(CO)_3$(R-DAB) with $Fe_2(CO)_9$ [N.B. Under the reaction conditions employed the same synthesis from $Fe(CO)_5$ also proceeds via $Fe_2(CO)_9$, see reaction (39) in Section III,D]. This reaction seems rather strange, but on closer inspection would become understandable, if $Fe(CO)_3$(R-DAB) coordinates to $Fe_2(CO)_n$, with the possible intermediate formation of $Fe_3(CO)_9$(R-DAB), after which $Fe_2(CO)_6$(R-DAB) is formed through loss of $Fe(CO)_3$. Another pathway is the equilibrium shown in reaction (40).

$$Fe_2(CO)_6(R\text{-}DAB) + R\text{-}DAB \leftrightarrows 2Fe(CO)_3(R\text{-}DAB) \qquad (40)$$

This equilibrium lies far to the right in the case of Fe (*110, 112*). In the case of Ru a similar equilibrium might exist, as was tentatively suggested on the basis of kinetic HPLC measurements (*173*), but this equilibrium lies then far to the left due to the greater $M-M$ bond strength for Ru.

An unexplained point is the rather astonishing virtually quantitative formation of $Ru_4(CO)_8(R\text{-}DAB)_2$ from $Ru_2(CO)_6(R\text{-}DAB)$ (*52*), which points to a relatively simple, kinetically easily available pathway (route d of Scheme 6). Bearing in mind the reaction conditions, it seems likely that reactive species $Ru_2(CO)_5(R\text{-}DAB)$ and also $Ru_2(CO)_4(R\text{-}DAB)$ may be formed through successive loss of CO from $Ru_2(CO)_6(R\text{-}DAB)$. Subsequent dimerization to a tetranuclear Ru_4 intermediate compound accompanied by migration of CO and of R-DAB may then lead to the formation of $Ru_4(CO)_8(R\text{-}DAB)_2$ when, as mentioned above, suitable R substituents are present.

3. The Influence of Geometric and Metal–Carbonyl Bond Strength Factors on the Formation of $\sigma\text{-}N,\mu^2\text{-}N',\eta^2\text{-}CN'$ (6e) Bonding

Because of the facile formation of $MnCo(CO)_6(R\text{-}DAB)$ (6e) from $MnCo(CO)_7(R\text{-}DAB)$ (4e) an attempted was made to prepare similar homobinuclear complexes of Mn and Co and the compounds $Mn_2(CO)_8(R\text{-}DAB)$, $Co_2(CO)_6(R\text{-}DAB)$, $Cr_2(CO)_4(R\text{-}DAB)_2$, and $Co_4(CO)_8(R\text{-}DAB)_2$ were isolated (*103, 140*). However, they all contain $\sigma,\sigma\text{-}N,N'$ chelated R-DAB ligands (*103, 140*) and neither heating nor irradiation led to CO substitution by a C=N group of the R-DAB ligand. In the case of $Co_2(CO)_6(R\text{-}DAB)$ this might be due to the fact that the R-DAB ligand is in a noninteractive position, since the five-membered $\overline{Co-N=C-C=N}$ ring is not perpendicular to but bent away from the $Co-Co$ bond. In the case of $Mn_2(CO)_8(R\text{-}DAB)$ the $\overline{Mn-N=C-C=N}$ ring is now in a position perpendicular to the $Mn-Mn$ bond. However, intramolecular attack of one C=N π bond on the $Mn(CO)_5$ unit does not occur, possibly due to the strength of the $Mn-CO$ bond. It is clear that more experimental work is needed to understand the formation of the 6e-R-DAB bonding mode.

B. Structural and Fluxional Features

1. X-Ray Structural Features

The various structures for the $\sigma\text{-}N$ (2e), $\sigma\text{-}N,\sigma\text{-}N'$ (4e), $\sigma,\sigma\text{-}N,N'$ (4e), $\sigma\text{-}N,\mu^2\text{-}N',\eta^2\text{-}C=N'$ (6e), and $\sigma\text{-}N,\sigma\text{-}N',\eta^2\text{-}C=N,\eta^2\text{-}C=N'$ (8e) bonded R-

DAB complexes have been discussed fairly broadly in the previous sections, while relevant details have been given in Table II.

Some important aspects that interest us here include the conformation of the R-DAB ligands and the $C=N$ and $C-C$ bond lengths. In all compounds the $N=C-C=N$ skeleton, whether in the E-s-trans-E conformation found for the monodentate and bridging situations, or in the E-s-cis-E conformation (chelate, 6e-bridging, 8e-bridging) is approximately planar. No doubt this is due to the fact that by virtue of this planarity the skeleton is better able to use its electron donating and accepting capabilities. Furthermore, steric factors also play a role, as has been pointed out before for the σ-N monodentate and σ-N,σ-N' bridging conformations.

In the last two bonding types the $N=C$ and central $C-C$ bond lengths are scarcely different. However, this situation changes for the other bonding modes. In the case of the σ,σ-N,N' chelate bonding there are small but significant differences between the $N=C$ and $C-C$ bond lengths. It is clear from Table II that generally speaking the low-valent electron-rich metal atoms have longer $C=N$ bond lengths and a shorter $C-C$ bond lengths due to population of the LUMO, which is antibonding in $C-N$ and bonding in $C-C$ (26). Larger changes, as have already been noted, occur in the case of η^2-$C=N$ bonding both for 6e- and 8e-bonding types, since the central C atoms of the R-DAB skeleton are also directly bonded to the metal atom. The η^2-$C=N$ bond length is somewhat shorter for $MnCo(CO)_6(t$-Bu-DAB) [1.358(16) Å; 6e] than for $Fe_2(CO)_6(c$-Hex-DAB) [1.397(4) Å; 6e] and for $Ru_2(CO)_4(i$-Pr-DAB)$_2$ [1.43(1) Å; 6e]. This might indicate less η^2-bonding interaction to Co than to Fe or Ru, which could change the bonding (formally) from σ-N,μ^2-N',η^2-$C=N'$ to σ-N,σ-N',η^2-$C=N'$ (see also Section III,E).

2. NMR Aspects

Most of the relevant NMR data have been dealt with in Section III but some observations should be made here. An important point is always the investigation into the correlation of NMR parameters with, e.g., charge density on ligands. Such systematic studies have, generally speaking, not given much insight since, for example, ^1H and certainly ^{13}C NMR are also dependent on other factors (3, 9, 10, 89, 120). In the case of diimine complexes relatively little systematic work has been carried out, though a large number of Ni[R-DAB(R',R'')] complexes (R = alkyl or aryl) have been studied in detail (9, 10). In general it was found that the ^1H signals of the imine-H substituents (2,3 positions) are downfield with respect to the free diimine and these shifts increase when the R groups show increased branching at C^α (9). However, the interpretation of these results remains difficult, since for Ni^0 complexes an energetically low-lying triplet level may strongly influence the shifts, with the added complication that the $Ni(R-DAB)_2$ complexes may have variable coordination geometries (9, 10).

Change of the electron configuration of the metal clearly has an influence on the ^1H NMR chemical shifts of the imine-H atoms, as is shown in Table VIII for a number of σ,σ-N,N' chelated diimines bonded to metal carbonyl fragments. It is clear that there is a high field shift on going to higher electronic configuration i.e., from $d^6 \rightarrow d^7 \rightarrow d^8 \rightarrow d^9$. In the case of the carbonyl free d^{10} complexes $Ni(R-DAB)_2$ a low field shift is again observed, no doubt due partly to an energetically low lying triplet state. Also the change in solvent may play a role by specific solvent–solute interactions, since the Ni compound (Table VIII) is the only compound in this series that was measured in C_6D_6. The trend in the high field shift of the carbonyl complexes might be due to increasing donation to the R-DAB ligand, but one should be particularly careful when comparing different geometries (120).

Much stronger upfield shifts for both the ^1H and ^{13}C resonances of imine-H and imine-C atoms, respectively, have been observed for η^2-C=N bonded groups. The ^1H shifts lie in the range of 3.3 to 5.5 ppm for σ-N,μ^2-N',η^2-C=N' (6e) bonding and in the range of 5.9 to 6.6 ppm for σ-N,σ-N',η^2-C=N,η^2-C=N' (8e) bonding (52, 163, 166). The smaller shifts for the 8e- compared to the 6e- bonding situation might be due to the fact that in the first case there is only one metal atom back donating to two C=N groups of the M—N=C—C=N ring instead of to one C=N group as for the 6e-bonding mode (Table V and Table VII).

Among the σ-N,μ^2-N',η^2-C=N' bonded R-DAB groups there are some interesting but unexplained differences in the Fe column, since the ^1H NMR chemical shifts go upfield from Fe to Os, while the corresponding ^{13}C shifts go downfield (163) (Table V). The analogous chemical shifts for the $MCo(CO)_6(R-DAB)$ (M = Mn,Re) (Table V) are downfield from these of the complexes of the Fe-triad metals and may be due to the smaller metal–η^2-C=N interaction for cobalt, as was suggested in Section IV,B,1, on the basis of the shorter C=N bond length of $MnCo(CO)_6(t$-Bu-DAB) (51).

3. Fluxional Processes Involving Carbonyl Groups

Fluxional processes connected, e.g., with olefin rotation in metal olefin complexes, with intramolecular exchange of σ-N (2e) bonded R-DAB groups and with cluster movements taking place for $Ru_4(CO)_8(R-DAB)_2$ have been discussed in previous sections. Here we shall restrict ourselves to CO scrambling processes.

TABLE VIII

^{1}H NMR Shifts of Imine H in Various σ,σ-N,N′ (4e) R-DAB Metal Carbonyl R-DAB Complexes[a,b]

RN=CH–CH=NR R	Cr(CO)$_4$(R-DAB) d^6	Mn$_2$(CO)$_8$(R-DAB) d^7	Fe(CO)$_3$(R-DAB) d^8	Co$_2$(CO)$_6$(R-DAB) d^9	Ni(R-DAB)$_2$[c] d^{10}
i-Pr	8.22	8.18	7.62	7.84	8.70
t-Bu	8.30	—	8.01	7.88	9.03
p-Tol	8.39	8.18	—	7.83	9.87

[a] From Refs. 9, 10, 140.
[b] In CDCl$_3$.
[c] In benzene-d_6 (9, 10).

Majunke *et al.* (*79*) discovered by means of stereospecific ^{13}CO labeling (i.e., one ^{13}CO cis to the σ,σ-N,N'-chelated R-DAB ligand) that an equilibrium exists between an axial and equatorial substituted tetracarbonyl. This equilibrium is fast on the NMR time scale for R = Ar at room temperature but slow for R = *t*-Bu. In the case of the aryl groups the $\Delta G\ddagger$ values for M = Mo increased in the order *p*-Tol < *o*-Tol < 2,6-Xyl < 2,4,6-tri-*i*-Pr-C$_6$H$_2$ (*79*). The interchange process proceeds according to an "umbrella mechanism" (*79*) ($C_{2v} \rightleftarrows C_{4v}$). This is relatively easy for aryl groups, since these groups can minimize the steric interaction with the equatorial CO groups by turning to a position perpendicular to the chelate plane.

These conformational interchanges have also been studied by Staal *et al.* (*83*) by means of electronic absorption spectra and resonance Raman spectroscopy in solutions or glasses at temperatures below $-73°C$ for compounds M(CO)$_4$(R-DAB) (M = Cr, Mo, W; R = *t*-Bu, *c*-Hex, *n*-Pent). It appeared that at low temperatures both conformations exist and interchange in a temperature dependent equilibrium.

Intramolecular CO exchange in five-coordinate complexes Fe(CO)$_3$(R-DAB) have been studied for R = *t*-Bu and *i*-Pr by ^{13}C NMR. The energy barrier for the process, which may be a Berry pseudorotation or a turnstile movement, is low, but is higher for *t*-Bu than for *i*-Pr, while both have lower energy barriers than measured for Fe(CO)$_3$(η^4-butadiene) (*106*).

Intramolecular CO site exchange processes have also been observed for M$_2$(CO)$_6$(R-DAB) (*51, 163*). In the case of MnCo(CO)$_6$[*c*-Pr-DAB(Me,Me)] the fluxional process involving the change of point of attachment of the R-DAB(Me,Me) ligand from η^2-C=N' to η^2-C=N and vice versa in the temperature range -40 to $+20°C$ has been discussed in Section III,D (*51, 163*) (Fig. 12). Independently, there is also an intramolecular site exchange, between the semibridging carbonyl group and the two terminal C$^\alpha$ carbonyl groups, which even at $-50°C$ is already fast on the NMR time scale. The terminal CO groups on Mn, however, remain rigid even at $+50°C$. Pseudorotational CO site exchange processes occurring at both metal centers have been observed for M$_2$(CO)$_6$(R-DAB) (M = Fe, Ru (*163*).

For the Ru$_2$(CO)$_6$(R-DAB) complexes the intramolecular site exchanges on each Ru atom are fast even at low temperatures ($-70°C$). In the case of Fe$_2$(CO)$_6$(*t*-Bu-DAB) it could be shown that the CO groups on the Fe atom containing the σ-N bonded N=C moiety already exchange locally at $-50°C$. However, the local scrambling of the CO groups on the other Fe atom containing the η^2-C=N bonded unit began on the NMR time scale at about room temperature. Obviously σ-N=C and η^2-C=N bonding types have a strong influence on the energy barriers of the intramolecular site exchange. No intermolecular site exchange was observed (*163*).

V

METAL-1,4-DIAZA-1,3-BUTADIENE RADICALS: ESR
SPECTROSCOPY AND REACTIVITY

Like 2,2'-bipyridine (174) the R-DAB ligands can be readily converted to stable paramagnetic radical anions by potassium in 1,2-dimethoxyethane (175) or THF (176). The radical anion t-Bu-DAB$^-$ exists, like the free molecule (see Section II,B) in the E-s-trans-E conformation (Fig. 19a) and displays a highly resolved ESR spectrum in which the two equivalent ^{14}N nuclei (5.62 G) ($I = 1$), the imino hydrogens (4.37 G), the 18 equivalent t-Bu hydrogen nuclei (0.15 G) ($I = \frac{1}{2}$) and the two pairs of ^{13}C nuclei ($I = \frac{1}{2}$) (176) couple. The absence of any further coupling indicates that no association with potassium occurs in solution ($175, 176$).

Direct treatment of complexes containing the R-DAB ligand in the E-s-cis-E conformation (i.e., σ,σ-N,N' bonded) such as CuCl(t-Bu-DAB) (177) or MoBr(η^3-C$_3$H$_5$)(CO)$_2$(t-Bu-DAB) (178) with potassium again generates R-DAB$^-$ in the E-s-trans-E conformation. However, the radical anion of MoBr(η^3-C$_3$H$_5$)(CO)$_2$(t-Bu-DAB) reacts with free t-Bu-DAB to

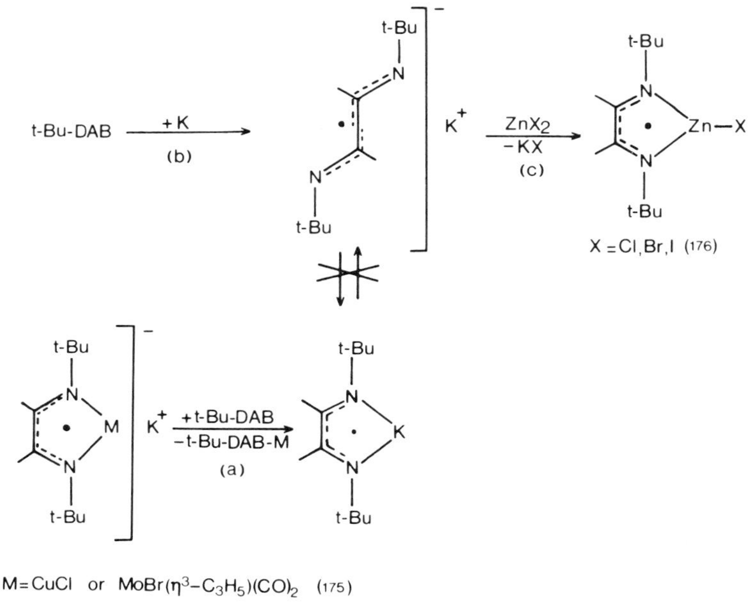

M=CuCl or MoBr(η^3–C$_3$H$_5$)(CO)$_2$ (175)

FIG. 19.   Reaction (a) of K[M(t-Bu-DAB)] with t-Bu-DAB [M = CuCl or MoBr(η^3-C$_3$H$_5$)(CO)$_2$], (b) of potassium with t-Bu-DAB, and (c) of ZnX$_2$ (X = Cl, Br, I) with K$^+$(R-DAB)$^-$.

give the (t-Bu-DAB)K complex in which the s-cis conformation of the N=C—C=N skeleton is retained and coordination of K$^+$ ($I = \frac{3}{2}$) is reflected in the ESR spectrum (*175*). No evidence was found for conversion of the s-trans to s-cis conformation (see Fig. 19) of the radical anionic species at temperatures up to +30°C (*175*) (see Fig. 19a, b).

The blocked conversion between the s-cis and s-trans conformations has been explained (*175, 176*) by occupation in the radical ion of a molecular orbital that has bonding interactions between the central C atoms (see Fig. 20) and hence an increased barrier to rotation around this central C—C bond. However, it is not clear yet to what extent the K—N interaction between K$^+$ and the s-cis conformer plays a stabilizing role. Reaction of t-Bu-DAB$^-$ with ZnX$_2$ affords an ESR spectrum in agreement with ZnX(t-Bu-DAB) radical revealing coupling with both the N nuclei (5.6 G), the imino hydrogens (5.6 G), and the chlorine nucleus (mean 0.58 G). Even using ^{67}Zn enriched samples the satellites arising from coupling with ^{67}Zn (4.4 G) could be established (*176*). The zinc atom in ZnX(t-Bu-DAB) is stable three-coordinate when X = halide (Fig. 19c). When X = CN or NCS, ZnX$_2$(t-Bu-DAB)$^-$ is obtained directly (*176*). Analogous MgX(t-Bu-DAB) was not only accessible via the t-Bu-DAB$^-$–MgX$_2$ route but also by reaction of MgCl$_2$ with t-Bu-DAB in the presence of metallic magnesium as the reducing reagent (*176*).

Reaction of [t-Bu-DAB]K with EtZnCl formed in quantitative yield the corresponding ZnEt(t-Bu-DAB) radical which, however, is in equilibrium with its C—C coupled dimer [see Eq. (41)].

$$\text{EtZnCl} + [t\text{-Bu-DAB}]^- \text{K}^+ \longrightarrow \text{KCl} + \qquad (41)$$

At low temperatures (< -50°C) the concentration of the radical is too low to be detected by ESR, indicating that the equilibrium is shifted completely to the side of the dimer. Above -50°C the solution displays a highly resolved ESR spectrum (a_N 4.87, $a_{H\alpha}$ 5.87, $a_{H\beta}$ 0.48 G), indicating a com-

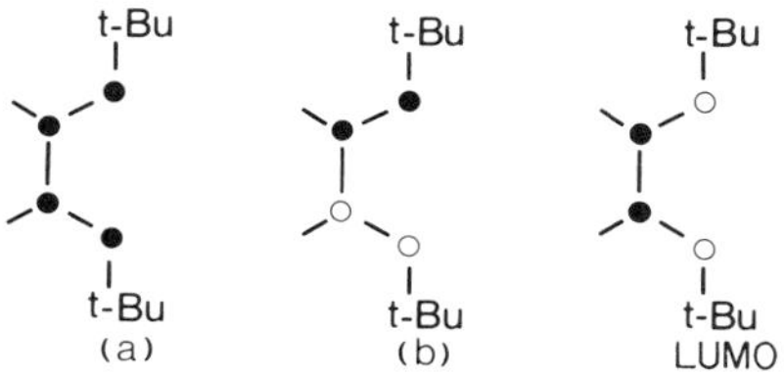

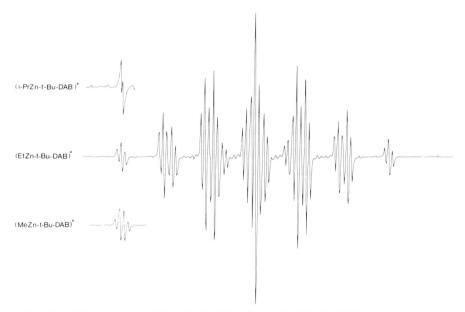

FIG. 20. Molecular orbitals, (a) and (b), of the free *t*-Bu-DAB (*s*-cis conformation shown) and the LUMO, which become partly occupied in the *t*-Bu-DAB radical anion.

plete delocalization of the unpaired electron density over the metallocycle and the neighboring nuclei (*2, 68*), see Fig. 21. NMR spectra reveal that the C—C coupling must occur with high stereospecificity while the fact that a sharp NMR resonance pattern was observed points to a low rate for the monomer association–dimer dissociation processes. Other ZnR′(R-DAB) (R = Me, *i*-Pr, *t*-Bu) radicals have also been prepared according to Eq. (41) (see Fig. 21 for the ESR patterns).

Also in the reaction of R_2Zn with *t*-Bu-DAB ZnR(*t*-Bu-DAB) radicals are formed, although in minor amounts, for R is a primary alkyl group (Et, *n*-Bu, or CH_2-*t*-Bu) (*68*). In these reactions the organo group does

FIG. 21. ESR spectra for R′Zn(*t*-Bu-DAB) radicals: R′ = Me, Et (fully shown) and *i*-Pr. Note the multiplicity of the outer multiplet arising from coupling with the R′ group α protons.

not escape from the $ZnR_2(t\text{-Bu-DAB})$ intermediate as a radical but is transferred either completely to the imino-N (for primary alkyls) or to the imino-C atoms (tertiary alkyl) or in some instances partially to the imino-N and the imino-C atom, see Section VI,C (179). An exception is the reaction for Me_2Zn with t-Bu-DAB which results in formation of the stable 1:1 complex $Me_2Zn(t\text{-Bu-DAB})$. Above room temperature this complex is converted according to Eq. (42) to the $MeZn(t\text{-Bu-DAB})$ radical (50%). Similarly, reaction of Et_2Zn with $C_5H_4NC(H){=}N\text{-}t\text{-Bu}$ afforded the equilibrium mixture of the three-coordinate radical and the C—C coupled dimer [Eq. (43)] via the four-coordinate $ZnEt_2[C_5H_4NC(H){=}N\text{-}t\text{-Bu}]$ complex (68). This equilibrium is shifted more to the side of the dimer than in the case of the t-Bu-DAB derivative in Eq. (41). Recently the structure of $[ZnEt\{C_5H_4NC(H)N\text{-}t\text{-Bu}\}]_2$ was established by an X-ray structure determination (179). Remarkable is the resemblance of this structure [schematically shown in Eq. (43)] to the structure of the C–C coupling product $Ru_2(CO)_5(IAE)$ (see Fig. 26; Section IV,C).

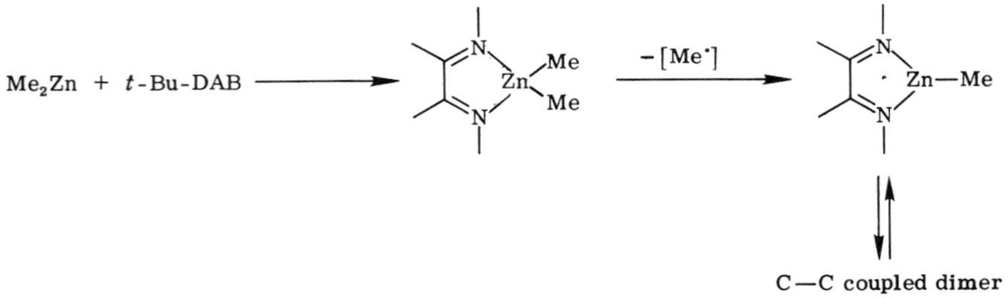

$$(42)$$

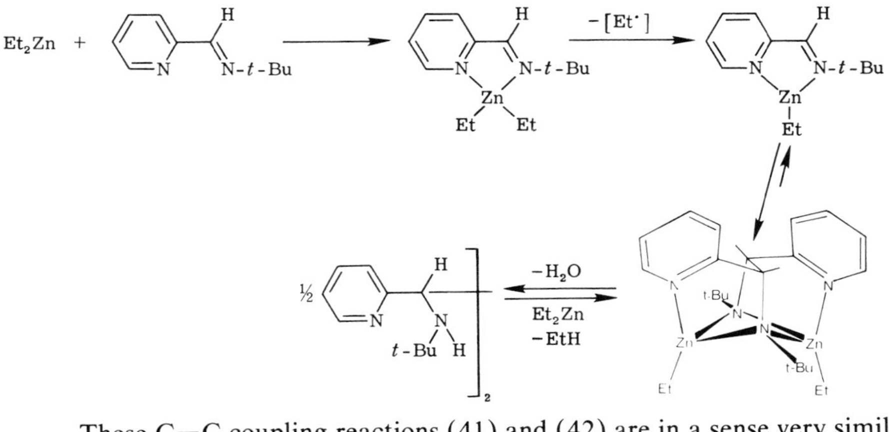

$$(43)$$

These C—C coupling reactions (41) and (42) are in a sense very similar to the ones occurring during the formation of $Mo_2(CO)_6(IAE)$ and $Ru_2(CO)_n(IAE)$ ($n = 4,5$) (see Section VI,C).

Extensive ESR investigations have been carried out on the radicals obtained by treating $M(CO)_{4-n}L_n[R\text{-}DAB(R',R'')]$ with potassium in 1,2-dimethoxyethane or THF (*81, 88, 145, 180*) in which M = Cr, Mo, W and R = Me, CH_2Ph, *i*-Pr, *c*-Hex, *c*-Pr for R′ = R″ = H or Me, and L = PBu_3 with *n* = 0, 1, or 2. Comparison of the ESR spin density at the coordination sites as well as the coupling data (in case of *n* = 1 or 2) for these complexes with those obtained for the corresponding 2,2′-bypyridine complexes showed that the π-acceptor capacity of 2,2′-bypyridine is about half that of the R-DAB ligand. This result corresponds with the higher stability of the zerovalent metal–R-DAB complexes and the lower reduction potential of many of these complexes as compared to those of the zerovalent metal–2,2′-bipyridine analogs (*84, 94*).

VI

CHEMICAL ACTIVATION OF METAL COORDINATED 1,4-DIAZA-1,3-BUTADIENE LIGANDS

Earlier we have mentioned the strong electron accepting properties of R-DAB ligands when coordinated in the σ,σ-N,N′ (4e) and particularly in the σ-N,μ^2-N′,η^2-C=N′ (6e) and σ-N,σ-N′,η^2-C=N,η^2-C=N′ (8e) bonding modes. In the latter two types one and two C=N groups are η^2-bonded, respectively. As a consequence one would expect chemical activation of the diimine itself, and also of the metal to which the diimine is coordinated.

First, we shall discuss activation by σ-bonded R-DAB groups and subsequently reactions involving the η^2-C=N metal linkages.

A. Complexes with (Intermediate) σ-N (2e) Bonded 1,4-Diaza-1,3-butadiene Ligands

An example is the stoichiometric reaction of $Fe_2(CO)_9$ with *t*-Bu-DAB which gives $Fe(CO)_5$, $Fe(CO)_3(DAB)$, and 2-imidazolinone in a 1:1:1 ratio (*30, 110*). A fairly detailed study of this formation of imidazolinone which, as far as we know, has not been observed for other metal carbonyl R-DAB complexes, showed that the reaction can be made catalytic (*110*). The proposed mechanism for both the stoichiometric and catalytic reactions (Fig. 22) starts with the attack of *t*-Bu-DAB on $Fe_2(CO)_9$ giving $Fe(CO)_5$ and $Fe(CO)_4(t\text{-Bu-DAB})$ in which the t-Bu-DAB is σ-N (2e) bonded to Fe and probably in the *E-s-trans-E* conformation. Rotation about the cen-

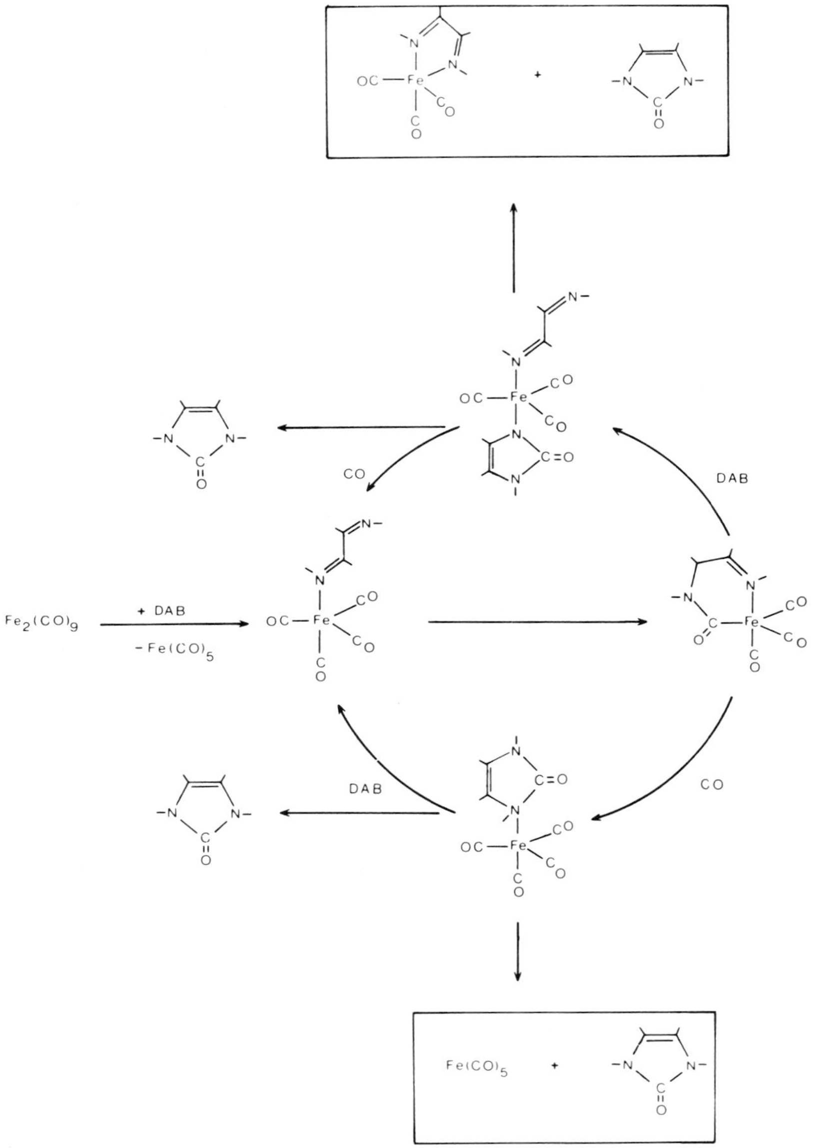

FIG. 22. Suggested (catalytic) reaction scheme for the formation of imidazolinone in the reaction of $Fe_2(CO)_9$ with t-Bu-DAB and CO (110).

tral C—C bond brings the diimine in this intermediate complex into the *E-s-cis-E* conformation and therefore in the right position to bind to one of the CO groups on Fe through intramolecular nucleophilic attack. Subsequent addition of *t*-Bu-DAB and extrusion of the weakly bonding imidazolinone then gives further Fe(CO)$_3$(*t*-Bu-DAB). [N.B. Recently some work was published on the possible coordination modes of imidazoline(1,3-*H*)-2-thione which is a sulfur analog of 2-imidazolinone (*181*) and it was shown that both S and N may be involved in metal coordination.] However, when the reaction was carried out under CO pressure, a catalytic process occurred possibly going via two pathways differing in the order of attack of R-DAB and CO, as is shown clearly in Fig. 22 (*110*). Unfortunately the catalyst was deactivated after about three cycles due to the unproductive side reactions that lead to Fe(CO)$_3$(R-DAB) and Fe(CO)$_5$.

B. *Complexes with σ,σ-N,N′ (4e) 1,4-Diaza-1,3-butadiene Ligands*

In a series of papers Krumholz and co-workers (*28, 115–117, 122, 130*) studied ligand oxidation mainly on iron diimine complexes Fe[Me-DAB(R′,R″)]$_3^{2+}$. These reactions using Ce^{IV} in strongly acidic environments were studied by means of potentiometric, photometric (*115, 116, 130*), and electrochemical techniques (*117*). For example Fe^{II}(Me-DAB)$_3^{2+}$ reacted in strong sulfuric acid with Ce^{IV} to yield Fe^{III}(Me-DAB)$_3^{3+}$, which disproportionated further as shown below (*115–117*). It is thought that the first step involves the oxidation of Fe^{II}(Me-DAB)$_3^{2+}$ [Eq. (44)]:

$$Fe(Me\text{-}DAB)_3^{3+} + Ce^{IV} \rightleftarrows Fe(Me\text{-}DAB)_3^{3+} + Ce^{III} \qquad (44)$$

Subsequently there is an intramolecular electron transfer assisted by nucleophilic attack of a water molecule [Eq. (45)]:

$$2Fe(Me\text{-}DAB)_3^{3+} \xrightarrow{\text{H}_2\text{O}} Fe(Me\text{-}DAB)_2[MeN=CH-C(OH)=NMe]^{2+}$$
$$+ Fe(Me\text{-}DAB)_3^{2+} + 2H^+ \quad (45)$$

which is followed by a fast intramolecular electron transfer reaction [Eq. (46)]:

$$Fe(Me\text{-}DAB)_2[MeN=CH-C(OH)=NMe]^{2+} + Fe(Me\text{-}DAB)_3^{3+} \xrightarrow{\text{fast}}$$
$$Fe(Me\text{-}DAB)_2[MeN=CH-C(OH)=NMe]^{3+} + Fe(Me\text{-}DAB)_3^{2+} \quad (46)$$

Parallel to reaction (45) reaction (47) occurs:

$$2Fe(Me\text{-}DAB)_3^{3+} \xrightarrow{\text{H}_2\text{O}} Fe(Me\text{-}DAB)_2(MeN=CHCHNCH_2OH)^{2+}$$

$$+ Fe(Me\text{-}DAB)_3^{2+} + 2H^+ \quad (47)$$

The intramolecular electron transfer by means of a nucleophilic reaction with H_2O is shown schematically [Eq. (48)]:

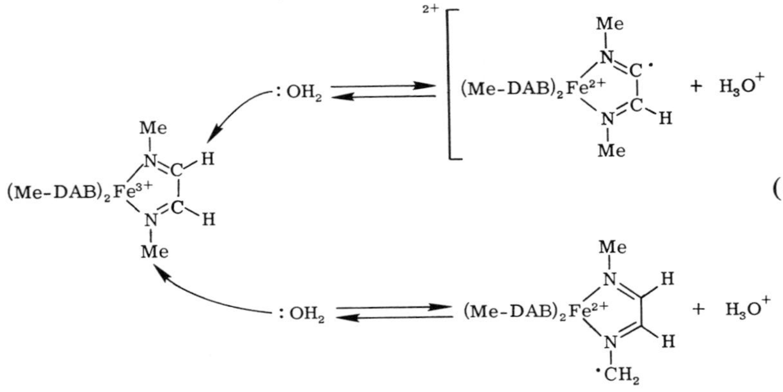

$$(48)$$

Subsequent reaction of the radical complex with water and Fe(Me-DAB)$_3^{3+}$ gives stable two-electron oxidation products, e.g., Eq. (49):

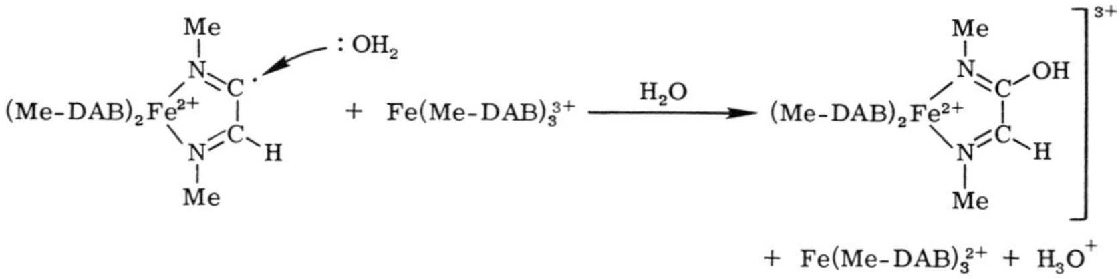

$$(49)$$

A rather similar reaction is the autoxidation of Fe[Me-DAB-(Me,Me)]$_3^{2+}$ in acidic aqueous media with oxygen, which was studied by spectrophotometric and NMR techniques (130). Also in this case a radical intermediate [Me-DAB(Me,Me)]$_2$Fe[MeN=C(Me)−C(Me)=N−ĊH$_2^{2+}$was proposed. The electron is mainly localized on the deprotonated Me group bonded to the N atom.

Asymmetric R-DAB ligands $C_5H_4N−CMe=NMe$ bonded to Fe^{II} are oxidized by Ce^{IV} in acidic environments to give via a radical intermediate an aldehyde group formed from the Me group bonded to the imine-C atom (122).

These studies therefore show that radical intermediates are of crucial importance in Fe–diimine reactions. The electron may be localized on the imino-C atom, on the substituent bonded to the imino-C atom and also on the substituent bonded to a N atom.

Other examples of activation of σ,σ-N,N' bonded R-DAB ligands include their reactions with RLi, RMgX, and R$_2$CuLi which gave after hydrolysis a large variety of products (159). However selective reactions were obtained between R-DAB and ZnR$_2$ and AlR$_3$, respectively (68, 182, 183). The four-coordinate complexes Zn(p-Tol)$_2$(R-DAB) are stable, but the compounds ZnEt$_2$[R-DAB(R',R'')] are stable only at low temperatures (R = t-Bu, i-Pr, CH$_2$(t-Bu), c-Hex, and i-Pr$_2$CH; R' = R'' = H and R = i-Pr with R' = H, R'' = Me and finally R = c-Hex and R' = R'' = Me). For example, ZnEt$_2$(t-Bu-DAB) converted at $-50°$C to EtZn$\overline{-}$N(Et) (t-Bu)CH=CHN-t-Bu via an intramolecular ethyl transfer from Zn^{II} to one of the σ-N coordinated N atoms (Fig. 23ab) (183). The reaction is

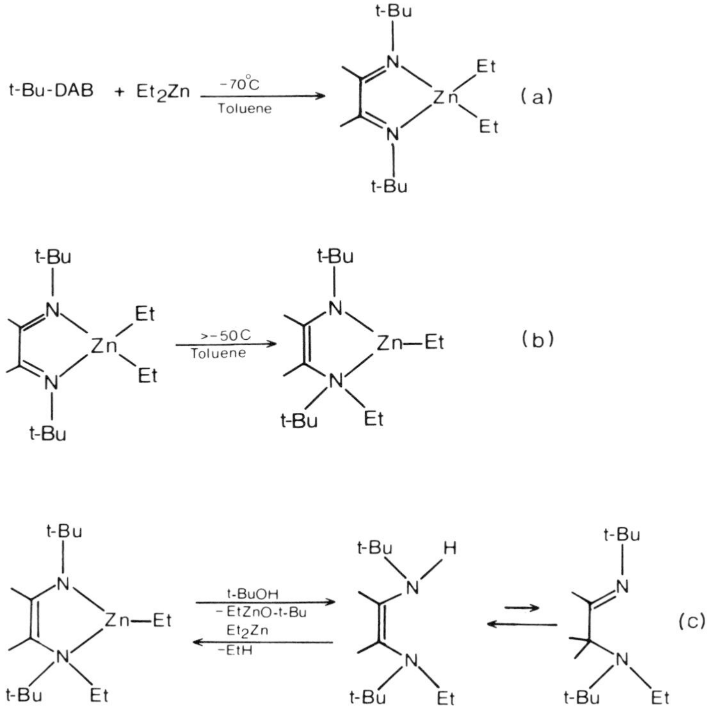

FIG. 23. Reaction (a) of ZnEt$_2$ with t-Bu-DAB; formation (b) of EtZnN(Et)-(t-Bu)CH=CHN-t-Bu; hydrolysis (c) of EtZnN(Et)(t-Bu)CH=CHN-t-Bu with t-BuOH to give a tautomeric mixture of enediamines (183).

almost quantitative, but a small amount ($<2\%$) of a persistent three-co-ordinated zinc radical species $EtZn(t\text{-}Bu\text{-}DAB)^{\cdot}$ was also obtained (see Section V). Both molecular weight and NMR measurements indicated the monomer nature of $EtZn\overline{N}(Et)(t\text{-}Bu)CH{=}CHN\text{-}t\text{-}Bu$. Careful protono-lysis with $t\text{-}BuOH$ produced $EtZnO\text{-}t\text{-}Bu$ and quantitatively a $4:1$ tau-tomeric mixture of the enediamine $(t\text{-}Bu)EtNCH{=}CHN(H)\text{-}t\text{-}Bu$ and an iminoaminoethane $(t\text{-}Bu)EtNCH_2CH{=}N\text{-}t\text{-}Bu$ (Fig. 23c). Reaction with $ZnEt_2$ then gave back the starting material in quantitative yield (Fig. 23).

The protonolysis with $t\text{-}BuOD$ showed that in a very fast reaction the monodeuterated enediamine is formed first, after which a slower reaction reaches equilibrium in about ten minutes and this leads to a number of monodeuterated enediamine and the monodeuterated iminoaminoethane compounds (Fig. 24) (183). After about one hour in a second process a statistical ratio of fully protonated, mono- and dideuterated enediamines and iminoaminoethanes is formed. It could be deduced from ^1H NMR that H_A in Fig. 24 is not participating in the isomerization and the H–D ex-change.

The ratio of the tautomers appeared to depend on the $R{-}N$ and $R'{-}C$ substituents, since branching at C^{β} in R shifted the equilibrium away from the enediamine and R' substitution of H for Me stabilized the enediamine tautomer (183). The mechanism of the unprecedented ethyl group transfer from Zn to N probably occurs via a mechanism involving initial homolytic cleavage of the $Et{-}Zn$ bond for which the formation of the radical $EtZn(R\text{-}DAB)^{\cdot}$ through loss of an ethyl radical is supporting evidence. In the case where the Et radical does not escape from the solvent cage it moves to the N atom via an intramolecular 1,2-shift. This can be rationalized by the character of the LUMO of the $\overline{Zn{-}N{=}C{-}C{=}N}$ entity, which is bonding in $C{-}C$ and antibonding in $C{-}N$ (183) (see Fig. 20 in Section V). Apparently, this activation of the chelated $N{=}C{-}C{=}N$ skeleton is a delicately tuned process because recent results have shown that in $R'_2Zn(R\text{-}DAB)$ ($R' = $ Me, $R = t\text{-}Bu$) complexes the $ZnMe(t\text{-}Bu\text{-}DAB)$ radical is formed exclusively (50%) [see Fig. 21 and Eq. (42) in Section V]. For $R' = $ primary alkyl a 1,2-shift is observed while for $R = t\text{-}Bu$ or $i\text{-}Pr$ a 1,3-shift to the imino-C atom is the predominant process.

In contrast to the Zn reactions, Al_2Ph_6 reacted with $p\text{-}Tol\text{-}DAB$ to pro-duce $Ph_2Al[(p\text{-}Tol)NCH(Ph)CH{=}N(p\text{-}Tol)]$ via the four coordinate in-termediate $Ph_2Al(p\text{-}Tol\text{-}DAB)$ (182). However, now $C{-}C$ coupling occurs instead of $N{-}C$ bond formation and the corresponding iminoamine can be isolated in virtually quantitative yields upon protonolysis. Of interest is the more complicated reaction of Al_2Me_6 with R-DAB, which afforded various reaction products in very selective ways depending on the R group

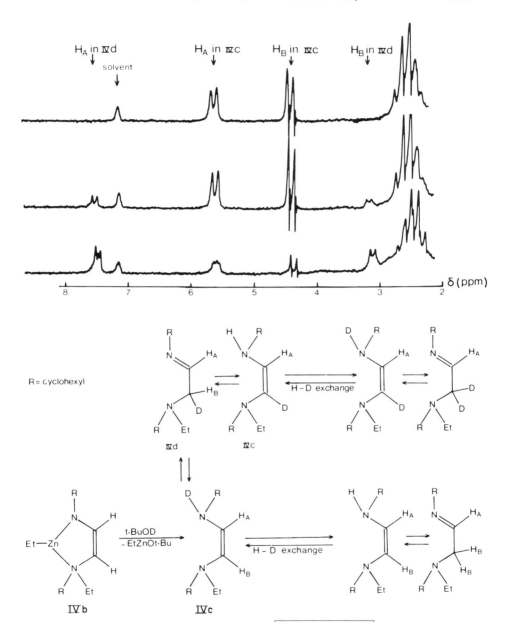

FIG. 24.   The hydrolysis of $EtZn\overline{N(Et)(R)CH{=}CH}NR$.

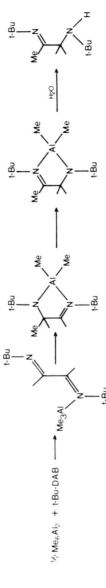

Fig. 25. The reaction of Me_6Al_2 with t-Bu-DAB (182).

(*182*). The first step (Fig. 25) yielded $Me_3Al(R\text{-}DAB)$ (R = 2,6-Xyl and 2,4,6-Mes) in which the R-DAB is σ-N (2e) bonded with the R-DAB probably in the *E-s-trans-E* conformation. In the subsequent step one of the methyl groups of Al is transferred to one of the imine-C atoms giving $Me_2\overline{Al(RN-CHMe-CH}=\overset{}{N}R)$, which could be isolated for R = *p*-XC_6H_4 with X = Me, OMe, Cl. In the final step this compound rearranges to $Me_2\overline{AlRN}=C(Me)-CH_2-\overset{}{N}R$ via a hydrogen migration (Fig. 25) (*182*). For R = 2,6-Xyl and Mes this last compound is directly formed from the first coordination product. Hydrolysis of the final product affords the particular iminoamine $RN(H)CH_2C(Me)=NR$. The rate of rearrangement increased in the order $p\text{-}ClC_6H_4 < p\text{-}MeC_6H_4 < p\text{-}MeO\text{-}C_6H_4$ and 2,6-Xyl < Mes $\ll$ *t*-Bu. For *t*-Bu the reactions take place very rapidly even in the solid state at $-20°C$. The proposed mechanism for the Me transfer involves an intramolecular attack of the nucleophilic methyl groups on the relatively less negative imine-C atoms.

Recent results show that in the case of Al_2Et_6 Et transfer from both Al to N and Al to C occurs in parallel reactions, while in the first reaction also a radical species $Et_2Al(RN=CHCH=NR)^{\cdot}$ was observed (*179*), i.e., a simple change from Me to Et may change the reaction picture completely as does also the change from Zn^{II} to Al^{III}.

In this respect it is of interest to mention that K(*t*-Bu-DAB) reacts with alkyl halides RX to give *t*-Bu-N=CHC(R)=N-*t*-Bu, while with Me_3SiCl, it produces in low yields cis-enediamines and bis-*N*-silylated trans-enediamines (*42*).

C. Stoichiometric C—C Coupling Reactions Involving σ-N,μ^2-N',η^2-C=N' Coordinated 1,4-Diaza-1,3-butadiene Ligands

The η^2-C=N bonded entity in complexes $M_2(CO)_6(R\text{-}DAB)$ is clearly activated as indicated by the lengthened C=N bond and the strong upfield ^1H and ^{13}C NMR chemical shifts of the relevant atoms (Table V). Investigations now show that it is possible to insert a number of unsaturated systems into the activated Ru—C bond of $Ru_2(CO)_6(R\text{-}DAB)$, compounds which have been most extensively studied (*2, 50, 165, 166, 173*). For example, $Ru_2(CO)_6(R\text{-}DAB)$ (R = *t*-Bu, *i*-Pr, *c*-Hex) afforded with R-DAB in the first instance $Ru_2(CO)_5(IAE)$ {IAE = bis[(alkylimino)(alkylamino)ethane]} (Fig. 26). This complex, which has a bridging CO group, but no metal–metal bond, contains two R-DAB ligands coupled together via one C—C bond. The structure is similar to that of $Mo_2(CO)_6(IAE)$ which shows the presence of a 10e-donating IAE ligand with a long C—C

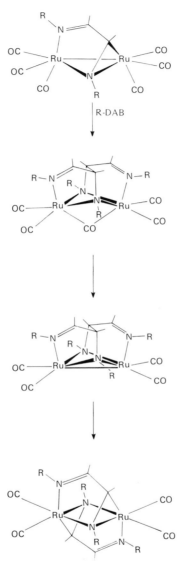

FIG. 26.   The reaction of $Ru_2(CO)_6(R\text{-}DAB)$ with R-DAB (R = i-Pr, c-Hex, t-Bu) to give via $Ru_2(CO)_5(IAE)$ and via $Ru_2(CO)_4(IAE)$ (only for R = i-Pr and c-Hex) the complex $Ru_2(CO)_4(R\text{-}DAB)_2$ (50).

bond of 1.61(2) Å connecting the two R-DAB groups (97). The C—C coupling reaction is completely regiospecific since in the case of asymmetric R-DAB(H,Me) it takes place uniquely between the C atoms bearing the H atoms (50, 165). The following step upon heating $Ru_2(CO)_5(IAE)$ is the loss of the bridging CO group and formation of $Ru_2(CO)_4(IAE)$. $Ru_2(CO)_4(IAE)$ has a metal–metal bond, as is indicated by the presence of a strong electronic absorption band at 290 nm whose intensity increases appreciably at lower temperatures and which is ascribed to a $\sigma \rightarrow \sigma^*$ transition localized on the Ru—Ru bond. $Ru_2(CO)_4(IAE)$ was isolated only for R = i-Pr and c-Hex since the reaction stopped at $Ru_2(CO)_5(IAE)$ for R = t-Bu. Further heating of $Ru_2(CO)_4(IAE)$ caused the C—C bond originally formed to break again. A point confirmed by the crystal structure of the product $Ru_2(CO)_4(R-DAB)_2$ (R = i-Pr) (50) (see also Section III,D). For R = Ar the analogous compound was formed rapidly from $Ru_3(CO)_{12}$ and R-DAB, although the intermediate complexes with IAE ligands could be identified in solution (50). From the above, it should be clear that $Ru_2(CO)_4(R-DAB)_2$ is not an intermediate, but the thermodynamically stable end product, while the $Ru_2(CO)_n(IAE)$ complexes are formed in a kinetically very favorable pathway, probably by coupling of an inserting R-DAB into the Ru—C bond activated by η^2-N=C bonding. In this respect it should be noted that there is a strong dependence not only on the R-group but on the metal too, since IAE complexes could not be formed from $Fe_2(CO)_6(R-DAB)$ (N.B. IAE complexes may be formed fleetingly but the C—C bond could then quickly break again), while $Os_2(CO)_6(R-DAB)$ reacted sluggishly and no "coupled" products have been isolated as yet (112). Coupling of C—C does occur in the reaction of N-α-methylbenzyl iminoacetate $PhCH(Me)N=CHCO_2Et$ (= L) with $Fe_2(CO)_6L$ (164) (see also Section III,E, Fig. 11). The resulting product $Fe_2(CO)_6L_2$, which was characterized by X-ray crystallography, shows the presence of a 2,3-diaminosuccinic acid derivative to which are linked a nitrogen bridged binuclear $Fe_2(CO)_6$ entity. Just as in IAE there is coupling between the η^2-N=C bonded C atom and the N=C C-atom of the inserting ligand (Fig. 27).

In order to obtain more information about the C—C coupling reaction kinetic measurements [using reversed phase HPLC analysis (high performance liquid chromatography)] were carried out on mixtures of $Ru_2(CO)_6(R-DAB)$ with R-DAB (173).

The reaction for R = i-Pr [Eq. (50)] turned out to have an overall order of one, being first order in complex and zero order in i-Pr-DAB (173). The reaction for the t-Bu case proceeded at a much lower rate and it was confirmed that the complex $Ru_2(CO)_5(IAE)$ was the end of the reaction sequence in Fig. 26.

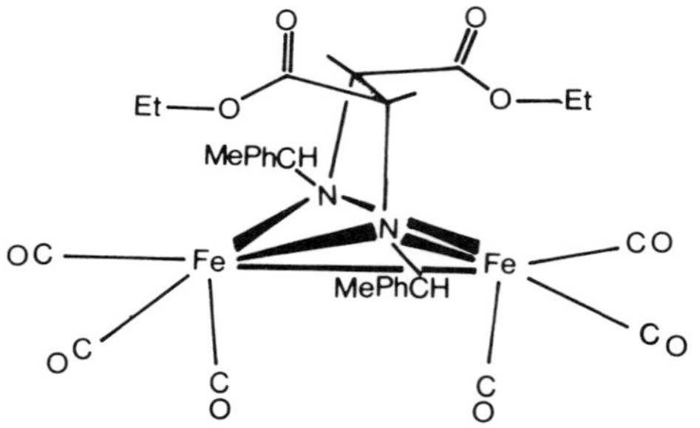

FIG. 27. Schematic structure of $Fe_2(CO)_6L_2$. L_2 is the C—C coupled dimer of $PhCH(Me)N=CHCO_2Et$ (164) (see also Fig. 11).

$$Ru_2(CO)_6(i\text{-Pr-DAB}) + i\text{-Pr-DAB} \rightarrow Ru_2(CO)_5(IAE) + CO \qquad (50)$$

Somewhat more complicated reactions appeared to take place for mixtures of $Ru_2(CO)_6(i\text{-Pr-DAB})$ with $t\text{-Bu-DAB}$ and of $Ru_2(CO)_6(t\text{-Bu-DAB})$ with $i\text{-Pr-DAB}$, since different products were characterized. The former reaction gave $Ru_2(CO)_6(t\text{-Bu-DAB})$, $Ru_2(CO)_n(i\text{-Pr-IAE})$, and $Ru_2(CO)_4(i\text{-Pr-DAB})_2$. However, for the second reaction mixture these last two compounds were present, but not $Ru_2(CO)_6(i\text{-Pr-DAB})$. There is some evidence that mixed IAE complexes containing $t\text{-Bu-DAB}$ C—C coupled to $i\text{-Pr-DAB}$ and furthermore a mixed compound $Ru_2(CO)_4(t\text{-Bu-DAB})(i\text{-Pr-DAB})$ might also be formed (173).

These various reactions may be rationalized by the following equilibrium equation

$$M_2(CO)_6(DAB) \underset{-DAB}{\overset{+DAB}{\rightleftharpoons}} 2M(CO)_3(DAB) \qquad (51)$$

which moves to the right for M = Fe and for R substituents doubly branched at C^α and C^β. Such a process would account for the presence of both mixed IAE and mixed R-DAB complexes. The intermediate complex $Ru_2(CO)_6(DAB)$ probably consists of a R-DAB ligand σ,σ-N,N' chelated to one M atom, while the other M atom is coordinatively unsaturated and therefore susceptible to attack of a second R-DAB. This would also be in line with the reaction scheme proposed for the formation of $MnCo(CO)_6(R\text{-DAB})$ (see also Sections III,E and IV,A and Fig. 13).

The mechanism of formation of the previously mentioned $Mo_2(CO)_6(IAE)$ is not very clear, since the preparations are different from those of Ru compounds (97). For example, the compound can be made by treating not only $MnBr(CO)_3(R\text{-DAB})$ but also $MnBr(CO)_5$ with $[Mo(CO)_4(R\text{-DAB})]K$ [Eqs. (52) and (53)]:

$$2MnBr(CO)_3(DAB) + 2[Mo(CO)_4(DAB)]K \rightarrow$$

$$Mo_2(CO)_6(IAE) + 2DAB + KBr + Mn_2(CO)_{10} \quad (52)$$

$$2MnBr(CO)_5 + 2[Mo(CO)_4(DAB)]K \rightarrow$$

$$Mo_2(CO)_6(IAE) + 2CO + 2KBr + Mn_2(CO)_{10} \quad (53)$$

It was further demonstrated that even though a Mo—Mn bonded complex is an intermediate, the IAE ligand is formed exclusively from the R-DAB bonded to Mo, indicating a bimetallic Mo_2 intermediate. The method giving the highest yields of $Mo_2(CO)_6(IAE)$ involves the treatment of $Hg[Mo(CO)_3(R\text{-}DAB)]_2$ with acids [Eq. (54)].

$$Hg[Mo(CO)_3(R\text{-}DAB)]_2 + 2H^+ \rightarrow Mo_2(CO)_6(IAE) + H_2 + Hg^{2+} \quad (54)$$

Using a similar reaction, only a few percent yield of $Cr_2(CO)_6(IAE)$ could be obtained, while the analogous tungsten complex could not be prepared. Use of asymmetric R-DAB(H,Me) gave quantitatively a regiospecific reaction in which only the C atoms bearing the H atoms were linked together. All these observations indicate a kinetically favorable pathway that is very sensitive to the various factors. The preferred reaction mechanism involves a dimerization of two coordinatively unsaturated $Mo(CO)_3(R\text{-}DAB)$ (R = t-Bu, i-Pr, c-Hex) units to form an Mo_2 intermediate on which the R-DAB ligands are held close together. Because of the observed regiospecificity and in view of the R groups that could be used [i.e., those also typical for the stabilization of $Ru_2(CO)_6(R\text{-}DAB)$] it seems likely that at least one R-DAB is σ-N,μ^2-N',η^2-C=N' (6e) bonded to the Mo_2 pair. Whether the subsequent C—C coupling reaction both here and with the Ru complexes proceeds via a radical mechanism [cf. the C—C coupling reactions found for Zn—Eqs. (42) and (43)] or via, e.g., a polar mechanism is open to question.

The molecular structure of $Mo_2(CO)_6(IAE)$ with R = i-Pr shows a Mo_2-unit [2.813(2) Å] bridged by the 10e-IAE ligand with C=N, C—N and C—C bond lengths of 1.28(2), 1.39(2) and 1.62(2) Å, respectively (97). The long C—C bond is a result of steric strain, and it is therefore not surprising that both heat and light lead to cleavage of this bond.

D. Catalytic C—C Coupling Reactions Involving σ-N,μ^2-N',η^2-C=N' Coordinated 1,4-Diaza-1,3-butadiene Ligands

A good example of C—C coupling with other substrates is the reaction of $Ru_2(CO)_6(R\text{-}DAB)$ (R = i-Pr, t-Bu, c-Hex) with acetylene, and mono- and disubstituted alkynes $R'C_2R''$ (R' = R'' = CH_3O_2C-, CD_3O_2C-; R'

= R″ = H; R′ = H, R″ = CH₃O₂C, Ph, p-Tol, t-Bu) (166). In this multistage reaction sequence the first step involves insertion of R′C≡CR″. The crystal structure of the insertion product with PhC₂H, i.e., Ru₂(CO)₅[(t-Bu-DAB)(PhC₂H)] shows that the DAB ligand and PhC₂H are coupled via a C—C bond. The 3-amino-4-imino-1-butene-1-yl (AIB) so formed is bonded to the Ru₂(CO)₅ unit as shown in Fig. 28. The length of the C—C bond formed is 1.546(10) Å. The Ru₂(CO)₅ entity shows four terminal CO groups and a fifth one bridging a Ru—Ru bond of 2.711(1) Å. The C=N and C—N bonds are 1.259(10) and 1.496(9) Å, respectively. The original C≡C bond is reduced to a double C=C bond of 1.346(10) Å. An interesting feature of this planar olefinic fragment is that it is bonded to the Ru₂ pair with the CH end, while the C—Ph end is coupled to the formally

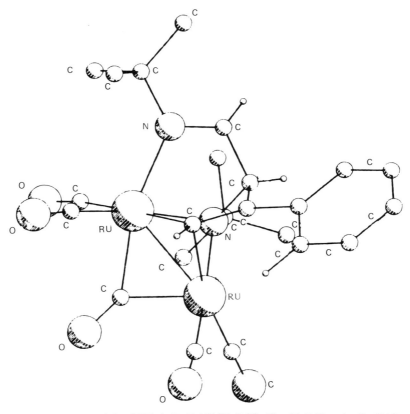

FIG. 28.   Structure of Ru₂(CO)₅(t-Bu-DAB)(PhC₂H). The PhC₂H and t-Bu-DAB are coupled via a C—C bond (166).

η^2-C=N bond imine-C atom. Here therefore we have coupling between a C—H and a C—Ph fragment in contrast to the C—H to C—H coupling in the case of insertion of R-DAB (50).

Subsequent reaction of $Ru_2(CO)_5(AIB)$ with another alkyne gave $Ru_2(CO)_5(AIB)$(alkyne) in which the second alkyne just adds to the first complex as a 2e donor without substitution, since $Ru_2(CO)_5(AIB)$ is electron deficient by 2e. This electron deficiency may be partly relieved by coordination either of the olefinic fragment or (if R is CH_3O_2C) by an oxygen atom of the carboxylate group to the Ru atom not bearing the σ-N=C bonded group. The CO groups in $Ru_2(CO)_5(AIB)$(alkyne) are all terminally bonded. Further heating produced $Ru_2(CO)_4(AIB)$(alkyne) with loss of CO, while the alkyne has changed its bonding from 2e-monodentate to 4e-bridging. Finally, further addition of alkynes at 110°C produced substituted benzenes catalytically (Fig. 29) (166). An example of such a reaction, monitored by HPLC, was the trimerization of $CH_3O_2CC\equiv CCO_2CH_3$ by $Ru_2(CO)_4(t$-Bu-DAB-R'C_2R')(R'C_2R') (R' = CH_3O_2C-) which gave rapidly in 380 cycles only the substituted benzene [$C_6(CO_2CH_3)_6$]. Use of mono substituted alkynes HC_2R' afforded with complete regiospecificity only 1,3,5-trisubstituted benzenes. This is very unusual since almost all catalysts produce mixtures of various products (184).

The mechanism of the catalytic reaction is not clear, but none of the present known mechanisms seem to explain the experimental data (185–189). Whatever the mechanism, the quantitative symmetry induction to give 1,3,5-substituted benzenes strongly indicates a binuclear intermediate as catalyst (166).

A final point is that involvement of HC_2H with $Ru_2(CO)_6(R$-DAB) did not give a catalytic reaction, but produced $Ru_2(CO)_4(\mu^2$-$HC_2H)(R$-DAB) via $Ru_2(CO)_5(R$-DAB)(HC_2H) which could be identified as an unstable intermediate. The μ^2-HC_2H complex is shown in the catalytic scheme of Fig. 29 while the structure was discussed in Section III,F (Fig. 14b). The complex shows no catalytic activity (31, 166).

E. Catalytic Reactions

Very little is known about catalytic reactions in which metal–R-DAB complexes are involved but for which on first sight of the R-DAB itself is not activated. Discussion is unfortunately hampered by the absence of any concrete data, since the available results have been laid down in preliminary discussions that have not been followed up by full papers (74, 105, 107,

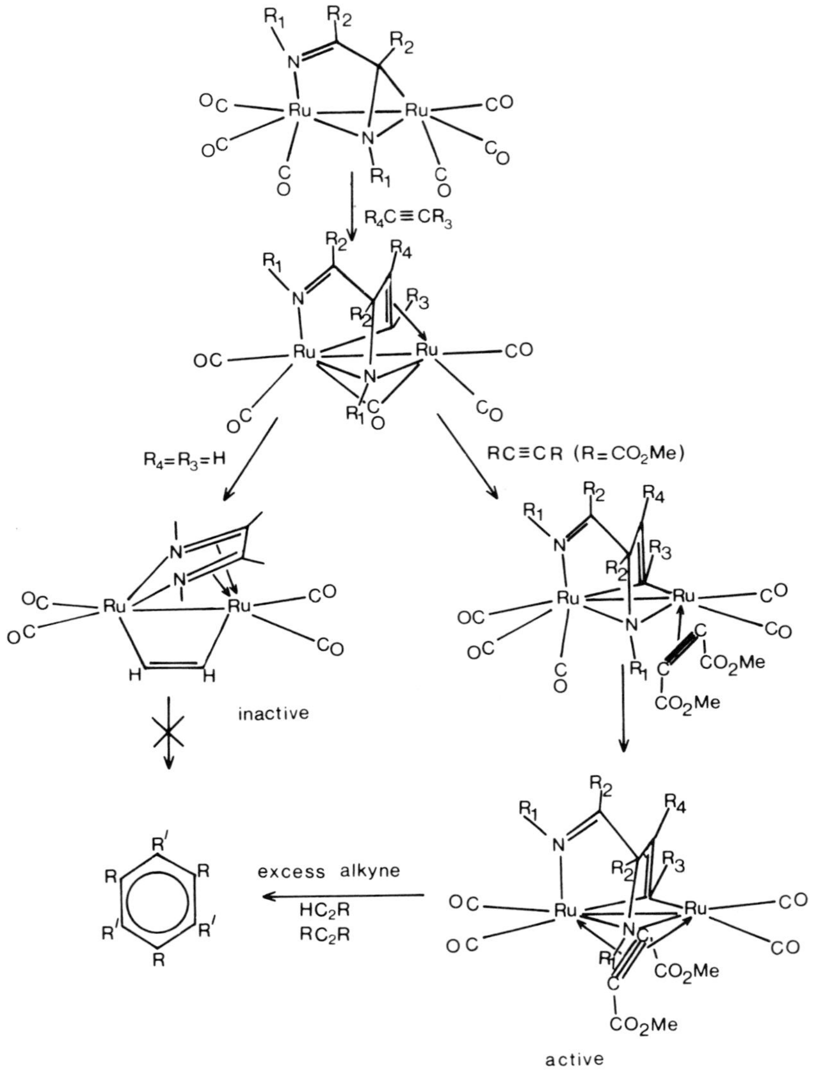

FIG. 29. The catalytic reaction of $Ru_2(CO)_6[R_1\text{-}DAB(R_2,R_2)]$ with RC_2R. Note that $Ru_2(CO)_4(R\text{-}DAB)(\mu^2\text{-}HC_2H)$ is the stable end product if HC_2H is used (31, 166). For asymmetric alkynes only 1,3,5-substituted benzenes are found.

148). We shall therefore restrict ourselves by necessity to some scarce details on Fe, Ni and Cr complexes.

For example, tetracoordinate Fe[R-DAB(R′,R″)]₂ (R = *t*-Bu, *c*-Hex, *i*-Pr, *i*-Pr₂CH, Ar; R′ = R″ = H or Me) which have a 16e configuration, show in the presence of Et₂AlOEt or AlEt₃ (2 mol equivalent) high catalytic activity in the dimerization of conjugated dienes in the temperature range 22–82°C (*105*). In the case of butadiene 1,5-cyclooctadiene (COD) and vinylcyclohexene were formed, while in the case of isopropene 1,6-dimethyl-COD and 1,5-dimethyl-COD could be isolated together with methylated vinylcyclohexenes and open chain products.

In the case of Fe(NO)₂(R-DAB(Me,Me) (R = Ph) the conversion of isoprene both with and without (AlEt₃)₂ led to formation of somewhat varying amounts of isoprene polymers as well as 1,4- and 2,4-dimethyl-4-vinyl-1-cyclohexene (*107*).

The complex Cr(R-DAB)₃ (R = *i*-Pr₂CH) activated by Et₂AlOEt yielded mainly 2,7-dimethyl-2-*trans*-4,6-octatriene through a selective tail-to-tail C—C coupling accompanied by a double H shift, in addition to some trimers and some higher molecular weight products. Change of R to *c*-Hex gave none of the first product, but instead another isomer, 2,7-dimethyl-1,3,6-octatriene and a head-to-tail dimer, 2,6-dimethyl-1,3,6-octatriene (*74*). A similar steric control was mentioned for V(R-DAB)₃–AlR₃ but no details were given (*74*).

Finally it has been mentioned in a short paragraph of Ref. *148* that the novel complexes NiMe₂(R-DAB) [R = 2,6-(*i*-Pr)₂C₆H₃] appear to be active in the catalytic oligomerization of butadiene at 120°C and at unstated pressures to give unidentified waxy polymers.

The steric control in some of these reactions seems to indicate the presence of R-DAB in the catalyst. However, in most cases the catalysts are activated by alkylaluminum compounds, which as we have seen are extremely active in the conversion of R-DAB itself (see Section VI,B, Fig. 25) (*182*). Furthermore, we must expect under these conditions the formation of active clusters.

VII

CONCLUDING REMARKS

In this review we have shown that the R-DAB species RN=C-(R′)C(R′)=NR are highly versatile ligands. The versatility is due to (i) the flexibility of the NCCN skeleton, (ii) the strong σ-donor and π-acceptor properties, and (iii) the possibility of changing both the R and R' substituents by which both the electronic and steric properties may be influenced appreciably.

It has been shown that due to this versatility the diimine ligands are able to bind as 2e, 4e, 6e, or 8e donor ligands. In the 2e and 4e bonding modes the N atoms are used for bonding while in the 6e and 8e bonding modes the π-C$=$N orbitals are also employed for metal-η^2-C$=$N bonding. In the last type of bonding the C$=$N bond length is increased appreciably by virtue of back donation into the low-lying LUMO of the NCCN skeleton.

In addition to the coordination types we have also surveyed the reactivity of coordinated diimine ligands. The most unusual aspect is the activation of the η^2-C$=$N bonding in the 6e bonding mode, which results in stoichiometric and catalytic C$-$C coupling reactions.

Considering what is now known about diimine chemistry, it is to be expected that much fruitful work should be carried out in the reactivity of coordinated R-DAB ligands.

ACKNOWLEDGMENTS

We wish to thank L. H. Staal, H. van der Poel, K. I. Cavell, L. H. Polm, J. Keijsper, R. van Vliet, and J. T. B. H. Jastrzebski for their experimental efforts and helpful discussions, D. M. Grove for the careful reading of the manuscript, H. Kraaykamp for assistance with the literature search, and L. C. Taylor for typing the manuscript.

REFERENCES

1. L. F. Lindoy and S. E. Livingstone, *Coord. Chem. Rev.* **2**, 173 (1967).
2. G. van Koten and K. Vrieze, *Recl. Trav. Chim. Pays-Bas* **100**, 129 (1981).
3. J. M. Kliegman and R. K. Barnes, *Tetrahedron* **26**, 2555 (1970).
4. J. M. Kliegman and R. K. Barnes, *J. Org. Chem.* **35**, 3140 (1970).
5. V. C. Barnay and P. W. D. Mitchell, *J. Chem. Soc.* p. 3610 (1953).
6. L. Horner and E. Jürgens, *Chem. Ber.* **90**, 2184 (1957).
7. J. F. Carson, *J. Am. Chem. Soc.* **75**, 4337 (1953).
8. H. van der Poel and G. van Koten, *Synth. Commun.* **8**, 305 (1978).
9. M. Svoboda, H. tom Dieck, C. Krüger, and Y.-H. Tsay, *Z. Naturforsch. B: Anorg. Chem., Org. Chem.* **36B**, 814 (1981).
10. H. tom Dieck, M. Svoboda, and T. Greiser, *Z. Naturforsch., B: Anorg. Chem., Org. Chem.* **36B**, 823 (1981).
11. H. tom Dieck, K. D. Franz, and W. Majunke, *Z. Naturforsch., B: Anorg. Chem., Org. Chem.* **30B**, 922 (1975).
12. J. M. Kliegman and R. K. Barnes, *Tetrahedron Lett.* **24**, 1953 (1969).
13. O. Exner and J. M. Kliegman, *J. Org. Chem.* **36**, 2014 (1971).
14. H. tom Dieck and I. W. Renk, *Chem. Ber.* **104**, 92 (1971).
15. O. Borgen, B. Mestvedt, and I. Skauvik, *Acta Chem. Scand., Ser. A* **30**, 43 (1976).
16. R. Benedix, P. Birner, F. Birnstock, H. Hennig, and H.-J. Hofmann, *J. Mol. Struct.* **51**, 99 (1979).

17. J. Keijsper, G. van Koten, and K. Vrieze, *Polyhedron*, in press (1982).
18. I. Hargittai and R. Seip, *Acta Chem. Scand. Ser. A* **30**, 540 (1976).
19. J. Tyrell, *J. Am. Chem. Soc.* **101**, 1766 (1979).
20. B. Crociani, G. Bandoli, and D. A. Clemente, *J. Organomet. Chem.* **184**, 269 (1980).
21. H. van der Poel and G. van Koten, *Polyhedron*, in press (1982).
22. L. L. Merritt, Jr. and E. D. Schroeder, *Acta Crystallogr.* **9**, 801 (1956).
23. K. Folting and L. L. Merritt, Jr., *Acta Crystallogr., Sect. B* **33**, 3540 (1977).
24. M. Lenner and O. Lindgren, *Act Crystallogr., Sect. B* **32**, 1903 (1976).
25. A. Almenningen, O. Bastiansen, and M. Traettenberg, *Acta Chem. Scand.* **12**, 1221 (1958).
26. J. Reinhold, R. Benedix, P. Birner, and H. Hennig, *Inorg. Chim. Acta* **33**, 209 (1979).
27. H. C. Barany, E. A. Braude, and M. A. Pianka, *J. Chem. Soc.* p. 1898 (1949).
28. P. Krumholz, *J. Am. Chem. Soc.* **75**, 2163 (1953).
29. H. van der Poel, G. van Koten, and K. Vrieze, *Inorg. Chem.* **19**, 1145 (1980).
30. H. W. Frühauf, A. Landers, R. Goddard, and C. Krüger, *Angew. Chem.* **90**, 56 (1978).
31. L. H. Staal, L. H. Polm, K. Vrieze, F. Ploeger, and C. H. Stam, *J. Organomet. Chem.* **199**, C13 (1980).
32. H. van der Poel, G. van Koten, K. Vrieze, M. Kokkes, and C. H. Stam, *Inorg. Chim. Acta* **39**, 197 (1980).
33. L. H. Staal, G. van Koten, and K. Vrieze, F. Ploeger, and C. H. Stam, *Inorg. Chem.* **20**, 1830 (1982).
34. H. van der Poel, G. van Koten, and K. Vrieze, *Inorg. Chim. Acta* **51**, 253 (1981).
35. H. van der Poel, G. van Koten, D. M. Grove, P. S. Pregosin, and K. A. O. Starzewski, *Helv. Chim. Act* **64**, 1174 (1981).
36. H. van der Poel and G. van Koten, *J. Organomet. Chem.* **217**, 129 (1981).
37. T. Mack, Ph.D. Thesis, University of Frankfurt, Frankfurt am Main (1979).
38. A. J. Graham, D. Akrigg, and B. Sheldrick, *Cryst. Struct. Commun.* **5**, 891 (1976).
39. A. J. Graham, D. Akrigg, and B. Sheldrick, *Cryst. Struct. Commun.* **6**, 253 (1977).
40. A. J. Graham, D. Akrigg, and B. Sheldrick, *Cryst. Struct. Commun.* **6**, 571 (1977).
41. A. J. Graham, D. Akrigg, and B. Sheldrick, *Cryst. Struct. Commun.* **6**, 577 (1977).
42. B. Bruder, Ph.D. Thesis, University of Frankfurt, Frankfurt am Main (1979).
43. B. Chaudret, H. Köster, and R. Poilblanc, *J. Chem. Soc., Chem. Commun.* p. 266 (1981).
44. V. Pank, J. Klaus, K. von Deuten, M. Feigel, H. Bruder, and H. tom Dieck, *Transition Met. Chem.* **6**, 185 (1981).
45. J. Kopf, J. Klaus, and H. tom Dieck, *Cryst. Struct. Commun.* **9**, 783 (1980).
46. H. D. Hausen and K. Krogmann, *Z. Anorg. Allg. Chem.* **389**, 247 (1972).
47. H. tom Dieck, M. Svoboda, and J. Kopf, *Z. Naturforsch. B: Anorg. Chem., Org. Chem.* **33B**, 1381 (1978); A. L. Balch and R. H. Holm, *J. Am. Chem. Soc.* **88**, 5201 (1966); D. Walther, *Z. Anorg. Allg. Chem.* **431**, 17 (1977).
48. H. van der Poel, G. van Koten, M. Kokkes, and C. H. Stam, *Inorg. Chem.* **20**, 2491 (1981).
49. L. Maresca, G. Natile, M. Calligaris, P. Delise, and L. Randaccio, *J. Chem. Soc., Dalton Trans.* p. 2386 (1976).
50. L. H. Staal, L. H. Polm, R. W. Balk, G. van Koten, K. Vrieze, and A. M. F. Brouwers, *Inorg. Chem.* **19**, 3343 (1980).
51. L. H. Staal, J. Keijsper, G. van Koten, K. Vrieze, J. A. Cras, and W. P. Bosman, *Inorg. Chem.* **20**, 555 (1981).
52. L. H. Staal, L. H. Polm, K. Vrieze, F. Ploeger, and C. H. Stam, *Inorg. Chem.* **20**, 3590 (1981).
53. R. D. Adams, *J. Am. Chem. Soc.* **102**, 7476 (1980).

54. J. T. B. H. Jastrzebski, G. van Koten, and K. Vrieze, to be published.

55. L. H. Staal, D. J. Stufkens, and A. Oskam, *Inorg. Chim. Acta* **26,** 255 (1978).

56. H. van der Poel, G. van Koten, K. Vrieze, M. Kokkes, and C. H. Stam, *J. Organomet. Chem.* **175,** C21 (1979).

57. B. Crociani, T. Boschi, and P. Uguagliati, *Inorg. Chim. Acta* **48,** 9 (1981).

58. K. J. Cavell, D. J. Stufkens, and K. Vrieze, *Inorg. Chim. Act* **47,** 67 (1980).

59. H. van der Poel, G. van Koten, and K. Vrieze, *J. Organomet. Chem.* **135,** C63 (1977).

60. B. Crociani and U. Belluco, *J. Organomet. Chem.* **177,** 385 (1979).

61. H. van der Poel, G. van Koten, and K. Vrieze, *Inorg. Chim. Acta* **51,** 241 (1981).

62. A. T. T. Hsieh and K. L. Ooi, *J. Inorg. Nucl. Chem.* **38,** 604 (1976).

63. B. Crociani, M. Nicolini, and R. L. Richards, *J. Organomet. Chem.* **133,** C22 (1976).

64. B. Crociani and R. L. Richards, *J. Organomet. Chem.* **154,** 65 (1978).

65. B. Crociani, M. Nicolini, and A. Mantovani, *J. Organomet. Chem.* **177,** 365 (1979).

66. P. L. Sandrini, A. Mantovani, and B. Crociani, *J. Organomet. Chem.* **185,** C13 (1980).

67. P. L. Sandrini, A. Mantovani, B. Crociani, and P. Uguagliati, *Inorg. Chim. Acta* **50,** 71 (1981).

68. J. T. B. H. Jastrzebski, J. M. Klerks, G. van Koten, and K. Vrieze, *J. Organomet. Chem.* **210,** C49 (1981); **224,** 107 (1982).

69. H. Bock and H. tom Dieck, *Chem. Ber.* **100,** 228 (1967).

70. D. Walther, *Z. Anorg. Allg. Chem.* **405,** 8 (1974).

71. A. T. T. Hsieh and B. O. West, *J. Organomet. Chem.* **112,** 285 (1976).

72. J. Matei, T. Lixandru, *Bul. Inst. Politeh. Iasi* **13,** 245 (1967); *Chem. Abstr.* **70,** 3626 (1969).

73. D. Walther, *Z. Chem.* **15,** 72 (1975).

74. H. tom Dieck and A. Kinzel, *Angew. Chem.* **91,** 344 (1979).

75. P. Overbosch, G. van Koten, and O. Overbeek, *J. Am. Chem. Soc.* **102,** 2091 (1980).

76. A. Kinzel, Ph.D. Thesis, University of Hamburg, Hamburg (1979).

77. L. H. Staal, A. Oskam, and K. Vrieze, *J. Organomet. Chem.* **170,** 235 (1979).

78. R. W. Balk, D. J. Stufkens, and A. Oskam, *Inorg. Chim. Acta* **34,** 267 (1979).

79. W. Majunke, D. Liebfritz, T. Mack, and H. tom Dieck, *Chem. Ber.* **108,** 3025 (1975).

80. R. W. Balk, T. Snoeck, D. J. Stufkens, and A. Oskam, *Inorg. Chem.* **19,** 3015 (1980).

81. H. tom Dieck, K.-D. Franz, and F. Hohmann, *Chem. Ber.* **108,** 163 (1975).

82. D. Walther and M. Teutsch, *Z. Chem.* **16,** 118 (1976).

83. L. H. Staal, A. Terpstra, and D. J. Stufkens, *Inorg. Chim. Acta* **34,** 97 (1979).

84. H. tom Dieck and I. W. Renk, *Chem. Ber.* **104,** 110 (1971).

85. H. tom Dieck and I. W. Renk, *Angew. Chem.* **82,** 805 (1970).

86. D. Walther, *J. Prakt. Chem.* **316,** 604 (1974).

87. R. W. Balk, D. J. Stufkens, and A. Oskam, *Inorg. Chim. Acta* **28,** 133 (1978).

88. H. tom Dieck, I. W. Renk, and K. D. Franz, *J. Organomet. Chem.* **94,** 417 (1975).

89. G. Häfelinger, R. G. Weissenhorn, F. Hack, and G. Wester-Mayer, *Angew. Chem.* **84,** 769 (1972).

90. D. Walther and U. Dinjus, *Z. Chem.* **15,** 196 (1975).

91. D. Walther and P. Hallpap, *Z. Chem.* **13,** 387 (1973).

92. I. W. Renk and H. tom Dieck, *Chem. Ber.* **105,** 1403 (1972).

93. F. Hohmann, H. tom Dieck, K. D. Franz, and K. A. Ostoja Starzewski, *J. Organomet. Chem.* **55,** 321 (1973).

94. H. tom Dieck and I. W. Renk, *Chem. Ber.* **105,** 1419 (1972).

95. H. Friedel, I. W. Renk, and H. tom Dieck, *J. Organomet. Chem.* **26,** 247 (1971).

96. R. G. Hayter, *J. Organomet. Chem.* **13,** P1 (1968); C. G. Hull and M. H. B. Stiddard, *ibid.* **9,** 519 (1967).

97. L. H. Staal, A. Oskam, K. Vrieze, E. Roosendaal, and H. Schenk, *Inorg. Chem.* **18,** 1634 (1979).
98. R. W. Balk, D. J. Stufkens, and A. Oskam, *J. Mol. Struct.* **60,** 387 (1980).
99. R. W. Balk, D. J. Stufkens, and A. Oskam, *J. Chem. Soc., Chem. Commun.* p. 1016 (1978).
100. H. Bock and H. tom Dieck, *Angew. Chem.* **18,** 159 (1966).
101. D. Walther, *Z. Chem.* **13,** 107 (1973).
102. R. W. Balk, D. J. Stufkens, and A. Oskam, *J. Chem. Soc., Chem. Commun.* p. 604 (1979).
103. L. H. Staal, G. van Koten, and K. Vrieze, *J. Organomet. Chem.* **175,** 73 (1979).
104. R. W. Balk, D. J. Stufkens, and A. Oskam, *J. Chem. Soc., Dalton Trans.* p. 1124 (1981).
105. H. tom Dieck and H. Bruder, *J. Chem. Soc., Chem. Commun.* p. 24 (1977).
106. D. Liebfritz and H. tom Dieck, *J. Organomet. Chem.* **105,** 255 (1976).
107. H. tom Dieck, H. Bruder, K. Hellfeldt, D. Liebfritz, and M. Fiegel, *Angew. Chem.* **92,** 395 (1980).
108. S. Otsuka, T. Yoshida, and A. Nakamura, *Inorg. Chem.* **6,** 20 (1967).
109. H. tom Dieck and A. Orlopp, *Angew. Chem.* **87,** 246 (1975).
110. L. H. Staal, L. H. Polm, and K. Vrieze, *Inorg. Chim. Acta* **40,** 165 (1980); M. W. Kokkes, D. J. Stufkens, A. Oskam, and C. H. Stam, to be published.
111a. H.-W. Frühauf, F.-W. Grevels, and A. Landers, *J. Organomet. Chem.* **178,** 349 (1979).
111b. H.-W. Frühauf and G. Wolnershäuser, *Chem. Ber.* **115,** 1070 (1982).
112. L. H. Staal, G. van Koten, and K. Vrieze, *J. Organomet. Chem.* **206,** 99 (1981).
113. D. H. Busch and J. C. Bailar, Jr., *J. Am. Chem. Soc.* **78,** 1137 (1956).
114. R. J. H. Clark, P. C. Turtle, D. P. Strommen, B. Streusand, J. Kincaid, and K. Nakamoto, *Inorg. Chem.* **16,** 84 (1977).
115. H. L. Chum and P. Krumholz, *Inorg. Chem.* **13,** 514 (1974).
116. H. L. Chum and P. Krumholz, *Inorg. Chem.* **13,** 519 (1974).
117. H. L. Chum, T. Rabockai, J. Phillips, and R. A. Osteryoung, *Inorg. Chem.* **16,** 812 (1977).
118. T. V. Harris, J. W. Rathke, and E. L. Muetterties, *J. Am. Chem. Soc.* **100,** 6966 (1978).
119. I. P. Evans, G. W. Everett, and A. M. Sargeson, *J. Am. Chem. Soc.* **98,** 8041 (1976).
120. E. Bayer, E. Breitmaier, and V. Schurig, *Chem. Ber.* **101,** 1594 (1968).
121. M. Tubino and E. J. S. Vichi, *Inorg. Chim. Acta* **28,** 29 (1978).
122. D. Soria, M. L. de Castro, and H. L. Chum, *Inorg. Chim. Acta* **42,** 121 (1980).
123. H. L. Chum, D. Koran, and R. A. Osteryoung, *J. Am. Chem. Soc.* **100,** 310 (1978).
124. H. L. Chum, A. M. G. da Costa, and P. Krumholz, *J. Chem. Soc., Chem. Commun.* p. 772 (1972).
125. P. Krumholz, O. A. Serra, and M. A. de Paoli, *Inorg. Chim. Acta* **15,** 25 (1975).
126. H. L. Chum and M. L. de Castro, *J. Am. Chem. Soc.* **96,** 5278 (1974).
127. P. Krumholz, H. L. Chum, M. A. de Paoli, and T. Rabockai, *Electroanal. Chem. Interfacial Electrochem.* **51,** 465 (1974).
128. D. W. Clack, L. A. P. Kane-Maguire, D. W. Knight, and P. A. Williams, *Transition Metal Chem.* **5,** 376 (1980).
129. R. D. Gillard, D. W. Knight, and P. A. Williams, *Transition Metal Chem.* **5,** 321 (1980).
130. A. M. G. de Costa Ferreira, P. Krumholz, and J. M. Riveros, *J. Chem. Soc., Dalton Trans.* p. 896 (1977).
131. V. Pauk, J. Klaus, K. von Deuten, M. Feigel, H. Bruder, and H. tom Dieck, *Transition Metal Chem.* **6,** 185 (1981).
132. B. Chaudret and R. Poilblanc, *J. Organomet. Chem.* **174,** C51 (1979).
133. B. Chaudret and R. Poilblanc, *J. Chem. Soc., Dalton Trans.* p. 539 (1980).

134. F. S. Hall and W. L. Reynolds, *Inorg. Chem.* **5**, 931 (1966).
135. C. Tänzer, R. Price, E. Breitmaier, G. Jung, and W. Voelter, *Angew. Chem.* **82**, 957 (1970).
136. K. Nakamoto, *in* "Advances in the Chemistry of the Coordination Compounds" (S. Kirschner, ed.). Macmillan, New York, 1961; K. Nakamoto, "Infrared Spectra of Inorganic and Coordination Compounds." Wiley, New York, 1963.
137. B. Chaudret and R. Poilblanc, *J. Organomet. Chem.* **204**, 115 (1981).
138. R. Benedix, J. Reinhold, and H. Hennig, *Inorg. Chim. Acta* **40**, 47 (1980).
139. J. Reinhold, R. Benedix, H. Zwanziger, and H. Hennig, *Z. Phys. Chem.* **5**, 989 (1980).
140. L. H. Staal, P. Bosma, and K. Vrieze, *Inorg. Chim. Acta* **43**, 125 (1980).
141. A. J. Graham, D. Akrigg, and B. Sheldrick, *Cryst. Struct. Commun.* **6**, 577 (1977).
142. G. Häfelinger, R. G. Weiszenhorn, F. Hack, and G. Westermayer, *Angew. Chem.* **84**, 769 (1972).
143. B. Crociani, M. Nicolini, and R. L. Richards, *J. Chem. Soc., Dalton Trans.* p. 1478 (1978).
144. A. L. Balch and R. H. Holm, *J. Am. Chem. Soc.* **80**, 520 (1966).
145. D. Walther, *Z. Chem.* **17**, 348 (1977).
146. H. Hoberg and C. Fröhlich, *J. Organomet. Chem.* **209**, C69 (1981).
147. H. tom Dieck and M. Svoboda, *Chem. Ber.* **109**, 1657 (1976).
148. M. Svoboda and H. tom Dieck, *J. Organomet. Chem.* **191**, 321 (1980).
148a. P. Overbosch, G. van Koten, and O. Overbeek, *Inorg. Chem.*, **21**, 2373 (1982).
149. V. I. Nefjedov, J. V. Salin, D. Walther, E. Uhlig, and E. Dinjus, *Z. Chem.* **17**, 190 (1977).
150. P. Overbosch, G. van Koten, D. M. Grove, and A. L. Spek, *Inorg. Chem.*, in press (1982); W. Beck and F. Holsboer, *Z. Natürforsch. B: Anorg. Chem., Org. Chem.* **28B**, 511 (1973).
151. P. Overbosch, G. van Koten, and K. Vrieze, *J. Organomet. Chem.* **208**, C21 (1981).
152. L. Cattalini, F. Casparrini, L. Maresca, and G. Natile, *J. Chem. Soc., Chem. Commun.* p. 369 (1973).
153. N. Chaudhury and R. J. Puddephatt, *Inorg. Chem.* **20**, 467 (1981).
153a. A. De Rentzi, B. Di Blasio, A. Saporito, M. Scalone, and A. Vitagliano, *Inorg. Chem.* **19**, 960 (1980).
154. H. van der Poel and G. van Koten, *Inorg. Chem.* **20**, 2950 (1981).
155. H. van der Poel and G. van Koten, *J. Organomet. Chem.* **187**, C17 (1980).
156. H. van der Poel, G. van Koten, and G. C. van Stein, *J. Chem. Soc., Dalton Trans.* p. 2164 (1981).
157. H. Aryanci, C. Daul, M. Zobrist, and A. von Zelewsky, *Helv. Chim. Acta* **58**, 1732 (1975).
158. G. van Koten, J. T. B. H. Jastrzebski, and J. G. Noltes, *J. Org. Chem.* **42**, 2047 (1977).
159. J. T. B. H. Jastrzebski and G. van Koten, to be published.
160. G. van Koten and J. G. Noltes, *J. Am. Chem. Soc.* **101**, 6593 (1979).
161. G. C. van Stein, G. van Koten, and C. Brevard, *J. Organomet. Chem.* **226**, C27 (1982).
161a. C. Brevard, G. C. van Stein, and G. van Koten, *J. Am. Chem. Soc.* **103**, 6746 (1981).
162. M. Dvolaitzky, presented by M. H. Normant, *C. R. Hebd. Seances Acad. Sci.* **270**, 96 (1970).
163. L. H. Staal, J. Keijsper, L. H. Polm, and K. Vrieze, *J. Organomet. Chem.* **204**, 101 (1981).
164. A. de Cian and R. Weiss, *J. Chem. Soc., Chem. Commun.* p. 249 (1976).
165. L. H. Staal, L. H. Polm, G. van Koten, and K. Vrieze, *Inorg. Chim. Acta* **37**, L485 (1979).

166. L. H. Staal, G. van Koten, K. Vrieze, B. van Santen, and C. H. Stam, *Inorg. Chem.* **20,** 3598 (1981).

167. F. A. Cotton, *Prog. Inorg. Chem.* **21,** 1 (1976).

168. A. A. Hock and O. S. Mills, *Acta Crystallogr.* **14,** 139 (1961).

169. H. B. Chin and R. Bau, *J. Am. Chem. Soc.* **95,** 5068 (1973).

170. Y. Degreve, J. Meunier-Piret, M. van Meersche, and P. Piret, *Acta Crystallogr.* **23,** 119 (1962).

171. M. R. Churchill, F. J. Hollander, and J. P. Hutchinson, *Inorg. Chem.* **16,** 2655 (1977).

172. K. Wade, *J. Chem. Soc., Chem. Commun.* p. 792 (1971); *Chem. Br.* p. 177 (1975); *Adv. Inorg. Chem. Radiochem.* **18,** 1 (1976).

173. C. H. Gast, J. C. Kraak, L. H. Staal, and K. Vrieze, *J. Organomet. Chem.* **208,** 225 (1981).

174. E. König and S. Herzog, *J. Inorg. Nucl. Chem.* **32,** 585, 601, 613 (1970).

175. H. tom Dieck and K.-D. Franz, *Angew. Chem.* **87,** 244 (1975); *Angew. Chem., Int. Ed. Engl.* **14,** 249 (1975).

176. P. Clopath and A. von Zelewsky, *Helv. Chim. Acta* **55,** 52 (1972).

177. K.-D. Franz, H. tom Dieck, U. Krynitz, and I. W. Renk, *J. Organomet. Chem.* **64,** 361 (1974).

178. K.-D. Franz, Ph.D. Thesis, University of Frankfurt (1975).

179. J. T. B. H. Jastrzebski, G. van Koten, and K. Vrieze, to be published.

180. K.-D. Franz, H. tom Dieck, K. A. Ostoja Starzewski, and F. Hohmann, *Tetrahedron* **31,** 1465 (1975).

181. E. S. Raper and P. H. Crackett, *Inorg. Chim. Acta* **50,** 159 (1981).

182. J. M. Klerks, D. J. Stufkens, G. van Koten, and K. Vrieze, *J. Organomet. Chem.* **181,** 271 (1979).

183. J. M. Klerks, J. T. B. H. Jastrzebski, G. van Koten, and K. Vrieze, *J. Organomet. Chem.* **224,** 107 (1982).

184. C. W. Bird, "Transition Metal Intermediates in Organic Synthesis." Academic Press, New York, 1967.

185. W. Hübel and C. Hoogzand, *Chem. Ber.* **93,** 103 (1960).

186. J. P. Collman and J. W. Kang, *J. Am. Chem. Soc.* **89,** 844 (1967).

187. S. Otsuka and A. Nakamura, *Adv. Organomet. Chem.* **14,** 245 (1976).

188. F. L. Bowden and A. B. P. Lever, *Organomet. Chem. Rev.* **3,** 227 (1968).

189. J. Browning, M. Green, J. L. Spencer, and F. G. A. Stone, *J. Chem. Soc., Dalton Trans.* p. 97 (1974).

ADVANCES IN ORGANOMETALLIC CHEMISTRY, VOL. 21

Multiply Bonded Germanium Species

JACQUES SATGÉ

Université Paul Sabatier
Toulouse, France

I

INTRODUCTION

In spite of the fact that intermediates with silicon doubly bonded to carbon, oxygen, sulfur, nitrogen, or silicon have been described and fully characterized for several years (*1–10*), no such germanium intermediate was described until 1973 (*2, 10, 11*). Intermediates with germanium doubly bonded to carbon, oxygen, sulfur, nitrogen, phosphorus, and germanium are, like their silaanalogs, very unstable (*2, 10, 11*). However, there are now many reactions in which such species have been "trapped" and characterized as well-established intermediates (*2, 10, 11*).

241

Until recently the second- and subsequent-row elements were considered unable to form $p\pi-p\pi$ bonds and the corresponding unsaturated molecules were classed as nonexistent compounds (12). The reasons given were the large size of the atoms reducing the sideways overlap of the p orbitals, the diffuse nature of (np) orbitals for $n > 2$, and their high energy (13–19).

Theoretical studies were carried out by Pitzer (20), who postulated that in second-row or heavier elements the repulsion between electrons of the bonding p orbital and the filled inner shell becomes important and is responsible for the relative absence of stable multiple bonds. Mulliken agreed in part with Pitzer's interpretation, especially for the heavier elements, on the basis of his overlap integral calculations (21). These theoretical studies were discussed by Gilman and Dunn (22), but other authors (23–27) proposed that the absence of stable $p\pi-p\pi$ bonding in the silicon, germanium, tin, and lead series was due to more effective use of the d orbitals in the polymer, making it more stable than the monomer.

Now, numerous experimental studies in silicon and germanium chemistry (1–11) have established the formation of these unsaturated species, but generally as very short-lived intermediates. However, very recently two stable species with $\mathrm{\ Si{=}C\ }$ (28) and $\mathrm{\ Si{=}Si\ }$ (29) units have been described.

This article is especially devoted to germanium doubly bonded species. The synthetic routes and the reactivity of each type of these intermediates will be presented in the following order:

$\mathrm{\ Ge{=}C\ }$ species of the germaethylene (or germene) type (e.g., $Et_2Ge{=}CH_2$, diethylgermaethylene);

$\mathrm{\ Ge{=}O\ }$ and $\mathrm{\ Ge{=}S\ }$ species of the germanone and germathione types, analogous to ketones and thiones (e.g., $Et_2Ge{=}O$, diethylgermanone);

$\mathrm{\ Ge{=}N{-}\ }$ species of the germaimine type analogous to imines (e.g., $Et_2Ge{=}NPh$, N-phenyldiethylgermaimine);

$\mathrm{\ Ge{=}P{-}\ }$ species of the germaphosphimine type analogous to phosphimines (e.g., $Et_2Ge{=}PPh$, diethylgermaphenylphosphimine);

$\mathrm{\ Ge{=}Metal\ }$ species including:

$\mathrm{\ Ge{=}Ge\ }$ species of digermaethylene (or digermene) type (e.g., $Ph_2Ge{=}GePh_2$, tetraphenyldigermaethylene);

$$\ce{\overset{\diagdown}{\underset{\diagup}{Ge}} = Bi -} \text{ species;}$$

$$\ce{\overset{\diagdown}{\underset{\diagup}{Ge}} = MnL}_n \text{ derivatives.}$$

Theoretical investigations and some spectroscopy results are also reported.

II

$\ce{\overset{\diagdown}{\underset{\diagup}{Ge}} = C \overset{\diagup}{\underset{\diagdown}{}}}$ SPECIES

In an attempt to explain its much greater acidity relative to triphenylgermane, $(p \cdot p)\pi$ bonding between germanium and carbon has been invoked in pentaphenylgermole (**1**) (pentaphenylgermole is at least 10^6 times more acidic than triphenylgermane) (*30*) [Eq. (1)].

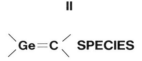

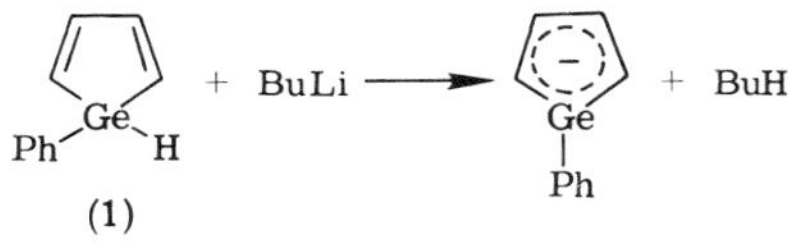

$$(1)$$

(1)

An attempt to generate a germanium–carbon $(p \cdot p)\pi$ double bond by pyrolysis of dimethylgermacyclobutane (**2**) proved unsuccessful (*31*). Both olefinic and carbenoid decomposition are observed [Eq. (2)].

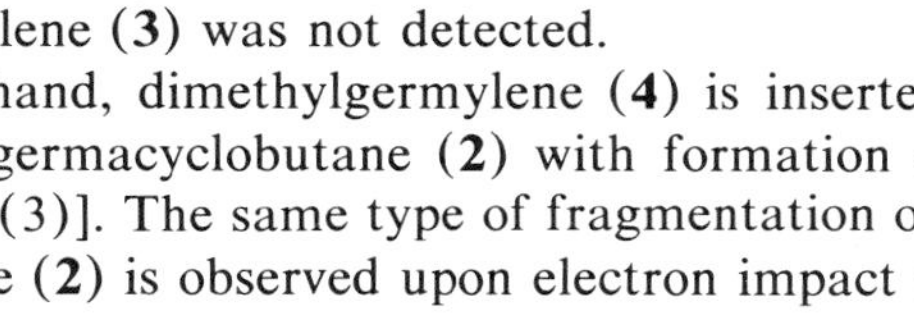

$$\ce{Me2Ge} \diagup\hspace{-2pt}\square\hspace{-2pt}\diagup \overset{?}{\nearrow} \begin{array}{l} \ce{(Me2Ge=CH2) + CH2=CH2} \\ \quad (3) \\ \ce{Me2Ge: + C3H6} \\ \quad (4) \end{array}$$

(2)

$$(2)$$

The elimination of ethylene is noted but the dimer (or polymer) of dimethylgermaethylene (**3**) was not detected.

On the other hand, dimethylgermylene (**4**) is inserted into the cyclic $\ce{Ge-C}$ bond of germacyclobutane (**2**) with formation of digermacyclopentane (**5**) [Eq. (3)]. The same type of fragmentation of 1,1-dimethyl-1-germacyclobutane (**2**) is observed upon electron impact (*31, 32*).

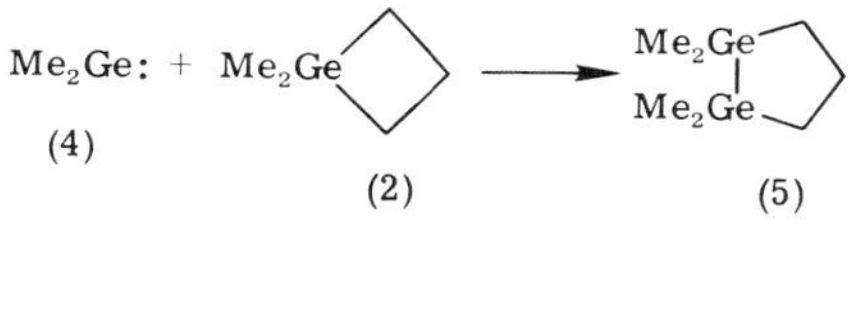

$$\ce{Me2Ge: + Me2Ge}\square \longrightarrow \ce{\underset{Me2Ge}{\overset{Me2Ge}{|}}}\pentagon$$

(3)

(4)   (2)   (5)

The first fully characterized intermediate with germanium doubly bonded to carbon was described by Barton (*33*) in the pyrolysis of the Diels–Alder adduct (**7**) of germacyclohexadiene (**6**) with perfluoro-2-butyne. 1,1-Diethyl-1-germaethylene (diethylgermene) (**8**) is characterized by formation of its dimer, 1,1,3,3-tetraethyl-1,3-digermacyclobutane (**9**), and by cycloaddition with dimethylbutadiene (Scheme 1).

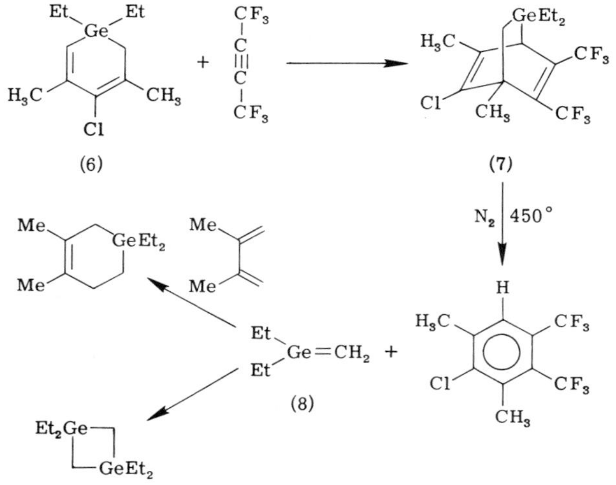

SCHEME 1

In a reaction of the same type, dimethylgermaethylene was characterized by addition to phenol (*34*) (Scheme 2).

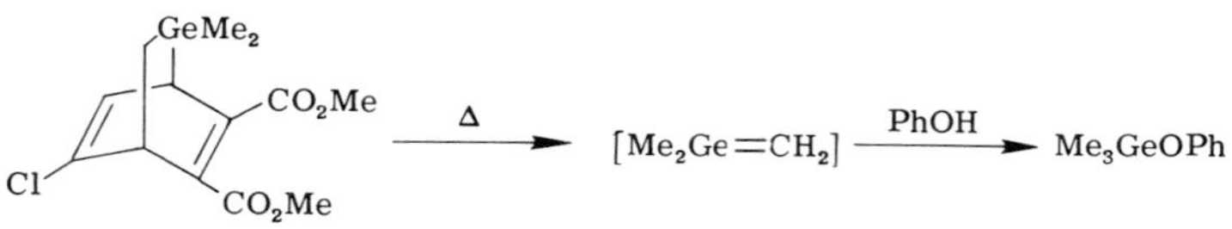

SCHEME 2

Barton and Hoekman (*8, 35, 36*) recently observed the formation of germene (**10**) from methyl migration to carbon on photolysis and pyrolysis of bis(trimethylgermyl)diazomethane (**11**). Pyrolysis of **11** produced **12** and **13**, each in 32% yield. Dimer **12** arises from head-to-head dimerization of germene (**10**) while germazene (**13**) probably is a result of reaction between **11** and **10**. Photochemical germene production is even more efficient, as irradiation of **11** produces dimer **12** in 89% yield and irradiation in the presence of MeOD affords adduct **14**, the product of methanol addition across the Ge=C bond (Scheme 3).

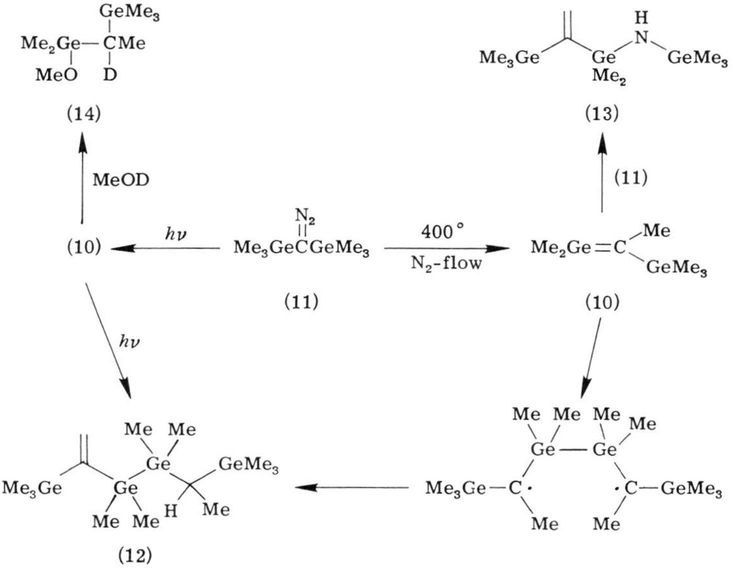

SCHEME 3

Thermolysis of trimethylsilyltrimethylgermyldiazomethane (**15**) produces only products from the silene intermediary, but photolysis and trapping with MeOD reveals a competition between silene (**16**) and germene (**17**) formation which favors the silene by a factor of four (*8, 35, 36*) (Scheme 4).

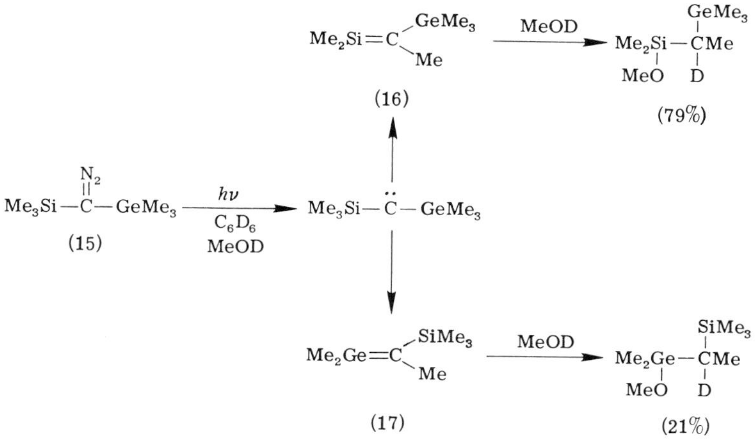

SCHEME 4

It is of interest to compare these results with those found by Jones *et al.*
(*37*) in the thermolysis of phenyl(trimethylgermyl)diazomethane (**18**).
Styrene can originate from germirane (**19**) and also from germene (**20**)
(Scheme 5).

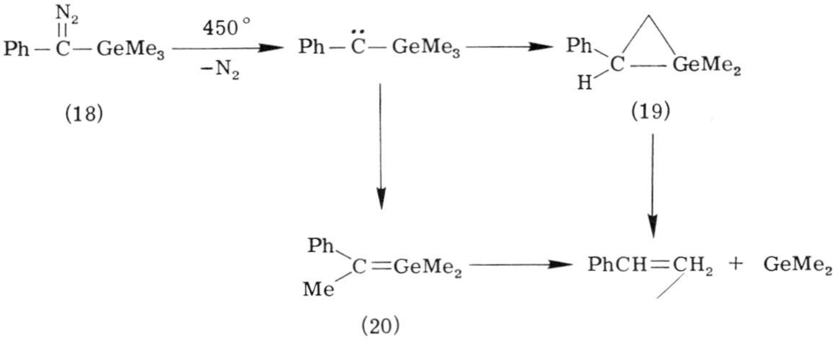

SCHEME 5

A new route to germaalkenes described recently is the reaction between
germylenes (*38*) and diazo derivatives such as diazophenylmethane and
ethyl diazoacetate (*39–41*). In this reaction type, the correlation of germylene reactivity with electrophilic character (*42*) permits postulation of
an initial nucleophilic attack of the germylene by the diazo compound,
which leads to an ylide type complex (**21**). Complex **21** proceeds with
evolution of nitrogen to germene (**22**) through a transient zwitterionic form.
Moreover, when a copper catalyst is used in these reactions germene (**22**)
can also be produced by direct interaction between germylene and the
generated carbene (Scheme 6).

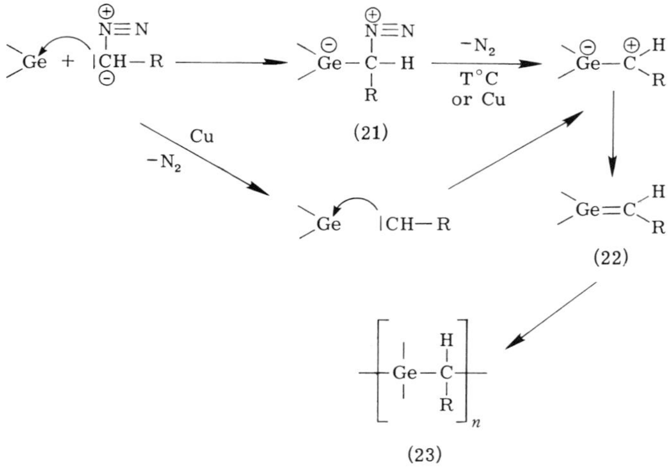

SCHEME 6

These intermediates are polycondensed quickly to lead to oligomers with germanium–carbon single bonds (**23**).

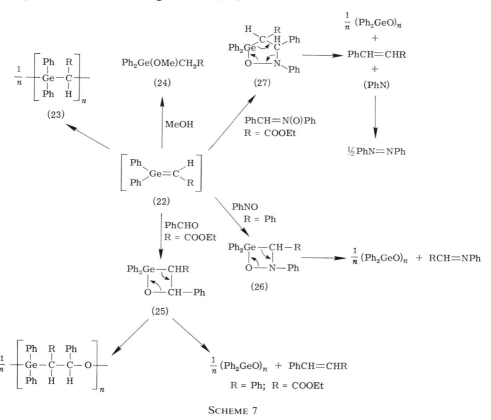

SCHEME 7

The intermediates (**22**) have been characterized by several trapping reactions (see Scheme 7). (1) Addition to alcohol (methanol) leads to the corresponding methoxygermanes (**24**). (2) Pseudo-Wittig reactions with benzaldehyde leading transitorily to germaoxetane (**25**) and finally to germanone (which is polycondensed to germoxane) and alkene. (3) Cycloaddition reaction with nitrosobenzene, formation of germaoxaazetidine (**26**) and finally, germoxane and imine. (4) A cycloaddition reaction is also seen with diphenylnitrone with transitory formation of germaoxaazolidine (**27**) decomposing to germanone, alkene, and phenylnitrene (characterized by the formation of azobenzene).

In the case of the reaction of halogermylenes (difluorogermylene and phenylchlorogermylene) with ethyl diazoacetate, the formed intermediate (**28**) develops in three different ways (*39*) (see Scheme 8).

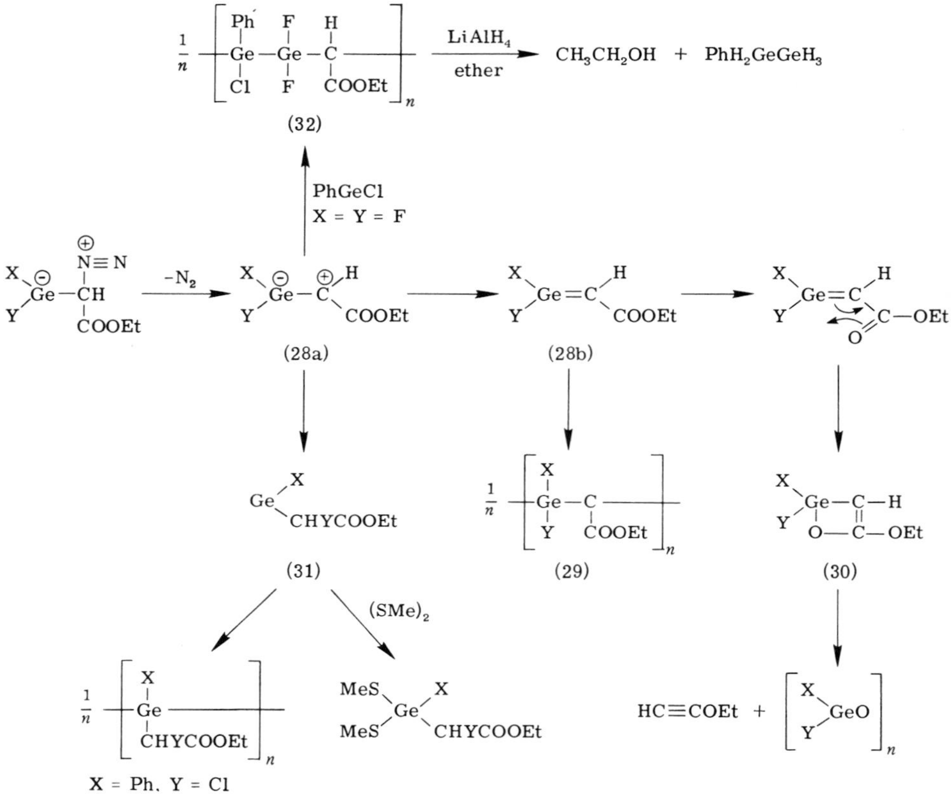

$$\frac{1}{n}\left[\begin{array}{c}\overset{Ph}{\underset{Cl}{Ge}}-\overset{F}{\underset{F}{Ge}}-\overset{H}{\underset{COOEt}{C}}\end{array}\right]_n \xrightarrow[\text{ether}]{LiAlH_4} CH_3CH_2OH + PhH_2GeGeH_3$$

(32)

PhGeCl
X = Y = F

$$\underset{(28a)}{\overset{X}{\underset{Y}{Ge}}-\overset{\oplus H}{\underset{COOEt}{C}}}$$

$$\underset{Y}{\overset{X}{Ge}}=\overset{H}{\underset{COOEt}{C}} \quad (28b)$$

$$\underset{Y}{\overset{X}{Ge}}=\overset{H}{\underset{O}{C}}\diagdown C-OEt$$

$$\overset{X}{\underset{CHYCOOEt}{Ge}} \quad (31)$$

$$\frac{1}{n}\left[\overset{X}{\underset{Y}{Ge}}-\overset{}{\underset{COOEt}{C}}\right]_n \quad (29)$$

$$\overset{X}{\underset{Y}{Ge}}\overset{C-H}{\underset{O-C-OEt}{}} \quad (30)$$

$$(SMe)_2$$

$$\frac{1}{n}\left[\overset{X}{\underset{CHYCOOEt}{Ge}}\right]_n$$

$$\overset{MeS}{\underset{MeS}{Ge}}\overset{X}{\underset{CHYCOOEt}{}}$$

$$HC\equiv COEt + \left[\overset{X}{\underset{Y}{Ge}}O\right]_n$$

X = Ph, Y = Cl

SCHEME 8

(1) The formed germene partially polycondenses to oligomers (**29**).

(2) A germanotropic rearrangement (*43*) is also seen in this case due to the electrophilic character of the germanium center. The germa-2-oxa-3-cyclobutene (**30**) thus formed quickly decomposes through a β-elimination process with formation of germoxane and ethoxyacetylene.

(3) Halogen migration from the metal to the carbonium ion in the α-position is also observed; it leads to a new functional germylene (**31**) which has been characterized in the polycondensed form and also by reaction with dimethyl disulfide (*42*).

Condensation reactions of these intermediates with an excess of germylenes is also observed by formation of polymers (**32**) with (—Ge—Ge—C—) units (Scheme 8).

The characterization of these halogenated germaalkenes with previously used trapping reagents (alcohol, nitrosobenzene, or nitrones) is impossible due to the high reactivity of the starting halogermylenes (*44*) and the germanium–halogen bonds (*43, 45*) with these reagents.

Dehydrobromination of 1,4-di-*tert*-butyl-1-germacyclohexa-2,4-diene (**33**) by the action of bases [*t*-BuLi, LiN(*i*-Pr)$_2$] gives, after nucleophilic substitution reaction and elimination, a new 1,4-di-*tert*-butylgermabenzene (**34**) which dimerizes or can be intercepted by Diels–Alder reaction with conjugated dienes (*46*) (see Scheme 9).

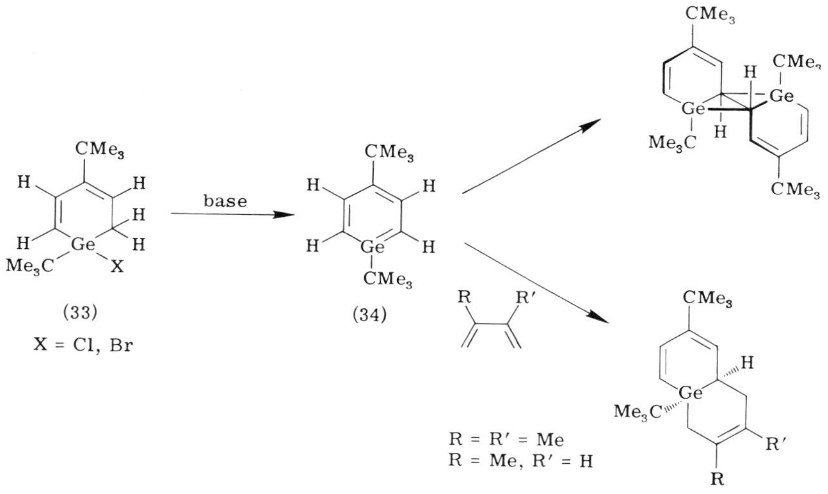

R = R′ = Me
R = Me, R′ = H

SCHEME 9

Thermolysis of germanium (and silicon) isologs of allyl aryl ethers (**35**) via transient metal doubly bonded intermediates (**36**) leads to aromatic Claisen-type rearrangements. The Claisen rearrangement leads to new oxametallacyclohexanes (**37**) by intramolecular trapping of the metal–carbon double bond (*34*) [Eq. (4)].

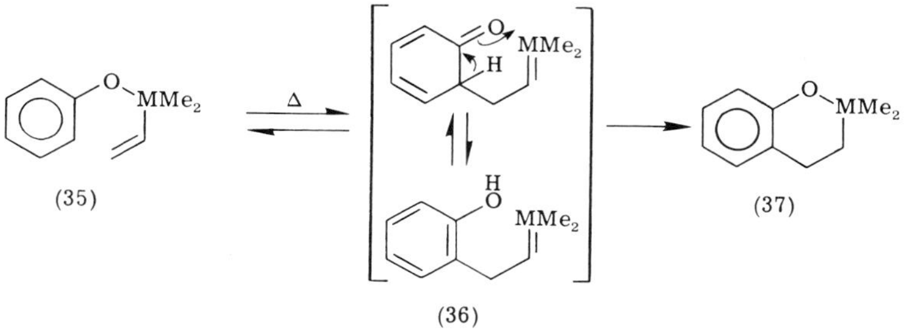

(4)

In the reaction of bis(trimethylsilyl)bromomethyllithium (**38**) with dimethyldihalo derivatives of silicon, germanium, and tin leading to formation of dimetallacyclobutanes (**39**), formation of the intermediate $(Me_3Si)_2C=MMe_2$ followed by dimerization has been postulated (*47, 48*) [Eq. (5)]. This route is nevertheless considered less likely than a process that proceeds entirely via organolithium derivatives.

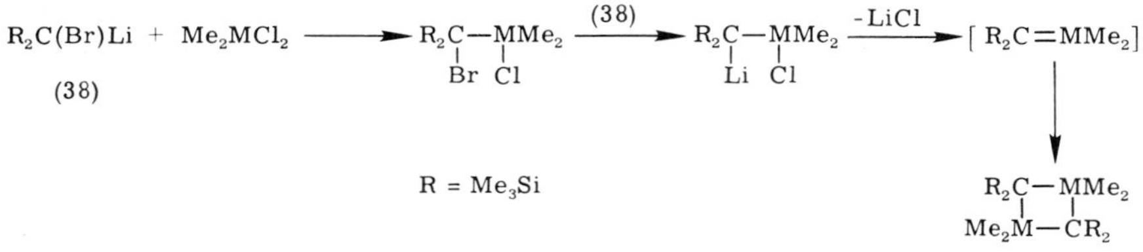

$$R_2C(Br)Li + Me_2MCl_2 \longrightarrow \underset{\underset{Br\ \ Cl}{|\ \ |}}{R_2C-MMe_2} \xrightarrow{(38)} \underset{\underset{Li\ \ Cl}{|\ \ |}}{R_2C-MMe_2} \xrightarrow{-LiCl} [\,R_2C=MMe_2\,] \quad (5)$$

(38)

$$R = Me_3Si$$

$$\underset{\underset{Me_2M-CR_2}{|\qquad\ \ |}}{R_2C-MMe_2}$$

(39)

Intermediate formation of diethylgermafulvene $\ \overset{\displaystyle\bigcirc}{}\!=GeEt_2\ $ is postulated in the reaction of diethylcyclopentadienylchlorogermane with triphenylmethylenephosphorane (*48a*).

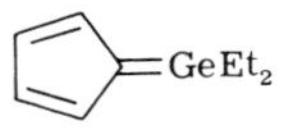

A comparative study of the pyrolysis of the tetramethyl derivatives of silicon, germanium, and tin using a wall-less reactor gives the following reaction sequence as a reasonable possibility (*49*) [Eq. (6)].

$$CH_3{}^{\cdot} + (CH_3)_4M \longrightarrow CH_4 + (CH_3)_3MCH_2{}^{\cdot}$$

$$(CH_3)_3MCH_2{}^{\cdot} \longrightarrow 3\ CH_3 + M=CH_2 \longrightarrow polymer$$

$$(6)$$

The formation of corresponding 1,3-dimetallacyclobutanes was not observed.

III

$\overset{\diagdown}{\underset{\diagup}{}}Ge=O$ AND $\overset{\diagdown}{\underset{\diagup}{}}Ge=S$ SPECIES

2-Germa-1-oxetanes (**40**) (*50–53*) and 2-germa-1,5-dioxanes (**41**) (*54*) as well as the adducts of 2-germa-1,3-dioxolanes, 2-germa-1,3-oxazolidines, and 2-germa-1,3-diazolidines with carbonyl compounds (aldehydes, ketones, and phenyl isocyanate) (**42**) (*55*) decompose thermally by β-elimination leading to transient dialkylgermanones which polymerize quickly to germoxanes $(R_2GeO)_n$ [Eqs. (7)–(9)].

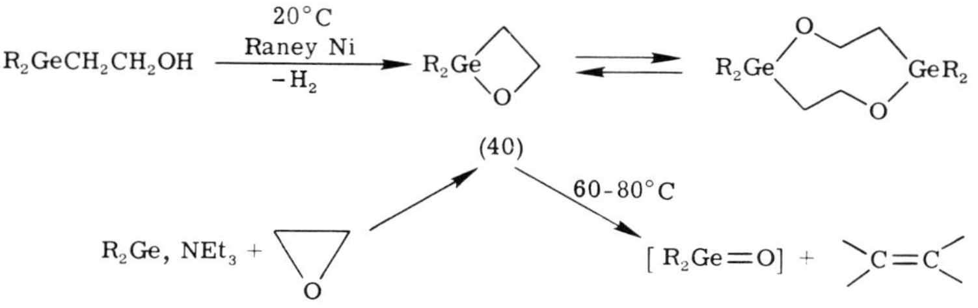

$$R_2GeCH_2CH_2OH \xrightarrow[\substack{-H_2}]{\substack{20°C \\ Raney\ Ni}} R_2Ge\!\!\bigtriangleup\!\!O \rightleftarrows R_2Ge\langle O \rangle GeR_2$$

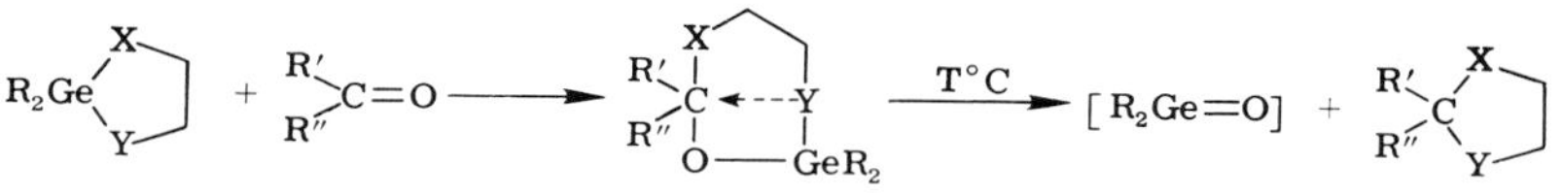

$$(40) \tag{7}$$

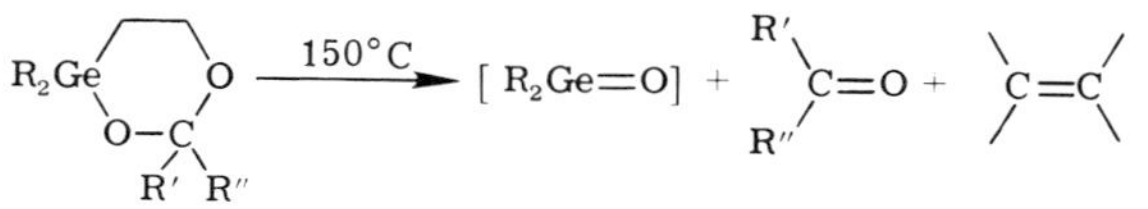

$$R_2Ge,\ NEt_3 + \!\!\bigtriangleup\!\!O \longrightarrow (40) \xrightarrow{60-80°C} [R_2Ge{=}O] + {>}C{=}C{<}$$

$$R_2Ge\langle O{-}C\langle\substack{R' \\ R''} \rangle \xrightarrow{150°C} [R_2Ge{=}O] + \substack{R' \\ R''}C{=}O + {>}C{=}C{<} \tag{8}$$

$$(41)$$

$$R_2Ge\langle\substack{X \\ Y} + \substack{R' \\ R''}C{=}O \longrightarrow \substack{R' \\ R''}C\!\!\langle\substack{X \\ O{-}GeR_2}\!\!{-}\!\!Y \xrightarrow{T°C} [R_2Ge{=}O] + \substack{R' \\ R''}C\langle\substack{X \\ Y} \tag{9}$$

$$(42)$$

The decomposition temperatures of the adducts in Eq. (9) depend on the nature of R′, R″, X, and Y (see Table I). The germanones thus generated were identified by various trapping reactions with methoxygermanes, oxagermacyclopentane, and ethylene oxide (55) [Eqs. (10)–(12)].

TABLE I

TEMPERATURE OF FORMATION OF GERMANONE BY
DECOMPOSITION OF ADDUCTS **42**[a]

R′	R″	X	Y	T (°C)
Me	H	O	O	160
Me	H	O	NMe	20
Me	H	NMe	NMe	20
Ph	H	O	O	160
Ph	H	O	NMe	20
Ph	H	NMe	NMe	20
Me	Me	O	O	160
Me	Me	O	NMe	20
Me	Me	NMe	NMe	20
$C_6H_5N{=}$		O	O	180
$C_6H_5N{=}$		NMe	NMe	140–150

[a] From Ref. 55.

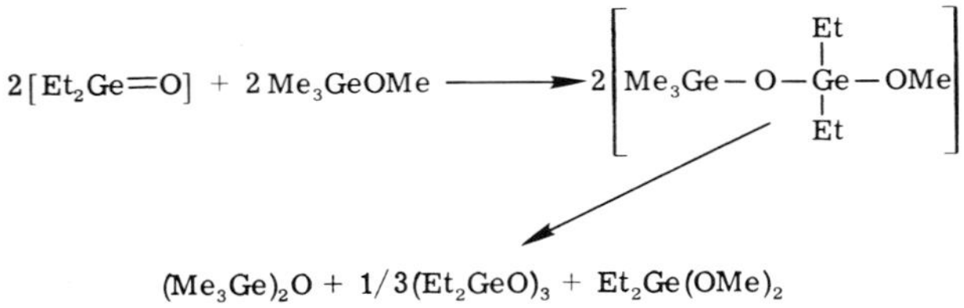

$$2[Et_2Ge{=}O] + 2\,Me_3GeOMe \longrightarrow 2 \left[Me_3Ge{-}O{-}\overset{\overset{\displaystyle Et}{|}}{\underset{\underset{\displaystyle Et}{|}}{Ge}}{-}OMe \right] \tag{10}$$

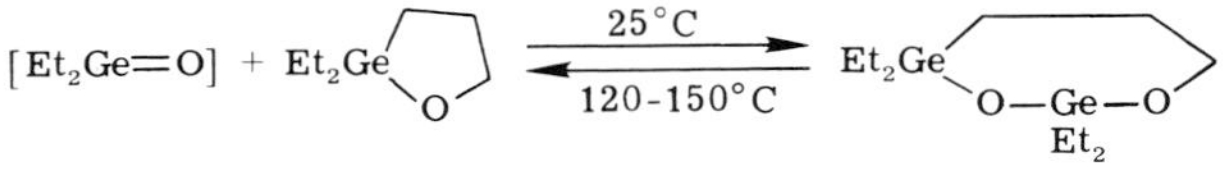

$$(Me_3Ge)_2O + 1/3\,(Et_2GeO)_3 + Et_2Ge(OMe)_2$$

$$[Et_2Ge{=}O] + Et_2Ge\!\!\!\diagup\!\!\!\!\diagdown\!\!\!\!\diagup_{O} \underset{120\text{-}150°C}{\overset{25°C}{\rightleftarrows}} Et_2Ge\!\!\!\diagup\!\!\!\!\diagdown\!\!\!\!\diagup_{O{-}\underset{Et_2}{Ge}{-}O} \tag{11}$$

$$[Et_2Ge{=}O] + \triangledown_{O} \longrightarrow Et_2Ge\!\!\diagup^{O}_{O} \tag{12}$$

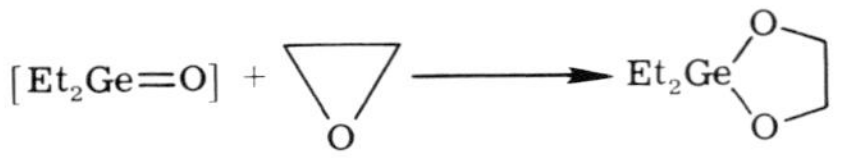

The action of benzaldehyde on cyclodigermazanes leads, after an insertion reaction to imines and germoxanes via the formation of transitory germanones (56) [Eq. (13)].

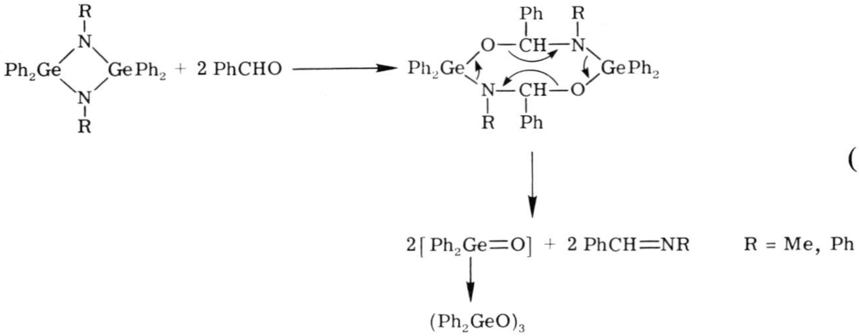

$$Ph_2Ge\overset{\overset{\displaystyle R}{\underset{\displaystyle |}{N}}}{\underset{\underset{\displaystyle R}{\underset{\displaystyle |}{N}}}{}}GePh_2 + 2\,PhCHO \longrightarrow Ph_2Ge\overset{O{-}\overset{Ph}{CH}{-}\overset{R}{N}}{\underset{N{-}\underset{R\quad Ph}{CH}{-}O}{}}GePh_2 \tag{13}$$

$$2[Ph_2Ge{=}O] + 2\,PhCH{=}NR \qquad R = Me,\ Ph$$

$$(Ph_2GeO)_3$$

The action of CS_2 on the same cyclodigermazanes leads to germathiones (56) [Eq. (14)].

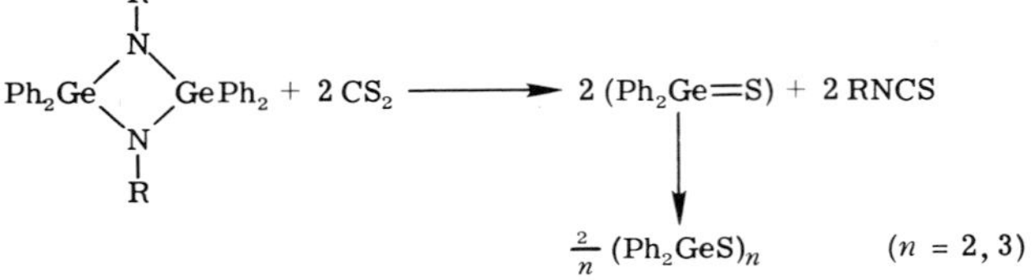

$$Ph_2Ge\overset{\overset{\displaystyle R}{\underset{\displaystyle |}{N}}}{\underset{\underset{\displaystyle R}{\underset{\displaystyle |}{N}}}{}}GePh_2 + 2\,CS_2 \longrightarrow 2\,(Ph_2Ge{=}S) + 2\,RNCS \tag{14}$$

$$\frac{2}{n}\,(Ph_2GeS)_n \qquad (n = 2, 3)$$

The bis(dialkylmethoxy)germanium oxides (**43**) and bis(dialkylmethylthio)germanium sulfides (**44**) are also precursors of germanones and germathiones, respectively, under mild conditions (*55, 57*) [Eqs. (15) and (16)].

$$[R_2GeCl]_2O \xrightarrow[-2Me_3GeCl]{+2Me_3GeOMe} \underset{(43)}{[R_2Ge(OMe)]_2O} \rightarrow R_2Ge(OMe)_2 + [R_2Ge{=}O] \qquad (15)$$

$$\underset{(44)}{[R_2Ge(SMe)]_2S} \rightarrow R_2Ge(SMe)_2 + [R_2Ge{=}S] \qquad (16)$$

Transient diethylgermathione is also obtained by thermal β-elimination reactions from adducts of germaoxazolidines and germadiazolidines with CS_2 or PhNCS (**45**) (*57*) (see Scheme 10).

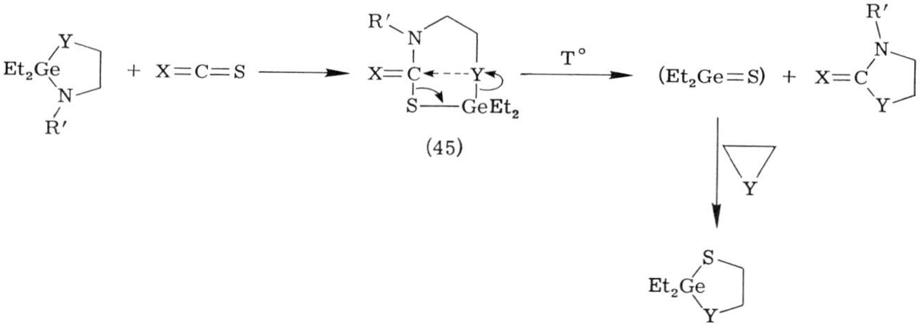

X = PhN Y = NMe R′ = Me T = 150 °C
X = S Y = NMe R′ = Me T = 150 °C
X = S Y = O R′ = H T = 0 °C

Y = O, S

SCHEME 10

Diethylgermathione undergoes insertion and ring expansion reactions with ethylene oxide or thiirane with formation of germaoxathiolane or germadithiolane (see Scheme 10) and adds to the germanium–sulfur bond of Et_3GeSMe (*57, 58*) [Eq. (17)].

$$2[Et_2Ge{=}S] + 2Et_3GeSMe \xrightarrow[2\ h]{150°C} 2\left[\begin{array}{c} Et \\ | \\ Et_3GeSGeSMe \\ | \\ Et \end{array}\right] \rightarrow$$

$$[(Et_2GeSMe)_2S] + (Et_3Ge)_2S$$
$$\rightleftharpoons$$
$$Et_2Ge(SMe)_2 + (Et_2Ge{=}S)$$
$$\searrow (Et_2GeS)_3 \qquad (17)$$

The formation of diethylgermathione is also observed in decomposition by β-elimination of germathiacyclobutane (**46**). The germathiacyclobutane arises from the dehydrocondensation reaction of hydrogermylthiol (**47**) (*58*), the condensation of germylene (**48**) with thiirane (*52*) or from the desulfuration of germadithiolane (**49**) by tributylphosphine (*57, 59*). The transient diethylgermathione is characterized by trimerization, by ring expansion with the starting germathiacyclobutane, and formation of digermadithiane (**50**) (*58*) and also by a condensation reaction with excess germylene leading to formation of a germanium sulfide with a Ge—Ge bond (**51**) (*60*) (Scheme 11).

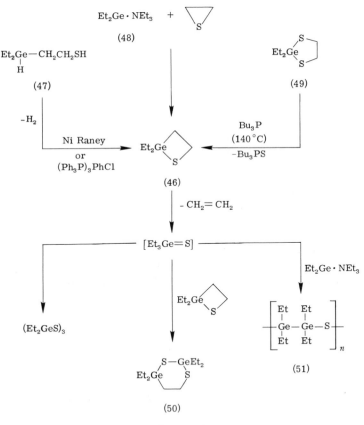

SCHEME 11

Another synthetic route to intermediates of the germanone type $\overset{R}{\underset{R'}{>}}Ge{=}O$ (R,R' = halogen, alkyl, aryl) is the direct oxidation of germylenes (via a bipolar intermediate) by various oxidizing agents (O_2, $KMnO_4$, DMSO, C_5H_5NO, $p\text{-}NO_2C_6H_4CO_3H$, etc.) (61) [Eq. (18)].

$$\overset{R}{\underset{R'}{>}}Ge \;+\; O{=}S(CH_3)_2 \longrightarrow \left[\overset{R}{\underset{R'}{>}}\overset{\ominus}{Ge}{-}O{-}\overset{\oplus}{S}(CH_3)_2\right] \longrightarrow \left[\overset{R}{\underset{R'}{>}}Ge{=}O\right] \;+\; (CH_3)_2S \quad (18)$$

RR'Ge=O intermediates, quickly polymerizable into germoxanes $[-RR'GeO-]_n$, have been characterized by various trapping reactions (ethylene oxide, methoxygermane, R_3GeOMe) and, like $>Ge{=}S$ intermediates, by condensation reactions with excess germylene leading to the formation of germoxanes with Ge—Ge bonds, $[-\overset{|}{\underset{|}{Ge}}-\overset{|}{\underset{|}{Ge}}-O-]_n$ (61).

The mechanism of the latter condensation reaction seems to correspond to nucleophilic attack of heteroelement Y on the germylene $(60, 61)$ in the sp^2 hybridized singlet state (62) [Eq. (19)].

$$>Ge{=}Y \;+\; >Ge \longrightarrow \;>\overset{\oplus}{Ge}{-}Y{-}\overset{\ominus}{Ge}< \longrightarrow \left[\overset{|}{\underset{|}{Ge}}-\overset{|}{\underset{|}{Ge}}-Y\right]_n \quad (19)$$

$$Y = O, S$$

Sulfuration (S_8) of germylenes leads to germathiones which quickly polymerize to cyclogermathianes (60). The reactivity of germylenes toward organic oxidizing agents is shown in the reactions of these bivalent species with nitrosobenzene, nitrones (and isomeric oxaziridines) $(44, 63)$.

The reactions of germylenes with nitrosobenzene lead to nitrene and germanone intermediates via the zwitterionic form of the corresponding germaoxaaziridines (**52**). Interactions between the germylenes and the nitrene generated in the reaction give new $>Ge{=}N-$ intermediates $(41, 44, 63)$ (see Scheme 12).

The formation of germaoxaazetidine (**53**) is observed in the insertion reaction of germylene into the oxazirane ring (**54**) in the 1,3-cycloaddition of germylene to the nitrone (**55**) which is an isomer of the oxazirane and

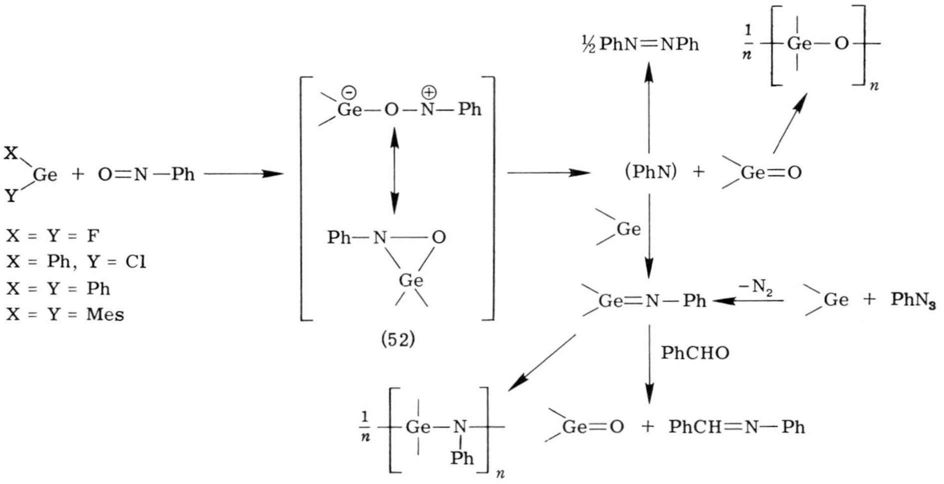

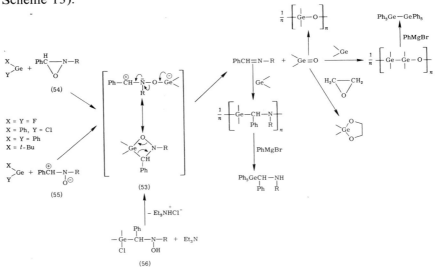

SCHEME 12

also in the dehydrochlorination reaction of C-germylated hydroxylamines

$\overset{|}{\text{Cl}}$GeCH(Ph)NOH(t-Bu) **(56)**. The β-elimination process from germa-

oxaazetidines leads to imine and $\diagup$Ge=O intermediates (44) (see

Scheme 13).

SCHEME 13

Reduction reactions of germanones and germathiones by trialkylphosphines have also been observed, leading to germylenes (*59*) [Eqs. (20) and (21)].

$$R_2Ge{=}S + Bu_3P \xrightarrow[3\,h]{170°C} Bu_3PS + \frac{1}{n}(R_2Ge)_n \qquad (20)$$

$$\updownarrow t°C$$

$$(R_2GeS)_3$$

$$Et_2Ge{=}O + Bu_3P \xrightarrow[3\,h]{170°C} Bu_3PO + \frac{1}{n}(Et_2Ge)_n + (Et_2GeO)_3$$

$$\underbrace{\phantom{Bu_3PO + \frac{1}{n}(Et_2Ge)_n}}_{70\%} \qquad 30\%$$

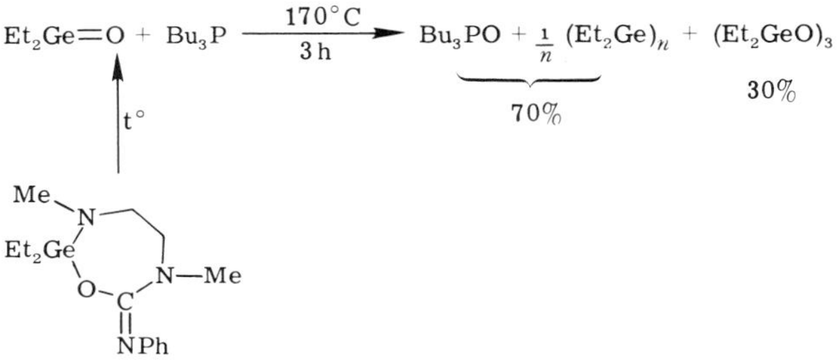

(21)

The thermal and catalytic depolymerization of cyclogermathianes with formation of a monomer/dimer to trimer equilibrium is induced by a temperature rise and by basic solvents such as triethylamine and HMPT (*57*) [Eq. (22)].

$$(Et_2GeS)_3 \underset{Et_3N,\ HMPT}{\overset{T°C}{\rightleftharpoons}} (Et_2GeS)_2 + Et_2Ge{=}S \qquad (22)$$

Monomer stabilization seems to exist with some transition metal complexes that strongly catalyze the reactions of germathione with small three- and four-membered rings (*64*) [Eq. (23)].

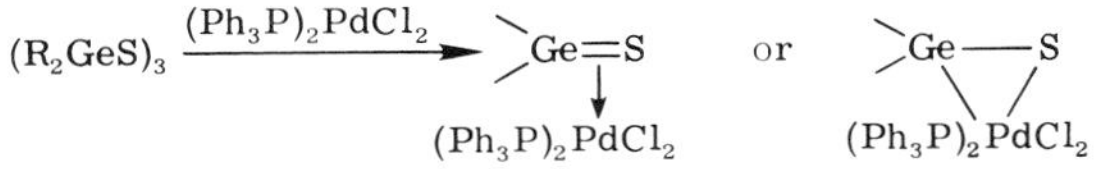

$$(R_2GeS)_3 \xrightarrow{(Ph_3P)_2PdCl_2} \underset{(Ph_3P)_2PdCl_2}{\overset{}{{>}Ge{=}S}} \quad or \quad \underset{(Ph_3P)_2PdCl_2}{{>}Ge{-}S} \qquad (23)$$

These germanones and germathiones with strongly polarized double bonds (*65*), are highly reactive and may therefore be used in organometallic synthesis in various types of reactions such as: (1) addition reactions to various σ bonds of acyclic and cyclic organometallic compounds (e.g., Ge—O and Ge—S, see above); (2) insertion reactions with ring expansion with small organic rings; and (3) cycloaddition reactions.

Certain results of insertion reactions with small organic rings are summarized (*64*) in Eqs. (24)–(28):

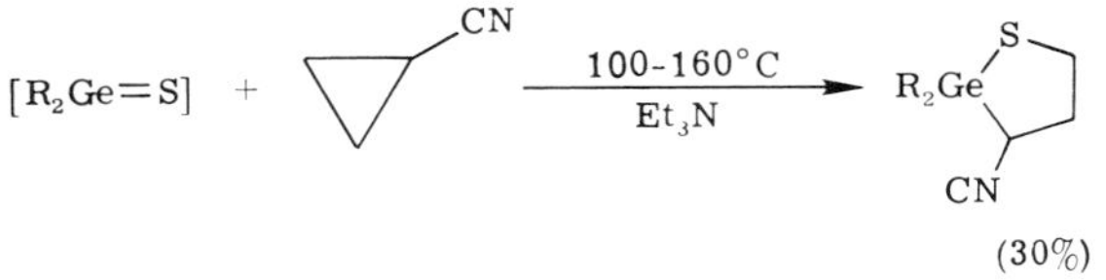

$$[R_2Ge{=}S] + \overset{CN}{\triangle} \xrightarrow[Et_3N]{100{-}160°C} R_2Ge\overset{S}{\underset{CN}{\diagdown}} \qquad (24)$$

$$(30\%)$$

$$[R_2Ge{=}Y] + \underset{O}{\triangle} \xrightarrow[Et_3N]{T^\circ} R_2Ge\underset{Y}{\overset{O}{\diagdown}} \quad Y = O, S, NR, PR \quad (25)$$

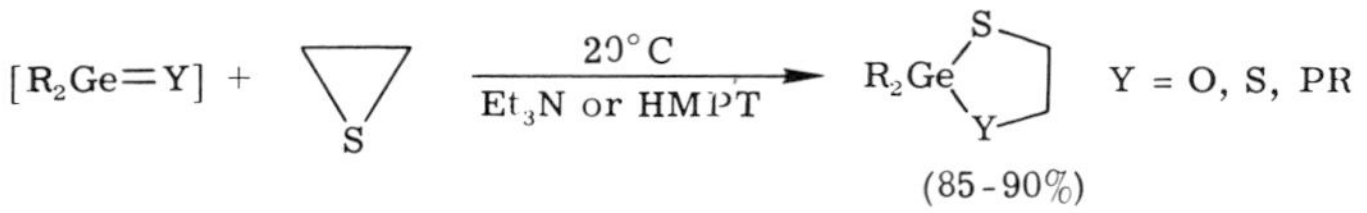

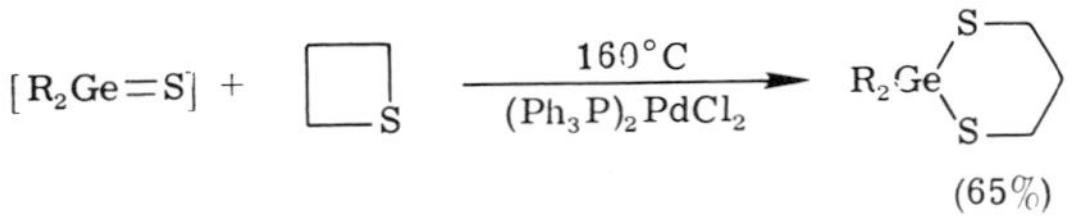

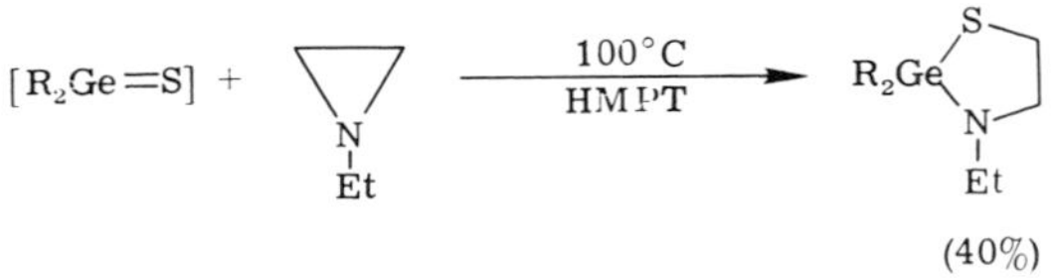

$$[R_2Ge{=}Y] + \underset{S}{\triangle} \xrightarrow[Et_3N \text{ or } HMPT]{20^\circ C} R_2Ge\underset{Y}{\overset{S}{\diagdown}} \quad Y = O, S, PR \quad (26)$$

$$(85\text{-}90\%)$$

$$[R_2Ge{=}S] + \underset{S}{\square} \xrightarrow[(Ph_3P)_2PdCl_2]{160^\circ C} R_2Ge\overset{S}{\underset{S}{\diagdown}} \quad (27)$$

$$(65\%)$$

$$[R_2Ge{=}S] + \underset{\underset{Et}{N}}{\triangle} \xrightarrow[HMPT]{100^\circ C} R_2Ge\overset{S}{\underset{\underset{Et}{N}}{\diagdown}} \quad (28)$$

$$(40\%)$$

The mechanism and the stereochemistry of some of these additions to small rings have been studied (64). The condensation reaction of germanone with oxirane begins by nucleophilic attack of oxygen on germanium followed by ring opening and cyclization. This nonconcerted mechanism is supported by the results of condensation of the germanone with cis or trans isomers of butene oxide. Each reaction leads to a mixture of cis and trans adducts: 45/55 from the *cis*-butene oxide and 52/48 from the *trans*-butene oxide (see Table II). However in the presence of triethylamine the percentage of the cis adduct increases, and with 300% excess of triethylamine almost exclusive formation of the cis adduct is observed from both *cis*- and *trans*-butene oxide, along with a maximum of 2% of the trans adduct (see table II) (64).

The high stereoselectivity in the presence of triethylamine seems to imply a germanium atom hexacoordinated by two molecules of triethylamine. The study of molecular models shows very strong steric hindrance between the methyl group and the triethylamino group in the equatorial position of the bipolar intermediate during the cyclization process (**57b**) (Scheme

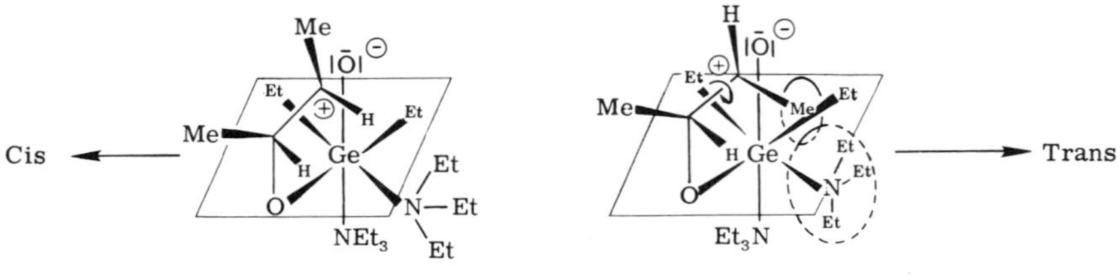

Cis ← (57a)　　SCHEME 14　　(57b) → Trans

TABLE II

INSERTION REACTIONS OF DIETHYLGERMANONE
INTO *CIS*- AND *TRANS*-BUTENE OXIDE[a]

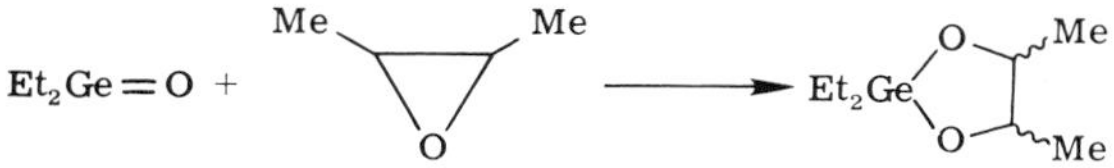

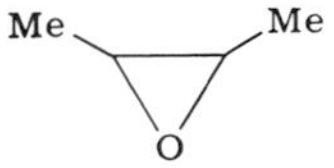

$$Et_2Ge = O \;+\; \underset{O}{\overset{Me \triangle Me}{}} \longrightarrow Et_2Ge \overset{O\text{—}Me}{\underset{O\text{—}Me}{}}$$

		Me⟍△⟋Me over O	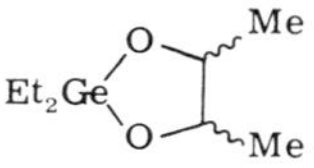 Et₂Ge with O–Me / O–Me cis/trans
		cis	45/55
		trans	52/48
		trans (100% excess Et₃N)	85/15
$Et_2Ge{=}O$		trans (300% excess Et₃N)	~100/0
		cis (100% excess Et₃N)	80/20
		cis (300% excess Et₃N)	~100/0

[a] Ref. *64*.

14). The position of least hindrance occurs when the two methyls are in the cis position opposite **(57a)** the triethylamino group position leading to the cis isomer.

On the other hand the condensation of germathione with *cis*- or *trans*-butene oxide gives a stereospecific reaction with inversion of configuration. *Cis*- and *trans*-butene oxides lead to trans and cis adducts, respectively, in proportions higher than 95% with or without triethylamine (*64*) (Table III). The mechanism of this reaction seems to proceed by nucleophilic attack of sulfur on the epoxide with configuration inversion (*64*) (Scheme 15).

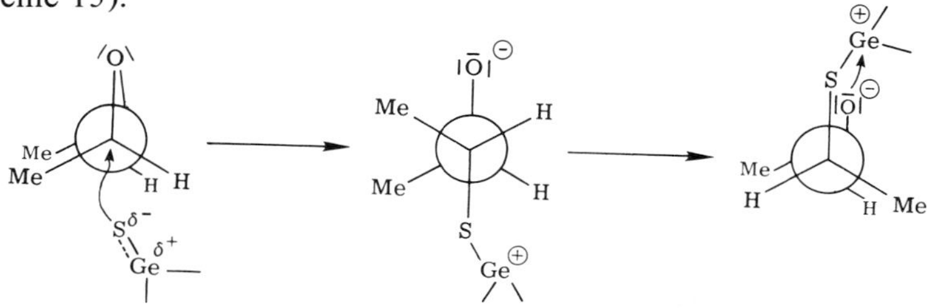

SCHEME 15

TABLE III

INSERTION REACTIONS OF DIMETHYLGERMATHIONE INTO *CIS*- AND
TRANS-BUTENE OXIDE[a]

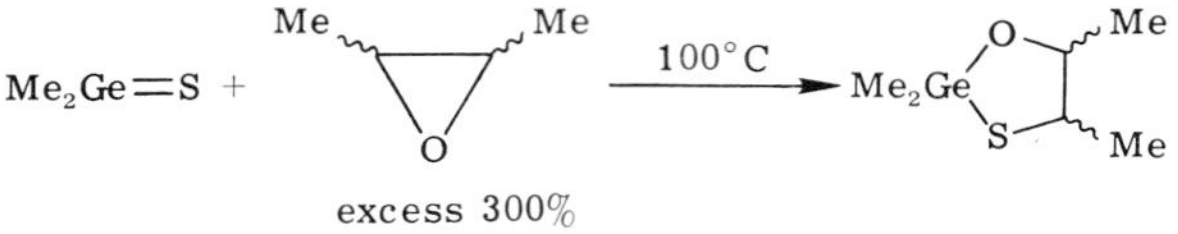

$$\text{Me}_2\text{Ge}{=}\text{S} \; + \; \underset{\text{excess } 300\%}{\text{(Me, Me butene oxide)}} \; \xrightarrow{100°\text{C}} \; \text{Me}_2\text{Ge}\begin{smallmatrix}\text{O} \\ \text{S}\end{smallmatrix}\begin{smallmatrix}\text{Me} \\ \text{Me}\end{smallmatrix}$$

	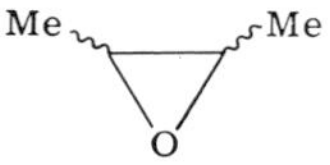 Me ⟍△⟋ Me O (butene oxide)	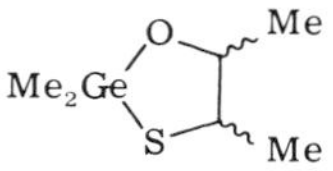 Me₂Ge⟨O,S⟩⟨Me,Me⟩	Yield (%)
With	cis	cis/trans ~0/100	85
Et₃N	trans	cis/trans 95/5	95
Without	cis	cis/trans ~0/100	40–50
Et₃N	trans	cis/trans 95/5	15

[a] Ref. *64*.

Cycloaddition Reactions

New cycloaddition reactions of germanone and germathione with 1,3-dipolar reagents (nitrilimines, nitrones, and nitrile oxides) have been reported (*53, 66*). The condensation reactions of doubly bonded germanium species precursors with diphenyl-2,5-tetrazole (precursor of nitrilimine) (**58**) at 160°C for 2 h lead to regioselective cycloaddition with formation of germaoxa- (**59a**) or germathiadiazolines (**59b**) [Eq. (29)].

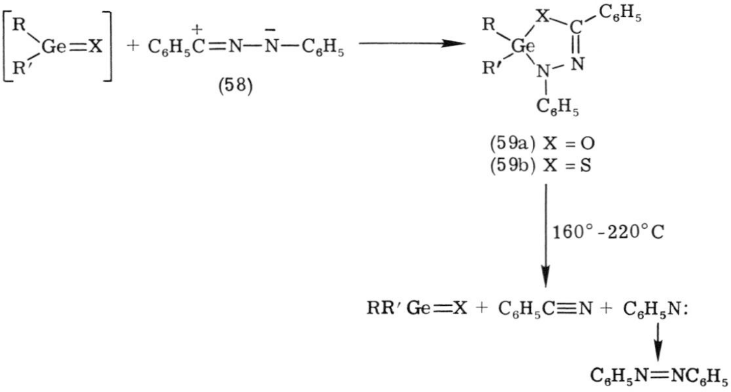

$$\left[\begin{smallmatrix} R \\ R' \end{smallmatrix}\!\text{Ge}{=}\text{X} \right] + \underset{(58)}{\text{C}_6\text{H}_5\overset{+}{\text{C}}{=}\text{N}{-}\overset{-}{\text{N}}{-}\text{C}_6\text{H}_5} \longrightarrow \begin{smallmatrix} R \\ R' \end{smallmatrix}\!\text{Ge}\begin{smallmatrix} \text{X}{\diagdown}\text{C}{-}\text{C}_6\text{H}_5 \\ \| \\ \text{N}{-}\text{N} \\ | \\ \text{C}_6\text{H}_5 \end{smallmatrix}$$

$$\begin{aligned}&(59\text{a}) \; \text{X} = \text{O}\\&(59\text{b}) \; \text{X} = \text{S}\end{aligned} \qquad\qquad (29)$$

$$\Big\downarrow 160°\text{-}220°\text{C}$$

$$\text{RR}'\,\text{Ge}{=}\text{X} + \text{C}_6\text{H}_5\text{C}{\equiv}\text{N} + \text{C}_6\text{H}_5\text{N}{:}$$

$$\Big\downarrow$$

$$\text{C}_6\text{H}_5\text{N}{=}\text{NC}_6\text{H}_5$$

Germaoxa- or germathiadiazolines are stable and can be characterized by NMR spectroscopy (^{1}H and ^{13}C) and also chemically; thermal decomposition of the type 5 = 2 + 2 + 1 can be seen at 160–220°C [Eq. (29)].

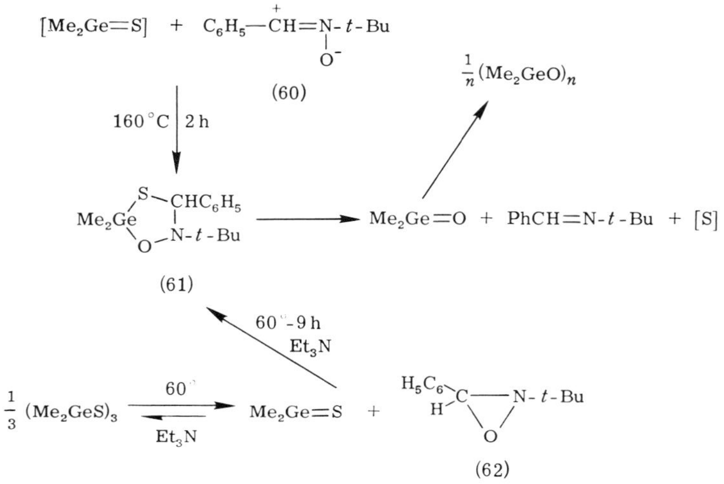

$$[Me_2Ge=S] \quad + \quad C_6H_5-\overset{+}{C}H=N\text{-}t\text{-}Bu$$
$$\underset{O^-}{\overset{|}{|}}$$

$$(60)$$

$$\frac{1}{n}(Me_2GeO)_n$$

$$160\,^\circ C \Big| 2\,h$$

$$Me_2Ge\overset{S-CHC_6H_5}{\underset{O-N\text{-}t\text{-}Bu}{\diagdown}} \quad \longrightarrow \quad Me_2Ge=O \;+\; PhCH=N\text{-}t\text{-}Bu \;+\; [S]$$

$$(61)$$

$$60\,^\circ\text{-}9\,h$$
$$Et_3N$$

$$\frac{1}{3}\,(Me_2GeS)_3 \quad \underset{Et_3N}{\overset{60\,^\circ}{\rightleftharpoons}} \quad Me_2Ge=S \quad + \quad \underset{H}{\overset{H_5C_6}{\diagdown}}\overset{}{C}\underset{O}{\overset{\diagup}{\diagdown}}N\text{-}t\text{-}Bu$$

$$(62)$$

SCHEME 16

The nitrones also give cycloaddition reactions with germathiones. Germathiones (generated from the adducts of germadiazolidines with PhNCS) react with diphenylnitrone and phenyl-*N*-*t*-butylnitrone (**60**) at 160°C in a sealed tube. The germaoxathiazolidines (**61**) formed have been characterized by ^{1}H NMR spectroscopy and by identification of their thermolysis products (*53, 66*) (see Scheme 16).

Dimethylgermathione [from hexamethylcyclotrigermathiane $(Me_2GeS)_3$ in the presence of triethylamine] reacts with 2-*t*-butyl-3-phenyloxazirane (**62**) (isomer of phenyl-*N*-*t*-butylnitrone) at 60°C by an insertion reaction in the oxazirane ring and formation of unstable germaoxathiazolidine, characterized after quantitative isolation of the decomposition products (decomposition [5 → 2 + 2 + 1]) (see Scheme 16).

Nitrile oxides (**63**) also give cycloadducts with germathiones (*53, 66*) [Eq. (30)].

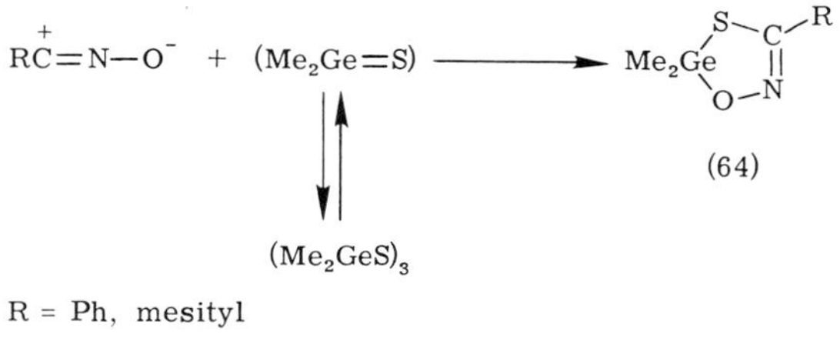

$$R\overset{+}{C}=N-O^- \;+\; (Me_2Ge=S) \quad \longrightarrow \quad Me_2Ge\overset{S-C\diagup R}{\underset{O-N}{\diagdown\,\|}}$$

$$\Big\updownarrow$$

$$(Me_2GeS)_3$$

$$(64) \hspace{3cm} (30)$$

R = Ph, mesityl

The 2-germa-1,3,5-oxathiazoles (**64**) formed with good yields (R = Ph, 80%; R = mesityl, 70%), are quite stable. They decompose at 120–160°C with formation of the corresponding germanone and phenyl isothiocyanate

(*66*) [Eq. (31)]. The regioselectivity of all these 1,3-cycloaddition reactions is analogous to that observed with ketones, thioketones, or CS_2 (*67*).

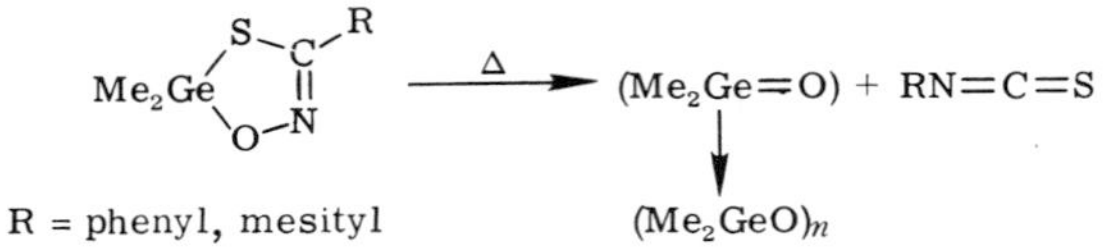
Me₂Ge
S
C
R
O
N
Δ
(Me₂Ge=O) + RN=C=S
(Me₂GeO)ₙ
R = phenyl, mesityl

(31)

IV

>Ge=N— SPECIES

The formation of new >Ge=N— intermediates and especially germaimines $R_2Ge{=}NR'$ has been observed in the reaction of germylenes with phenyl or methyl azide (*41*, *68*). The reactivity of azide increases with the electrophilic character of the germylenes. This fact is consistent with a nucleophilic attack of azide on germylene leading to bipolar intermediate (**65**) which decomposes generating nitrogen and forming the germaimine (**66**) (Scheme 17).

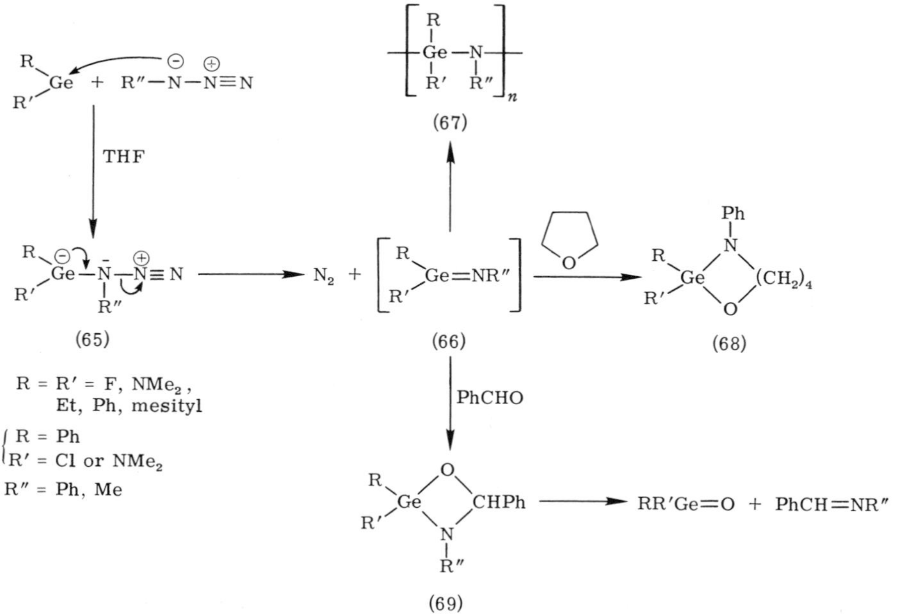
R
R′
Ge + R″—N—N≡N
THF
(65)
R = R′ = F, NMe₂, Et, Ph, mesityl
R = Ph
R′ = Cl or NMe₂
R″ = Ph, Me
N₂ +
Ge=NR″
(66)
(67)
(68)
PhCHO
CHPh
RR′Ge=O + PhCH=NR″
(69)

SCHEME 17

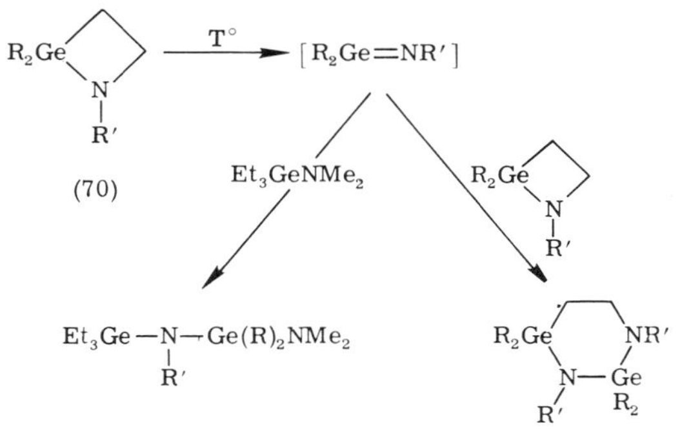

Scheme 18

Germaimines are characterized, in the form of polygermazanes (**67**), by insertion into a tetrahydrofuran ring and formation of a seven-membered heterocycle (**68**) as well as by pseudo-Wittig reaction with benzaldehyde and formation of unstable germaoxaazetidine (**69**) (Scheme 17). The direct interaction of a germylene with a nitrene also leads to germaimines (*41*, *44*, *68*) (see Section III, Scheme 12).

The rather unstable germaazetidines (**70**) lead, through a β-elimination process, to the corresponding germaimines which have mainly been characterized by insertion into the germanium–nitrogen bond of both the starting germaazetidines and diethyl(triethylgermyl)amine (*53*, *60*, *69*, *70*) (Scheme 18).

Like silaimines (*2*, *71*) the germaimines can be generated by photolysis of triorganogermanium azides (**71a**, **71b**). Germaimines have been quantitatively trapped by pinacol (*72*) (Scheme 19).

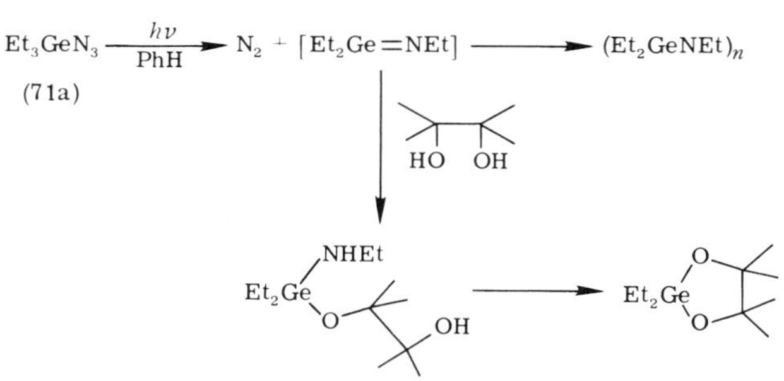

Scheme 19

Photolysis of triphenylgermanium azide (71b) is more complex and leads to four products [Eq. (32)].

$$Ph_3GeN_3 \xrightarrow[PhH]{h\nu} [Ph_2Ge{=}NPh] + Ph_2Ge + PhNH_2 + Ph{-}Ph$$

(71b)

$$(Ph_2GeNPh)_{2,n} \qquad (Ph_2Ge)_{4,n}$$

(32)

Two possible mechanisms might explain these experimental results [Eq. (33)].

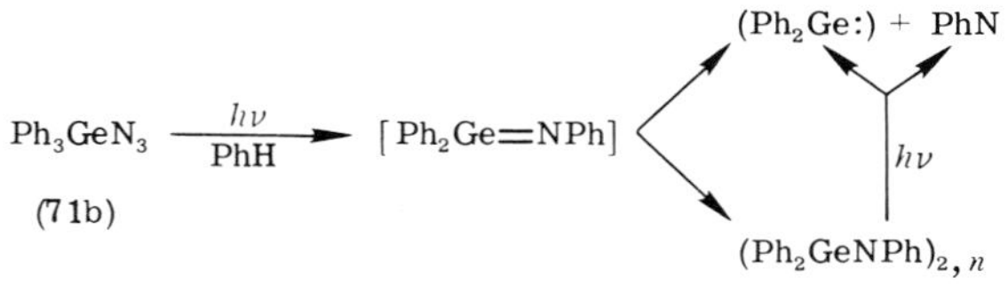

$$Ph_3GeN_3 \xrightarrow[PhH]{h\nu} [Ph_2Ge{=}NPh]$$

(71b)

(Ph$_2$Ge:) + PhN

$h\nu$

(Ph$_2$GeNPh)$_{2,n}$

(33)

One mechanism involves hydrogen abstraction from benzene by phenyl-nitrene giving aniline and biphenyl. The photolytic decomposition of germaimine (72) is the reverse of the reaction between germylenes and nitrenes (or organic azides) cited above (41, 44, 63, 68).

Another possible route to germaimines is the reaction of two moles of phenyl isocyanate with one mole of cyclodigermazane leading to a diadduct, which decomposes through successive eliminations to a germaimine stabilized by dipole–dipole interaction with phenyl isocyanate and thence to equilibrium with an unstable germaoxodiazetidine (73) (53, 56) (Scheme 20).

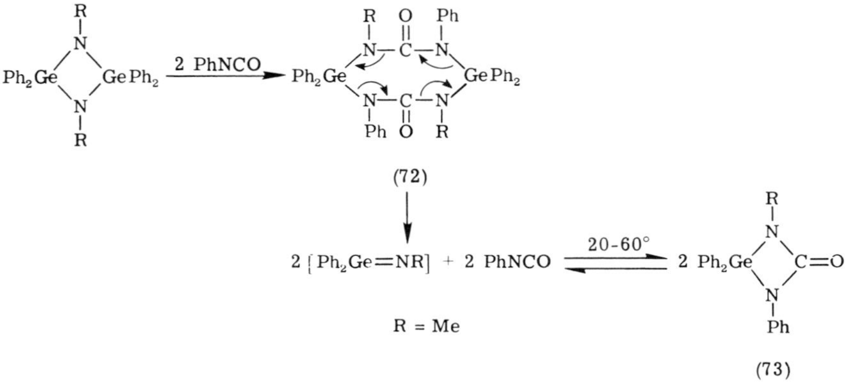

(72)

$$2\,[Ph_2Ge{=}NR] + 2\,PhNCO \underset{20\text{-}60°}{\rightleftharpoons} 2\,Ph_2Ge\diagdown C{=}O$$

R = Me

(73)

SCHEME 20

The thermal depolymerization of cyclogermazanes $(R_2GeNR')_{2,3}$ with formation of monomer–dimer equilibrium is induced, as in the cyclogermathiane series, by heat and by basic solvents such as triethylamine and HMPT (73). Probably nucleophilic assistance on germanium aids the opening of the germazane cycle and enhances its reactivity as a monomer species [Eq. (34)].

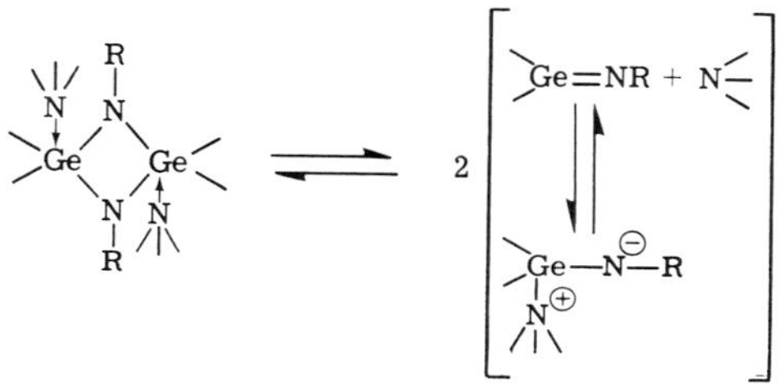

$$ \text{(34)} $$

Lewis acids such as $ZnCl_2$ in catalytic amounts also seem to induce depolymerization of cyclodigermazane (74) [Eq. (35)]:

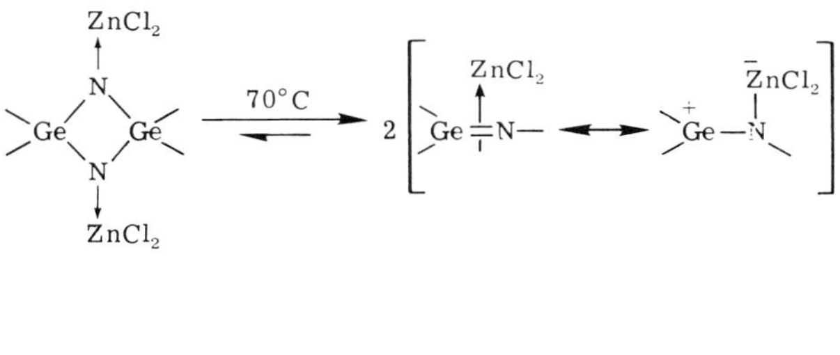

$$ \text{(35)} $$

Complexes of the same type have been observed with BF_3 and cyclodigermazanes (74).

It is possible to characterize the germaimine intermediates in the ammonolysis of diphenyldichlorogermane by a rapid trapping reaction (under conditions in which the cyclogermazane does not react) of the germaimine with CS_2 and formation of a germathiaazetidine (74) which decomposes to $Ph_2Ge=S$ and MeNCS. Subsequent reaction of MeNCS on germaimine is also observed with formation of germathiaazetidine (75) (Scheme 21). The presence of CS_2 prevents the formation of the cyclogermazane $(Ph_2GeNMe)_{2,3}$. In the absence of CS_2 the formation of the cyclogermazane is quantitative (73).

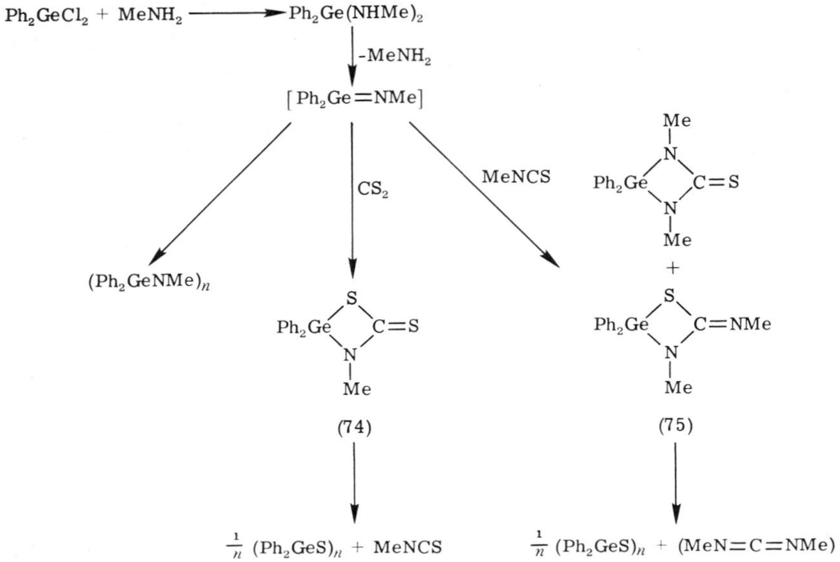

$Ph_2GeCl_2 + MeNH_2 \longrightarrow Ph_2Ge(NHMe)_2$

$\downarrow -MeNH_2$

$[Ph_2Ge{=}NMe]$

$(Ph_2GeNMe)_n$

CS_2

MeNCS

(74)

(75)

$\frac{1}{n}(Ph_2GeS)_n + MeNCS$

$\frac{1}{n}(Ph_2GeS)_n + (MeN{=}C{=}NMe)$

SCHEME 21

A. *1,2-Cycloadditions*

1,2-Cycloadditions of germaimines with phenyl or methyl isocyanates produce unstable germaoxodiazetidines (**76**) which lead to 2-germa-4,6-dioxo-1,3,5-perhydrotriazine (**77**) (*74*) (Scheme 22).

In the case of isothiocyanates the formation of adducts from C=N and C=S addition are observed. Addition across the C=N bond leads to 2-

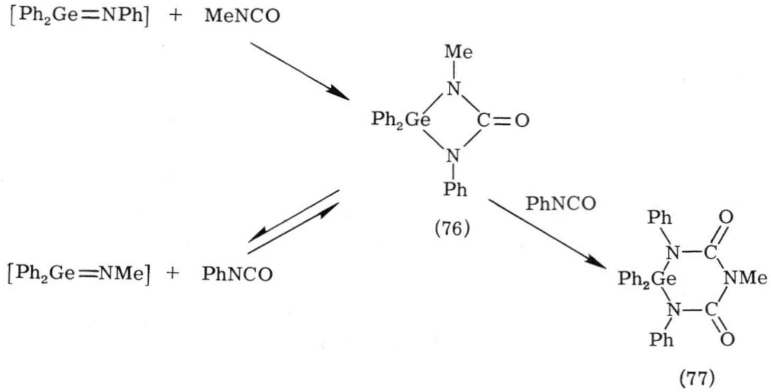

$[Ph_2Ge{=}NPh] + MeNCO$

$[Ph_2Ge{=}NMe] + PhNCO$

PhNCO

(76)

(77)

SCHEME 22

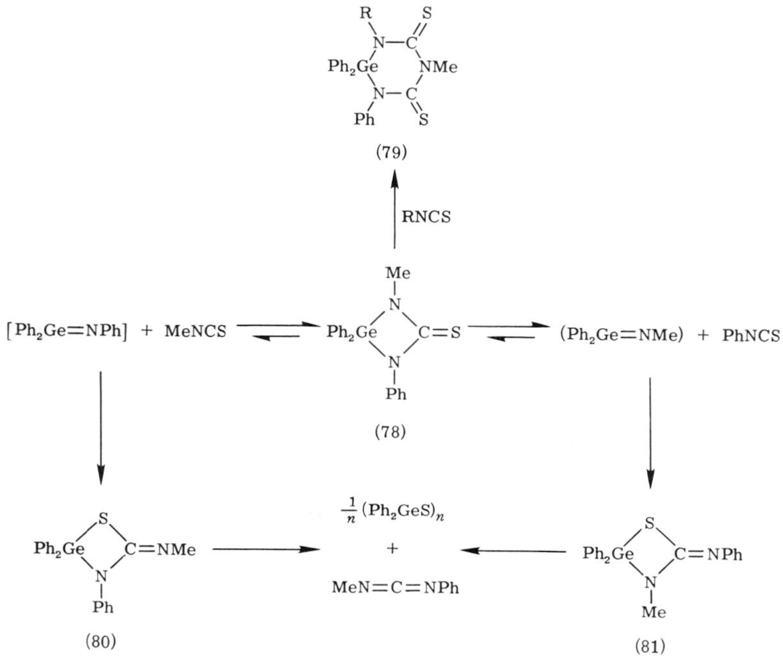

SCHEME 23

germa-4-thioxo-1,3-diazetidine **(78)** and 2-germa-4,6-thioxo-1,3,5-perhydrotriazine **(79)**. Addition across the C=S bond leads to unstable 2-germa-4-methylimino-1,3-thiazetidine **(80)** and 2-germa-4-phenylimino-1,3-thiazetidine **(81)** which decompose to diphenylgermanium sulfide dimer and trimer and carbodiimide (Scheme 23) *(74)*.

N-Methyldiphenylgermaimine [from $Ph_2Ge(NMe)_2$ + PhNCO] adds to CS_2 with formation of unstable 2-germa-4-thioxo-1,3-thiaazetidine **(82)** *(74)* [Eq. (36)].

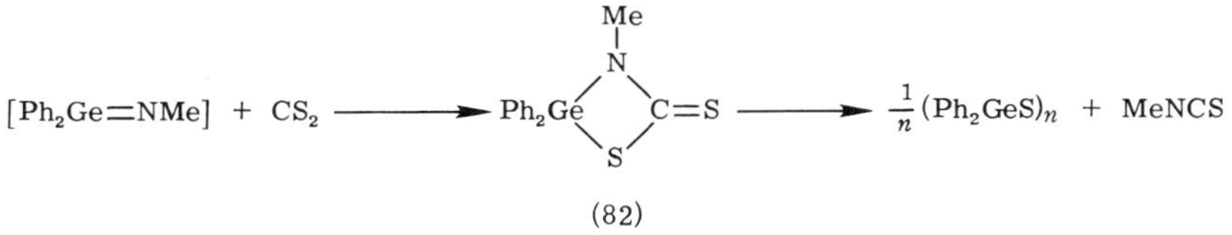

In the reaction of germylenes with phenyl azide in benzaldehyde, the formation of benzilideneaniline and germoxane via germaoxaazetidine **(83)** is noted *(68)*. Besides this the formation of condensation products of germylenes with benzaldehyde **(84)** is also observed for GeF_2 and PhGeCl

(68, 75) [Eq. (37)]. In the case of Ph_2Ge and Mes_2Ge, the main reaction is the 1,2-cycloaddition of the transient germaimine to benzaldehyde (68).

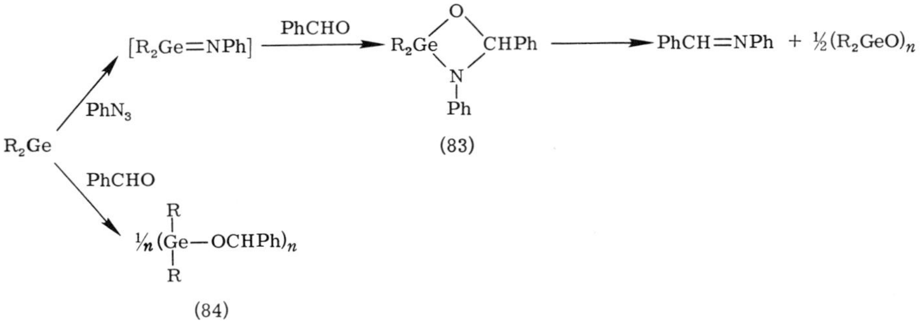

$$(37)$$

B. *1,3-Cycloadditions*

The 1,3-cycloadditions of germaimines $RR'Ge=NR''$ (R = Ph, R' = Ph or Cl; R'' = Me, Ph, t-Bu) with nitrones (**85**) (R''' = Ph, t-Bu) as well as their insertion reactions on oxaziridines (**86**) lead to germanones ($RR'Ge=O$), imines $PhCH=NR''$ and nitrenes probably via transient 2-germa-1-oxa-3,5-diazolidines (**87**). The imines formed contain the NR'' group of the initial germaimines (76) (Scheme 24).

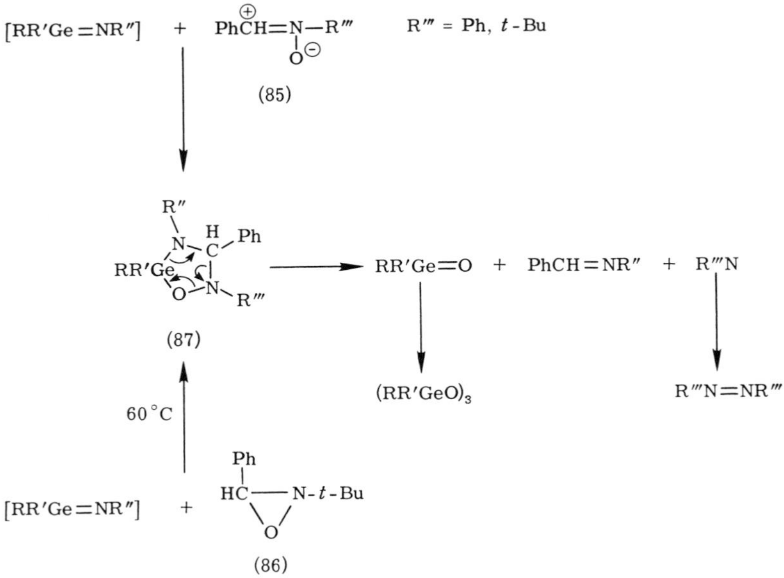

SCHEME 24

Germaimines (from catalytic depolymerization of cyclodigermazane by $ZnCl_2$, Et_3N) are inserted into ethylene oxide with formation of 2-germa-1,3-oxazolidine (**88**) (*74*) [Eq. (38)].

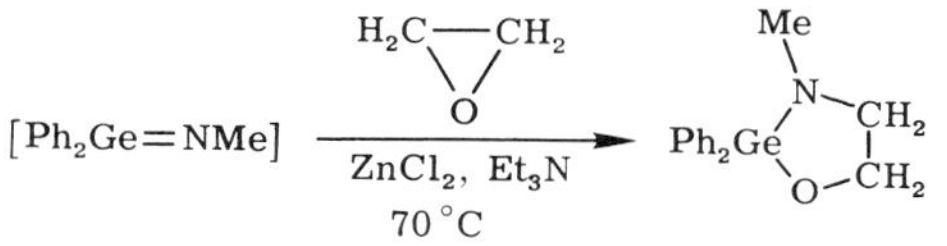

$$[Ph_2Ge{=}NMe] \xrightarrow[\substack{ZnCl_2,\ Et_3N \\ 70\,°C}]{\substack{H_2C{-}CH_2 \\ \diagdown\,\diagup \\ O}} Ph_2Ge\begin{array}{c} Me \\ | \\ N\diagdown \\ \diagup \quad \diagdown CH_2 \\ \diagdown O{-}CH_2 \end{array} \quad (\text{Yield } 43\%) \qquad (38)$$

(88)

No reaction was observed with cyclodigermazane without catalyst or basic solvents at 120°C (*74*).

V

$\diagup\!\!\diagdown$Ge$=$P$-$ SPECIES

Germaphosphimine $R_2Ge{=}PR'$ (*53, 77–79*) and silaphosphimine $R_2Si{=}PR'$ (*53, 77–80*) species with germanium or silicon doubly bonded to a dicoordinated phosphorus can be obtained from 2-germa- (**88a**) or 2-silaphosphetanes (**88b**) by a thermal β-decomposition reaction [Eq. (39)].

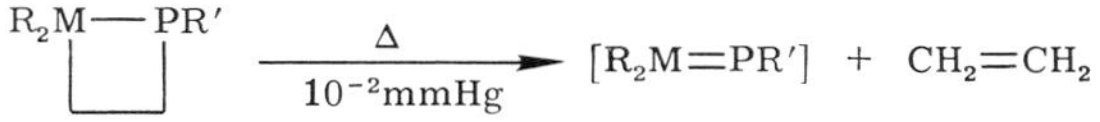

$$\begin{array}{c} R_2M{-}PR' \\ |\quad\quad| \\ \rule[0.5ex]{2.2em}{0.4pt} \end{array} \xrightarrow[10^{-2}\mathrm{mmHg}]{\Delta} [R_2M{=}PR'] + CH_2{=}CH_2 \qquad (39)$$

(88a) M = Ge
(88b) M = Si

Germaphosphimines and silaphosphimines add to 2-germa- (**88a**) or 2-silaphosphetanes (**88b**) leading to a 2,4-digerma- (**89**) and 2,4-disila-1,3-diphospholane (**90**), then after loss of germylene or silylene to a 3-germa-(**91**) or 3-sila-1,2-diphospholane (**92**) [Eqs. (40) and (41a)].

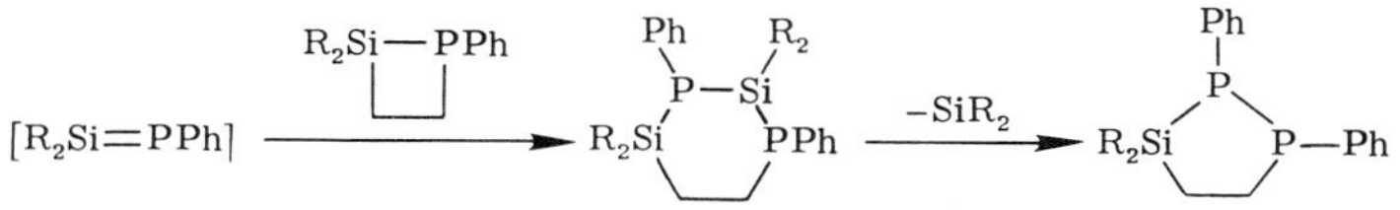

$$[R_2Si{=}PPh] \longrightarrow \begin{array}{c} R_2Si{-}PPh \\ |\quad\quad| \\ \rule[0.5ex]{2.2em}{0.4pt} \end{array} \longrightarrow \begin{array}{c} Ph\quad R_2 \\ \diagdown\diagup \\ P{-}Si \\ \diagup\quad\quad\diagdown \\ R_2Si\quad\quad PPh \\ \diagdown\rule[0.3ex]{1em}{0.4pt}\diagup \end{array} \xrightarrow{-SiR_2} \begin{array}{c} Ph \\ | \\ P \\ \diagup\ \diagdown \\ R_2Si\quad\quad P{-}Ph \\ \diagdown\rule[0.3ex]{1em}{0.4pt}\diagup \end{array} \quad (40)$$

(90)   (92)

In the case of condensation of germaphosphimines with 2-germaphosphetanes the formation of *P*-germylated 3-germa-1,2-diphospholanes (**92**) is observed. The unexpected formation of this derivative seems to indicate that germaphosphimines, unlike silaphosphimines, give a *P*-germylated phosphinidene (93) by a thermal transposition reaction. The *P*-germylated phosphinidene is inserted into the Ge—P bond of germaphosphetane (78) [Eq. (41b)]. Insertion of phosphinidenes into a Ge—P bond is a well-known reaction (81).

$$
\begin{array}{ccc}
\underset{(89)}{\overset{\displaystyle \underset{\text{R}_2\text{Ge}}{\overset{\text{Ph}}{\text{P—Ge}}}\ \overset{\text{R}_2}{\underset{\text{PPh}}{}}}{\Big\rfloor}} & \xrightarrow{\ -\text{R}_2\text{Ge}\ } & \underset{(91)}{\overset{\displaystyle \underset{\text{R}_2\text{Ge}}{\overset{\text{Ph}}{\text{P—PPh}}}}{\Big\rfloor}}
\end{array}
\qquad (41a)
$$

$$
[\text{R}_2\text{Ge}{=}\text{Ph}]
$$

$$
\text{R}_2\text{Ge—PPh}
$$

$$
\underset{(93)}{\overset{\text{Ph}}{\underset{\displaystyle \text{R}_2\text{Ge—P:}}{|}}} \xrightarrow{\ \text{R}_2\text{Ge—PPh}\ } \underset{(92)}{\overset{\text{R}_2\text{GePh}}{\underset{\displaystyle \text{R}_2\text{Ge}\ \ \text{PPh}}{\overset{|}{\text{P}}}}}
\qquad (41b)
$$

The formation of germaphosphimines has also been observed in the exchange reaction between 2,5-disilaphospholanes (94) and dialkyldihalogermanes (82). These transient species have been clearly characterized by formation of dimeric and trimeric cyclic germylphosphine (95) and by insertion and ring expansion reactions with ethylene oxide and sulfide with formation of new heterocycles 2-germa-3-oxaphospholanes (96) and 2-germa-3-thiophospholanes (97) (77, 82) [Eq. (42)].

$$(\text{Me}_2\text{GePPh})_{2,3}$$
$$(95)$$

$$
\underset{(94)}{\overset{\displaystyle \underset{\text{Me}_2}{\underset{\overset{|}{\text{Si}}}{\overset{\text{Me}_2}{\overset{|}{\text{Si}}}}}\ \text{PhP}}{\Big\rfloor}} + \text{Me}_2\text{GeCl}_2 \xrightarrow[20\,°\text{C}]{\text{THF}} \overset{\displaystyle \underset{\text{Me}_2}{\underset{\text{Cl—Si}}{\overset{\text{Me}_2}{\text{Cl—Si}}}}}{\Big\rfloor} + [\text{Me}_2\text{Ge}{=}\text{PPh}]
\qquad (42)
$$

$$
\underset{\substack{(96)\ \text{Y} = \text{O} \\ (97)\ \text{Y} = \text{S}}}{\overset{\displaystyle \underset{\text{Me}_2\text{Ge}}{\overset{\text{Ph}}{\underset{\text{Y}}{\text{P}}}}}{\Big\rfloor}}
$$

These exchange reactions can be extended to the tin and phosphorus series. Dimethylstanna(phenyl)phosphimine (**98**) has been obtained by exchange reactions between dimethyltin dichloride (or diorganotindiamines) and 2,5-disilaphospholanes (**94**) (*77, 79, 83*) [Eq. (43)].

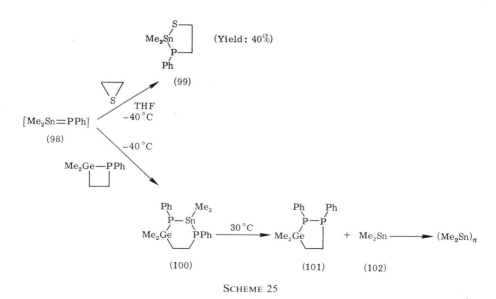

$$Me_2SnCl_2 \ + \ PhP \underset{Si}{\overset{Si}{\Big\langle}} \begin{array}{c} Me_2 \\ \\ \\ Me_2 \end{array} \quad \xrightarrow[-40\,°C]{THF} \quad [Me_2Sn=PPh] \ + \ \begin{array}{c} Me_2 \\ Cl-Si \\ Cl-Si \\ Me_2 \end{array} \qquad (43)$$

(94)   (98)

$(Me_2SnPPh)_3$

Dimethylstanna(phenyl)phosphimine (**98**) has been characterized, like germaphosphimines, by formation of its trimer and by insertion reactions into strained heterocycles such as thiirane and 2-germaphosphetanes with formation of stannathiaphospholane (**99**) and germastannadiphosphorinane (**100**), respectively. Compound **100** loses dimethylstannylene (**102**) with formation of germadiphospholane (**101**) (*77, 79, 83*) (Scheme 25).

Exchange reactions of the same type have been observed between 2,5-disilaphospholanes or bis(trimethylsilyl)alkyl- or arylphosphines ($Me_3Si)_2PR$ and organodichlorophosphines leading to intermediates with phosphorus–

SCHEME 25

phosphorus double bond (diphosphenes) characterized by cycloaddition to dimethylbutadiene (*77, 84*) [Eq. (44)].

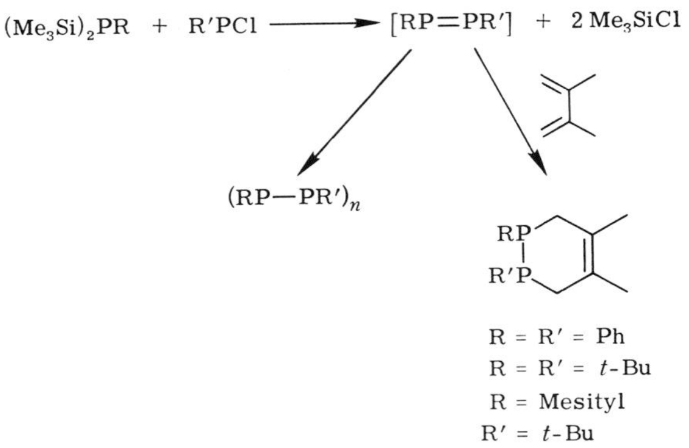

$$(Me_3Si)_2PR \; + \; R'PCl \longrightarrow [RP{=}PR'] \; + \; 2\,Me_3SiCl$$

$$(RP{-}PR')_n$$

(44)

$$\underset{R'P}{\overset{RP}{\diagdown}}$$

R = R' = Ph
R = R' = *t*-Bu
R = Mesityl
R' = *t*-Bu

VI

$\diagdown$Ge=METAL SPECIES

A. $\diagdown$Ge=Ge$\diagup$ Species

In the study of GeH$_2$ Kraus (*85*) claimed it likely that the dihydride GeH$_2$ is a dimer with doubly bonded germanium atoms, and that upon addition of sodium one bond is broken to form the salt NaH$_2$GeGeH$_2$Na.

Attempts to synthesize intermediates with a $\diagdown$Ge=Ge$\diagup$ bond were made by Curtis and Triplett (*86*) by the action of *t*-BuLi on Me$_2$Ge—Cl

Ge—Me$_2$ [Eq. (45)]
|
Cl

$$Me_2ClGeGeClMe_2 + t\text{-BuLi} \rightarrow Me_2Ge{=}GeMe_2\,(?) + t\text{-BuX} + LiX \qquad (45)$$
$$\downarrow$$
$$(Me_2Ge)_n$$

(*t*-Bu)Me$_2$GeGeMe$_2$(*t*-Bu), HMe$_2$GeGeMe$_2$(*t*-Bu) were also characterized in the same reaction. The production of large quantities of polygermanes

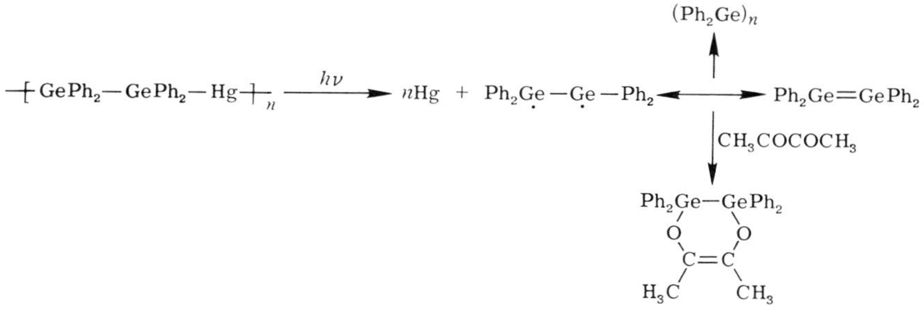

$$\underset{(104)}{} $$

Scheme 26 area:

$$(Ph_2Ge)_n$$

$$\displaystyle +GePh_2-GePh_2-Hg+_n \xrightarrow{h\nu} n Hg + Ph_2Ge-\overset{\cdot}{Ge}-Ph_2 \rightleftarrows Ph_2Ge=GePh_2$$

$$\downarrow CH_3COCOCH_3$$

$$\begin{array}{c} Ph_2Ge-GePh_2 \\ \diagup \quad \diagdown \\ O \qquad O \\ \diagdown \; C=C \; \diagup \\ H_3C \qquad CH_3 \end{array}$$

(103)

SCHEME 26

in this reaction may indicate the formation of $Me_2Ge=GeMe_2$ which immediately polymerizes even in the presence of a trapping reagent (1,3-dienes). However other mechanisms can also explain the formation of polymers.

The photolysis of polynuclear germylmercury compounds of the type $[-GeR_2GeR_2Hg-]_n$ leads to a mixture of cyclopolygermanes probably by intermediate formation of digermyl diradical $R_2Ge\dot{G}eR_2$ (87, 88) [Eq. (46)]

$$[-GeR_2-GeR_2-Hg-]_n \xrightarrow[-Hg]{h\nu\ 80°C} (R_2Ge)_4 + (R_2Ge)_5 + (R_2Ge)_m \tag{46}$$

These digermyl diradicals, which can be considered as the limiting form of $\diagdown\!\!\!\diagup Ge=Ge\diagup\!\!\!\diagdown$ intermediates, have been partially trapped with biacetyl forming a 2,3-digerma-1,4-dioxoline (103) (87, 88) (Scheme 26). However the preponderant formation of polymeric $(Ph_2Ge)_n$ seems to indicate the short life of these intermediates.

The tetraethyldigermyl diradical formed in the photolysis of the corresponding tetraethyldigermylmercury derivative (104) condenses on 1,3-dienes (isoprene and dimethylbutadiene) with formation of digermacyclohexadiene (105) (89) [Eq. (47)]

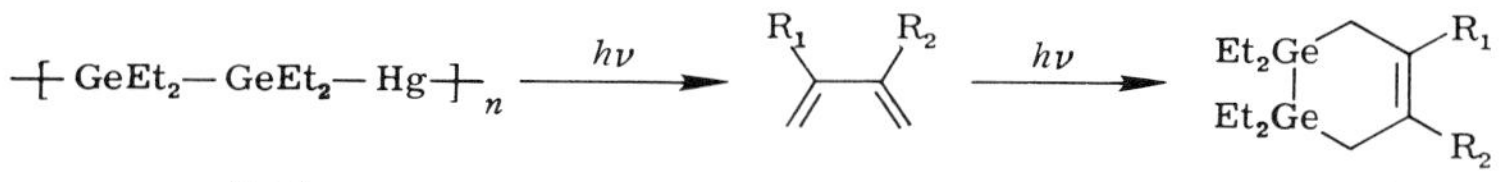

$$\underset{(104)}{+GeEt_2-GeEt_2-Hg+_n} \xrightarrow{h\nu} \underset{\substack{R_1 = H,\ R_2 = CH_3 \\ R_1 = R_2 = CH_3}}{\overset{R_1\ \ R_2}{\diagup\!\!\diagup\ \diagdown\!\!\diagdown}} \xrightarrow{h\nu} \underset{(105)}{\begin{array}{c} Et_2Ge \diagdown \quad \diagup R_1 \\ | \qquad \| \\ Et_2Ge \diagup \quad \diagdown R_2 \end{array}} \tag{47}$$

B. $\diagup\hspace{-0.5em}\diagdown Ge{=}Bi{-}$ Species

The intermediate formation of ylide species $\left[\diagup\hspace{-0.5em}\diagdown Ge{=}Bi{-} \leftrightarrow \diagup\hspace{-0.5em}\diagdown \overset{+}{Ge}{-} \right.$

$\left. \overset{-}{Bi}{-} \right]$ **(106)** is cited by Razuvaev *et al.* (*90, 91*) in the reaction of

bis(pentafluorophenyl)germane **(107)** with triethylbismuth [Eq. (48)]

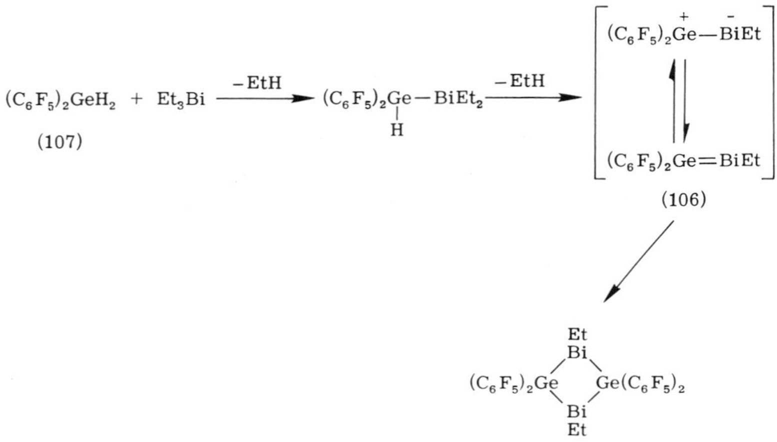

$$(C_6F_5)_2GeH_2 + Et_3Bi \xrightarrow{-EtH} (C_6F_5)_2\underset{H}{Ge}{-}BiEt_2 \xrightarrow{-EtH} \left[\begin{array}{c} (C_6F_5)_2\overset{+}{Ge}{-}\overset{-}{BiEt} \\ \updownarrow \\ (C_6F_5)_2Ge{=}BiEt \end{array} \right]$$

$$\begin{array}{c} \text{(106)} \end{array} \qquad\qquad (48)$$

$$(C_6F_5)_2Ge \overset{\overset{\displaystyle Et}{\displaystyle Bi}}{\underset{\underset{\displaystyle Et}{\displaystyle Bi}}{\diagup\hspace{-0.5em}\diagdown}} Ge(C_6F_5)_2$$

C. $\diagup\hspace{-0.5em}\diagdown Ge{=}MnL_n$ Derivatives

New stable species with Mn=Ge double bonds have been recently described by Gäde and Weiss (*92*). Treatment of $K[(\eta^5{-}CH_3C_5H_4){-}Mn(CO)_2GeH_3]$ with acetic acid gives $[(\eta^5{-}CH_3C_5H_4)Mn(CO)_2]_2Ge$ **(108)**. An X-ray analysis shows a linear Mn=Ge=Mn system. The molecule is centrosymmetric and the ligands on the Mn atoms are in the trans position. The reaction of $K[(\eta^5{-}CH_3C_5H_4)Mn(CO)_2GeH_3]$ with Hg^{2+} ions yields small amounts of $[(\eta^5{-}CH_3C_5H_4)Mn(CO)_2]_3Ge$ **(109)** which also contains a $Ge{=}Mn(CO)_2Cp$ moiety. The $Cp(CO)_2MnMn(CO)_2Cp$ unit is linked to the Ge atom through two Ge—Mn single bonds, thus forming a triangular Mn_2Ge ring (*92*). The molecular structures of **108** and **109** are shown in Figs. 1 and 2, respectively.

(108)

(109)

VII

THEORETICAL STUDIES

The first theoretical studies on germanium doubly bonded intermediates were carried out by Gowenlock and Hunter (*93, 94*). The results of the CNDO/2 calculations indicate that the germanium–carbon bond in ger-

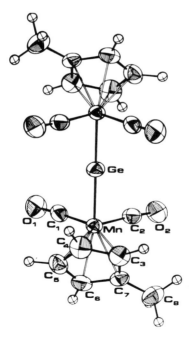

FIG. 1.   Molecular structure of $[(\eta^5\text{-CH}_3C_5H_4)Mn(CO)_2]_2Ge$ from W. Gäde and E. Weiss (*92*).

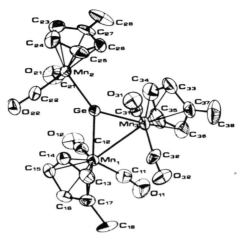

FIG. 2. Molecular structure of $[(\eta^5\text{-}CH_3C_5H_4)Mn(CO)_2]_3Ge$ from W. Gäde and E. Weiss (*92*).

maethylene $H_2Ge=CH_2$ and fluorinated germaethylenes $H_2Ge=CF_2$, $F_2Ge=CH_2$, and $F_2Ge=CF_2$ has a somewhat unusual double bond-like character with a relatively weak σ bond and much stronger π bond and the polarity $\overset{\delta-}{\underset{}{\diagup}}Ge=\overset{\delta+}{\underset{}{\diagdown}}C\diagdown$ (*93*). CNDO/2 calculations have also been performed to predict dipole moments for germaethylene and fluorinated germaethylenes (*94*).

More recently valuable *ab initio* calculations using pseudopotentials (*95*) have been carried out on the germanium doubly bonded intermediate series (*65, 96, 97*). The geometry of germaethylene has been established; $H_2Ge=CH_2$ is a planar molecule with C_{2v} symmetry (*96, 97*). The optimized geometry is the following at the SCF level: double $\zeta + d$ (Ge) basis sets (Scheme 27).

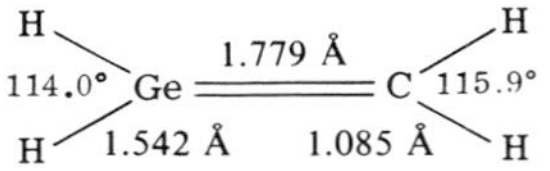

$$\begin{array}{c}
H \diagdown \qquad\qquad 1.779\ \text{Å} \qquad\qquad \diagup H \\
114.0° \quad Ge ======== C \quad 115.9° \\
H \diagup 1.542\ \text{Å} \qquad 1.085\ \text{Å} \quad \diagdown H
\end{array}$$

SCHEME 27

The $Ge=C$ bond length is shorter than the $Ge-C$ σ bond in CH_3GeH_3 ($d_{Ge-C} = 1.945$ Å (*19, 98*).

Calculations with configuration interactions [double $\zeta + d$ (Ge and C) basis sets] give the following results:

$$d_{Ge=C} = 1.812\ \text{Å}, \quad k_{Ge=C} = 4.99\ \text{mdyn/Å}, \quad \nu_{Ge=C} = 847\ \text{cm}^{-1}$$

The methylgermylene isomer $H\dot{G}eCH_3$ is more stable than germaethylene

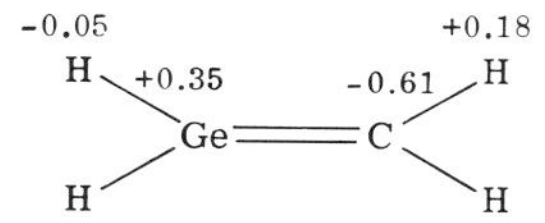

$\mu = 0.89$ D

bi ζ + d(Ge and C) basis sets

SCHEME 28

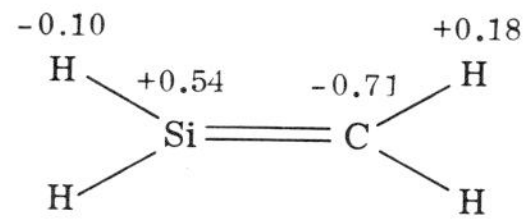

$\mu = 0.84$ D

bi ζ + d(Si and C) basis sets

SCHEME 29

by 15 kcal/mol, unlike the silicon analog (*96, 97*). The charge distribution is shown in Scheme 28.

On the same basis the silaethylene diagram is shown in Scheme 29. The charge difference in $\underset{/}{\overset{\backslash}{}} \overset{\delta+}{Si} = \overset{\delta-}{C} \underset{\backslash}{\overset{/}{}}$ is predicted to be greater than the charge difference in $\overset{\delta+}{Ge} = \overset{\delta-}{C}$ in agreement with the Allred–Rochow electronegativity scale (*99, 100*), according to which germanium is more electronegative than silicon. The π orbitals follow the same pattern:

$$\underset{Si=C}{+0.18 \quad -0.18} \qquad \underset{Ge=C}{+0.16 \quad -0.16}$$

Isodensity curves of the π orbital in $H_2Ge=CH_2$ are shown in Fig. 3.

Very recently an attempt to examine the properties of the germanium–carbon double bond in $H_2Ge=CH_2$ by *ab initio* SCF methods has been carried out by Kudo and Nugase (*101*), who discuss the geometry, proton affinity, and thermodynamic stability of $H_2Ge=CH_2$ in comparison with previous data on $H_2C=CH_2$ and $H_2Si=CH_2$ (*101*).

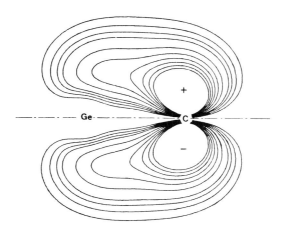

FIG. 3.   Isodensity curves of the π orbital in $H_2Ge=CH_2$. The values correspond to ψ^2 = 0.004, 0.006, 0.008, 0.010, 0.015, 0.020, 0.025, 0.030, 0.035, and 0.040. From Ref. *96*.

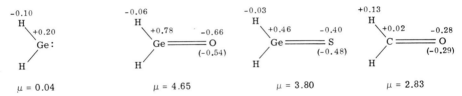

$\mu = 0.04$ $\mu = 4.65$ $\mu = 3.80$ $\mu = 2.83$

FIG. 4. Net atomic charges (π charges in parentheses) and calculated dipole moments (D). From Ref. 65.

Ab initio quantum calculations using pseudopotentials and including electron correlation were performed on $H_2Ge=O$ and $H_2Ge=S$ with double ζ ($+d$ orbitals) basis sets (65). Full geometry optimization performed at the SCF level led to planar structures

$$Ge=O = 1.63 \text{ Å}, Ge-H = 1.55 \text{ Å} \quad \text{and} \quad \sphericalangle HGeH = 112° \quad \text{for} \quad H_2GeO$$

and

$$Ge=S = 2.02 \text{ Å}, Ge-H = 1.55 \text{ Å} \quad \text{and} \quad \sphericalangle HGeH = 110° \quad \text{for} \quad H_2GeS$$

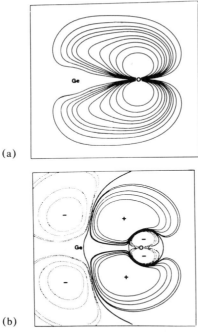

(a)

(b)

FIG. 5. Charge–density (a) and density–difference (b) contour maps for the π MO of germanone. The isodensity curves correspond to the values $\psi^2 = 0.002, 0.004, 0.006, 0.008, 0.01, 0.015, 0.02, 0.03, 0.05,$ and 0.1. The differential density curves correspond to $\psi_\pi^2 - (\psi_{4p_zGe}^2 + \psi_{2p_zO}^2) = -0.005, -0.003, -0.001,$ and -0.0008 (dashed lines), 0 (thick lines), 0.0008, 0.001, 0.002, 0.003, and 0.005 (solid lines). From Ref. 65.

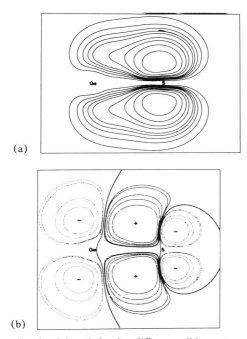

(a)

(b)

FIG. 6. Charge–density (a) and density–difference (b) contour maps for the π MO of germathione. The lines plotted correspond to the same values as in Fig. 5. From Ref. *65*.

The force constants were calculated as well as the theoretical vibrational frequencies $\nu(\text{Ge}=\text{O}) = 1038$ cm^{-1}, $\nu(\text{Ge}=\text{S}) = 586$ cm^{-1} at SCF level.

As expected, the Ge=X bond is strongly polarized, especially in $H_2\overset{\delta+}{\text{Ge}}=\overset{\delta-}{\text{O}}$. The σ and π Ge$-$O bond polarities suggest that the bonding is intermediate between $\pi(H_2\text{Ge}=\text{O})$ and semipolar $H_2\text{Ge}:\rightarrow\bar{\text{O}}|$ bonding. Extended C.I. was used to compute the Ge=X bond energies such as the $H_2\text{GeX} \rightarrow H_2\text{Ge}(^1A_1) + X(^3P)$ reaction enthalpy. They were predicted to be about 108 kcal/mol for Ge=O and 83 kcal/mol for Ge=S. $H_2\text{Ge}=\text{O}$ is found to be less stable than its germylene isomer HGeOH by 18 kcal/mol (*65*) and 19.9 kcal/mol in more extended calculations (*97*).

The following figures give the net atomic charges in germylene (GeH$_2$), germanone $H_2\text{Ge}=\text{O}$, germathione $H_2\text{Ge}=\text{S}$, and formol with corresponding dipole moments (Fig. 4), the charge density and density difference contour maps for the π molecular orbital of germanone $H_2\text{Ge}=\text{O}$ (Fig. 5) and germathione $H_2\text{Ge}=\text{S}$ (Fig. 6).

Germaimine, $H_2\text{Ge}=\text{NH}$, is predicted to be a planar molecule. Its optimized geometry at the SCF level is shown in Scheme 30 (*96, 97*).

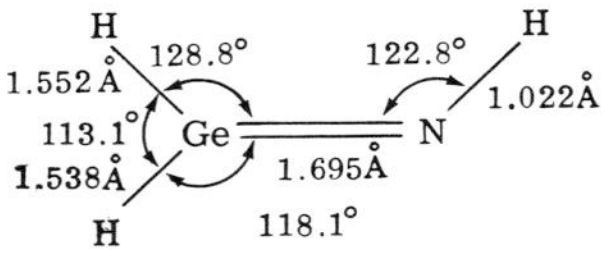

H   128.8°   122.8°   H
1.552Å \   / 1.022Å
         \   /
113.1° (Ge ═══════════════ N
1.538Å /   1.695Å   \
H   118.1°

SCHEME 30

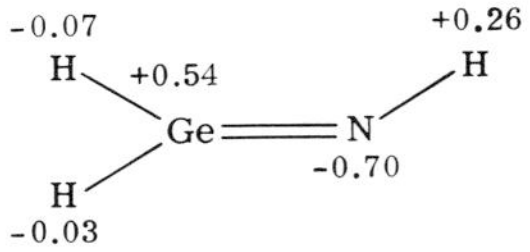

-0.07   +0.26
H \   +0.54   / H
   \   /
    Ge ═══════════ N
   /   -0.70
H
-0.03

μ = 2.99 D

SCHEME 31

The bond length Ge=N (1.695 Å) is shorter than Ge—N σ (e.g., in N(GeH₃)₃ (1.836 Å) (*102*) or in (Cl₂GeNMe)₃ 1.78–1.81 Å (*103*). The length, force constant, and absorption frequency of Ge=N (noncoupled) bond are at C.I. level

$$d_{Ge=N} = 1.727 \text{ Å}, \; k_{Ge=N} = 5.37 \text{ mdyn/Å}, \; \nu_{Ge=N} = 854 \text{ cm}^{-1}$$

In all cases the germylene isomer form HGeXH is more stable than the doubly bonded form H₂Ge = X. The energy differences between HGeXH (more stable form) and H₂Ge = X with X = O, NH, CH₂ [double ζ + d (Ge, O, N, C) basis sets] are: X = O 19.9 kcal/mol, X = NH 32 kcal/mol, X = CH₂ 15 kcal/mol (*96, 97*). The charge distribution and dipole moment of the germaimine [double ζ + d (Ge and N) basis sets] is shown in Scheme 31.

The geometry of germaphosphimine H₂Ge=PH at the SCF level is the following in the ground state (*104*) (Scheme 32). Calculations with configuration interactions give the following results:

$$d_{Ge=P} = 2.169 \text{ Å}, \; k_{Ge=P} = 3.05 \text{ mdyn/Å}, \; \nu_{Ge=P} = 481 \text{ cm}^{-1}$$

The distribution of net atomic charges and the calculated dipole moment are shown in Scheme 33. The $\diagdown$Ge=P— bond length is shorter than Ge—P σ bond in trigermylphosphine [(H₃Ge)₃P]d_{Ge-P} = 2.308 Å (*105*).

Ab initio calculations using pseudopotentials have been carried out on singlet digermene (H₂Ge=GeH₂) and its germylgermylene isomer.

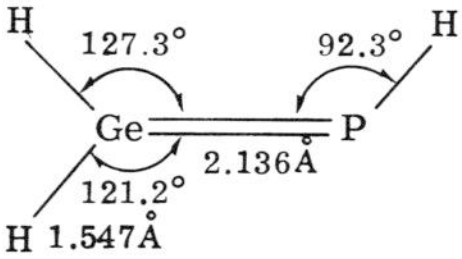

H   127.3°   92.3°   H
 \   /
  \   /
  Ge ═══════════ P
 /   2.136Å
121.2°
H 1.547Å

SCHEME 32

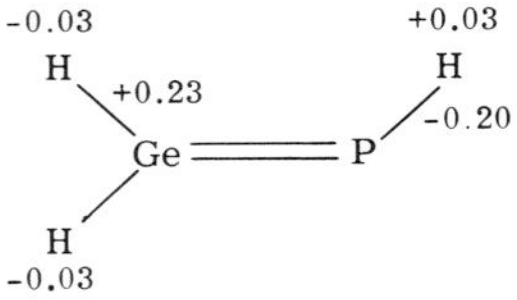

-0.03   +0.03
H \   +0.23   / H
   \   /
    Ge ═══════════ P
   /   -0.20
H
-0.03

μ = 2.34 D

SCHEME 33

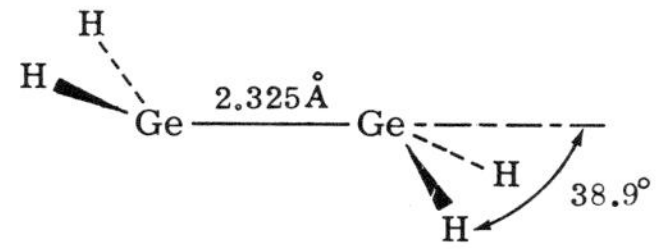

SCHEME 34

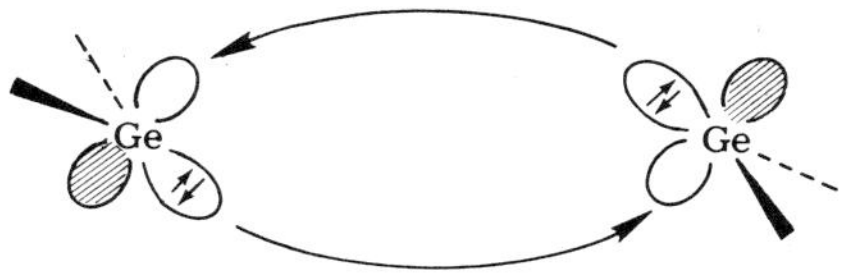

SCHEME 35

HGeGeH$_3$ at both SCF (double ζ + d basis set) and C.I. levels. H$_2$Ge=GeH$_2$ is 5 kcal/mol more stable than HGeGeH$_3$ (*106*).

Digermene is predicted to have a trans-bent geometry that does not depend strongly on correlation effects: the wagging angle of the GeH$_2$ groups is 39°. The planar form lies 3–4 kcal/mol higher in energy (Scheme 34).

The new type of bonding occurring in H$_2$Ge=GeH$_2$ can be described as two semipolar bent bonds between two singlet germylenes (*106*) (Scheme 35).

VIII

SPECTROSCOPIC STUDIES

A. Infrared and Photoelectron Spectroscopy

The instability of germanium doubly bonded intermediates makes the spectroscopic observation and measurement very difficult. However, vapor-phase germanium monoxide GeO (with Ge$_2$O$_2$ and Ge$_2$O$_3$) has been isolated in nitrogen and argon matrices. In a nitrogen matrix infrared absorptions are observed at 973.4 cm^{-1} (^{74}Ge^{16}O) (*107*).

A theoretical study of photoelectron spectra of GeS, GeSe, and GeTe has been published recently (*108*).

In the reaction of difluorogermylene with phenyl azide leading to F$_2$Ge=NPh, the infrared spectra of the reaction mixtures show two bands, at 1230 and 970 cm^{-1}, which disappear slowly while the absorption Ge—N—Ge (820–860 cm^{-1}) increases. It would seem that the 970 cm^{-1} absorption could be attributed to the $\diagdown$Ge=N— bond (*109*) in accordance with theoretical studies (*96, 97*). The difference of ~100 cm^{-1} between the experimental and calculated values may be attributed to substituent effects. The same bands appear in the thermal depolymerization of cyclo-digermazane (*109*).

B. *Mass Spectrometry: Formation of Ions or Neutral Fragments with $p\pi$–$p\pi$ Doubly Bonded Intermediates of Germanium*

The mass spectra of germacyclobutanes (1,1-dimethylgermacyclobutane **(2)** (*31, 32*), 1,1-dibutylgermacyclobutane (*110*), 1,1-dibutyl-3-methyl-1-germacyclobutane (*110*) and 4-germa-spiro-3,4 octane (*110, 111*) show olefinic and carbenoid decomposition in dissociative ionization such as in thermal dissociation [Eq. (49)].

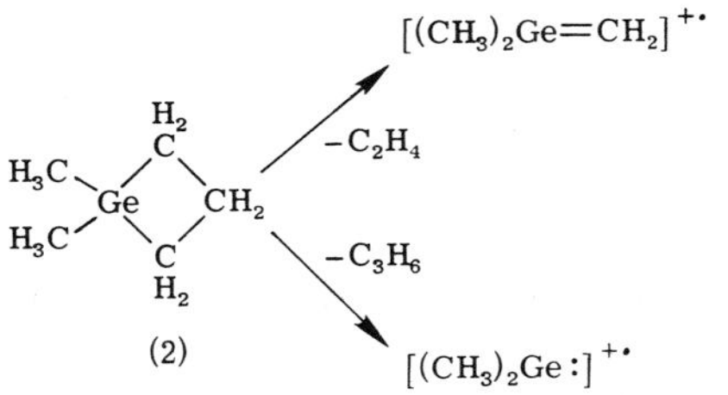

$$[(CH_3)_2Ge{=}CH_2]^{+\cdot}$$

$$(49)$$

The formation of $Me_2Ge{=}GeMe_2^{+\cdot}$ **(111)** in the fragmentation of 1,1,2,2-tetramethyl-1,2-digermacyclopentane **(110)** is also consistent with the results of thermolysis of the same compound (*32*) [Eq. (50)].

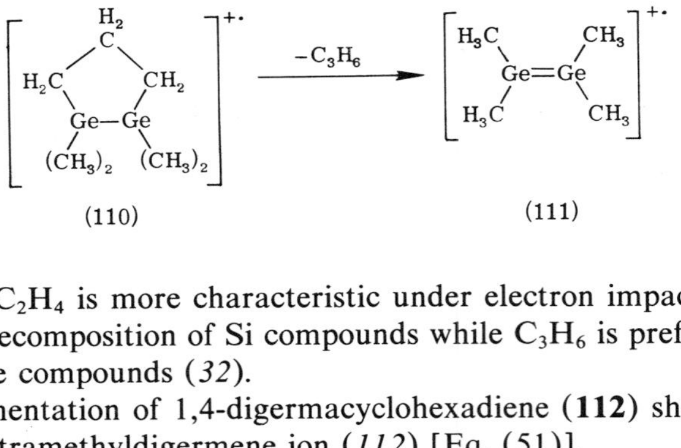

$$(50)$$

The loss of C_2H_4 is more characteristic under electron impact as well as in thermal decomposition of Si compounds while C_3H_6 is preferably eliminated by Ge compounds (*32*).

The fragmentation of 1,4-digermacyclohexadiene **(112)** shows the formation of tetramethyldigermene ion (*112*) [Eq. (51)].

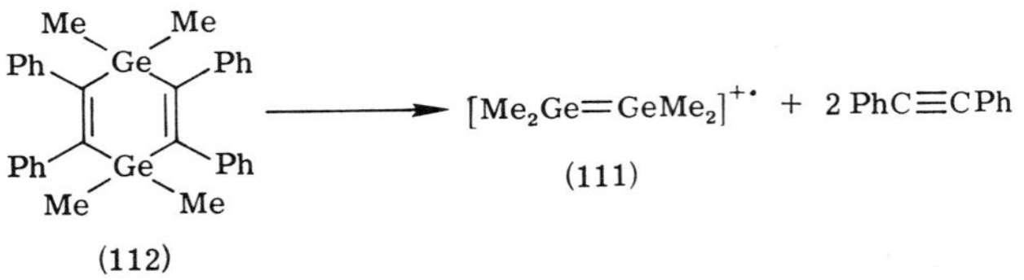

$$[Me_2Ge{=}GeMe_2]^{+\cdot} + 2\,PhC{\equiv}CPh \qquad (51)$$

Results of the same type are reported for $Ge_2(C_2H_2)_2Cl_4$ [tetrachloro-digermacyclohexadiene dimer of 1,1-dichlorogermirene (*113*)].

In a detailed study of mass spectra of organogermanium compounds Glockling and Light (*114*) reported that hexamethyldigermoxane shows an unusual transition in which Me_2GeO is eliminated as a neutral fragment [Eq. (52)].

$$Me_3GeOGeMe_2^+ \rightarrow Me_3Ge^+ + Me_2GeO \qquad (52)$$

IX

CONCLUSION

Doubly bonded germanium species, as well as germylenes (*38*), are of great fundamental and practical interest. These intermediates are powerful synthetic reagents in organometallic chemistry. The polar character of their multiple bonds leads to numerous types of reactions (such as insertion, addition, cycloaddition, transposition, etc.). Moreover, the symmetrical structure of some species (e.g., $\diagdown Ge{=}Ge\diagup \leftrightarrow \diagdown \dot{G}e{-}\dot{G}e\diagup$) and the weaker polar character of other species leads to radical activity, which is also very convenient in synthesis.

One of the major remaining problems is stabilization of these short-lived species. Attempts are being made to stabilize them by stereoelectronic effects of substituents on the metal and heteroelements, by matrix isolation techniques, and by complexing with Lewis acids or transition metals. These studies are now in progress. This stabilization would allow spectroscopic measurements in connection with theoretical studies as well as the study of additional chemical properties of these interesting doubly bonded species.

Acknowledgments

The author wishes to express gratitude and thanks to his collaborators who have carried out a great part of the germanium multiply bonded intermediate chemistry described above: J. D. Andriamizaka, J. Barrau, A. Castel, C. Couret, G. Dousse, J. Escudié, H. Lavayssière, S. Richelme, P. Rivière, M. Rivière-Baudet, and B. Saint-Roch.

The author also thanks Drs. Trinquier, Barthelat, and Malrieu for their work and collaboration in the theoretical studies of germanium doubly bonded species.

References

1. L. E. Gusel'nikov, N. S. Nametkin, and V. M. Vodvin, *Acc. Chem. Res.* **8**, 18 (1975).
2. L. E. Gusel'nikov and N. S. Nametkin, *Chem. Rev.* **79**, 533 (1979).

3. I. Fleming, *Compr. Org. Chem.* **3**, Part 13, 671 (1979).
4. B. Coleman and M. Jones, *Rev. Chem. Intermed.* **4**, 297 (1981).
5. G. Bertrand, G. Trinquier, and P. Mazerolles, *J. Organomet. Chem. Libr.* **12**, 1 (1981).
6. M. Ishikawa and M. Kumada, *Adv. Organomet. Chem.* **19**, 51 (1981).
7. M. Ishikawa and M. Kumada, *Rev. Silicon Germanium, Tin Lead Compds.* **4**, 7 (1979).
8. T. J. Barton, *Pure Appl. Chem.* **52**, 615 (1980).
9. T. J. Barton, *Compr. Organomet. Chem.* **1** (in press).
10. P. Jutzi, *Angew. Chem.* **87**, 269 (1975); *Angew. Chem., Int. Ed. Engl.* **14**, 232 (1975).
11. P. Rivière, J. Satgé, and M. Rivière-Baudet, *Compr. Organomet. Chem.* **1** (in press).
12. W. E. Dasent, "Nonexistent Compounds." Dekker, New York, 1965.
13. L. D. Pettit, *Q. Rev., Chem. Soc.* **25**, 1 (1971).
14. C. Eaborn, "Organosilicon Compounds." Butterworth, London, 1960.
15. A. D. Petrov, V. F. Mironov, V. A., Ponomarenko, and E. A. Chernyshev, "Synthesis of Organosilicon Monomers." Acad. Sci. USSR, 1961 (Engl. Transl. Consultants Bureau, New York, 1964).
16. V. Bazant, V. Chvalovsky, and J. Rathousky, "Organosilicon Compounds," Vol. 1. Academic Press, New York, 1965.
17. F. G. A. Stone, "Hydrogen Compounds of the Group IV Elements," Prentice-Hall, Englewood Cliffs, New Jersey, 1962.
18. E. A. V. Ebsworth, "Volatile Silicon Compounds." Pergamon, Oxford, 1963.
19. E. A. V. Ebsworth, *in* "Organometallic Compounds of the Group IV Elements" (A. G. MacDiarmid, ed.), Vol. 1, Part 1 p. 1. Dekker, New York, 1968.
20. K. S. Pitzer, *J. Am. Chem. Soc.* **70**, 2140 (1948).
21. R. S. Mulliken, *J. Am. Chem. Soc.* **72**, 4493 (1950).
22. H. Gilman and G. E. Dunn, *Chem. Rev.* **52**, 77 (1953).
23. I. R. Beattie and T. Gilson, *Nature (London)* **193**, 1041 (1962).
24. C. J. Attridge, *Organomet. Chem. Rev., Sect. A* **5**, 323 (1970).
25. F. G. A. Stone and D. Seyferth, *J. Inorg. Nucl. Chem.* **1**, 112 (1955).
26. F. Rijkens and G. J. M. Van der Kerk, "Organogermanium Chemistry." T. N. O. Utrecht, 1964.
27. H. Burger and R. Eujen, *Top. Curr. Chem.* **50**, 1 (1973).
28. A. G. Brook, F. Abdesaken, B. Gutekunst, G. Gutekunst, and R. K. Kallury, *J. Chem. Soc., Chem. Commun.* p. 191 (1981).
29. R. West, M. J. Fink, and J. Michl, *Science* **214**, 1343 (1981).
30. M. D. Curtis, *J. Am. Chem. Soc.* **91**, 6011 (1969).
31. N. S. Nametkin, L. E. Gusel'nikov, R. L. Ushakova, V. Yu. Orlov, O. V. Kuz'min, and V. M. Vdovin, *Dokl. Akad. Nauk SSSR* **194**, 1096 (1970).
32. V. Yu. Orlov, L. E. Gusel'nikov, N. S. Nametkin, and R. L. Ushakova, *Org. Mass Spectrom.* **6**, 309 (1972).
33. T. J. Barton, E. A. Kline, and P. M. Garvey, *J. Am. Chem. Soc.* **95**, 3078 (1973).
34. G. Bertrand, P. Mazerolles, and J. Ancelle, *Tetrahedron* **37**, 2459 (1981).
35. T. J. Barton and S. K. Hoekman, *J. Am. Chem. Soc.* **102**, 1584 (1980).
36. S. K. Hoekman, Thesis, Iowa State University, Ames (1980); *Diss. Abstr. Int.* **40**, 5672-B (1980).
37. E. B. Norsoph, B. Coleman, and M. Jones Jr., *J. Am. Chem. Soc.* **100**, 994 (1978).
38. J. Satgé, M. Massol, and P. Rivière, *J. Organomet. Chem.* **56**, 1 (1973).
39. P. Rivière, A. Castel, and J. Satgé, *J. Am. Chem. Soc.* **102**, 5413 (1980).
40. P. Rivière, A. Castel, and J. Satgé, *Organomet. Coord. Chem. Germanium, Tin Lead, Int. Conf., 3rd, 1980* Abstracts, p. 15 (1980).
41. P. Rivière, M. Rivière-Baudet, A. Castel, and J. Satgé, *Bull. Soc. Chim. Fr.* No. 11-12, 34 (1980).

42. P. Rivière, J. Satgé, and A. Castel, *C. R. Hebd. Seances Acad. Sci., Ser. C* **281**, 835 (1975).
43. M. Lesbre, P. Mazerolles, and J. Satgé, "The Organic Compounds of Germanium." Wiley (Interscience), New York, 1971.
44. P. Rivière, J. Satgé, A. Castel, and A. Cazès, *J. Organomet. Chem.* **177**, 171 (1979).
45. P. Rivière, M. Rivière-Baudet, and J. Satgé, *J. Organomet. Chem.* **96**, C7 (1975).
46. V. G. Märkl and D. Rudnik, *Tetrahedron Lett.* p. 1405 (1980).
47. D. Seyferth and J. L. Lefferts, *J. Am. Chem. Soc.* **96**, 6237 (1974).
48. D. Seyferth and J. L. Lefferts, *J. Organomet. Chem.* **116**, 257 (1976).
48a. N. N. Zemlyanskii, I. V. Borisova, Yu. N. Luzikov, N. D. Kolosova, Yu. A. Ustynyuk, I. P. Beletskaya, *Izv. Akad. Nauk SSSR, Ser. Khim.* 2668 (1980); *C. A.* **94**, 139907k (1981).
49. J. E. Taylor and T. S. Milazzo, *J. Phys. Chem.* **82**, 847 (1978).
50. M. Massol, D. Mesnard, J. Barrau, and J. Satgé, *C. R. Hebd. Seances Acad. Sci., Ser. C* **272**, 2081 (1971).
51. J. Barrau, M. Massol, D. Mesnard, and J. Satgé, *J. Organomet. Chem.* **30**, C67 (1971).
52. J. Barrau, M. Bouchaut, H. Lavayssière, G. Dousse, and J. Satgé, *Helv. Chim. Acta* **62**, 152 (1979).
53. J. Satgé, J. Barrau, A. Castel, C. Couret, G. Dousse, J. Escudié, H. Lavayssière, P. Rivière, and M. Rivière-Baudet, *Organomet. Coord. Chem. Germanium, Tin Lead, Int. Conf., 3rd, 1980* Abstracts, p. 14 (1980).
54. J. Barrau, M. Massol, D. Mesnard, and J. Satgé, *Recl. Trav. Chim. Pays-Bas* **92**, 321 (1973).
55. H. Lavayssière, J. Barrau, G. Dousse, J. Satgé, and M. Bouchaut, *J. Organomet. Chem.* **154**, C9 (1978).
56. M. Riviere-Baudet, G. Lacrampe, and J. Satgé, *C. R. Hebd. Seances Acad. Sci., Ser. C* **289**, 223 (1979).
57. H. Lavayssière, G. Dousse, J. Barrau, J. Satgé, and M. Bouchaut, *J. Organomet. Chem.* **161**, C59 (1978).
58. J. Barrau, M. Bouchaut, H. Lavayssière, G. Dousse, and J. Satgé, *Synth. React. Inorg. Met.-Org. Chem.* **10**, 515 (1980).
59. J. Barrau, H. Lavayssière, G. Dousse, C. Couret, and J. Satgé, *J. Organomet. Chem.* **221**, 271 (1981).
60. J. Satgé, J. Barrau, A. Castel, C. Couret, G. Dousse, J. Escudié, H. Lavayssière, P. Rivière, and M. Rivière-Baudet, *Abstr., Int. Conf. Organomet. Chem., 9th, 1979* p. A35 (1979).
61. J. Barrau, M. Bouchaut, A. Castel, A. Cazès, G. Dousse, H. Lavayssière, P. Rivière, and J. Satgé, *Synth. React. Inorg. Met.-Org. Chem.* **9**, 273 (1979).
62. J. C. Barthelat, B. Saint-Roch, G. Trinquier, and J. Satgé, *J. Am. Chem. Soc.* **102**, 4080 (1980).
63. A. Castel, A. Cazès, P. Rivière, and J. Satgé, *Abstr. Int. Conf. Organomet. Chem., 9th, 1979* p. A37 (1979).
64. J. Barrau, H. Lavayssière, G. Dousse, and J. Satgé, unpublished results.
65. G. Trinquier, M. Pelissier, B. Saint-Roch, and H. Lavayssière, *J. Organomet. Chem.* **214**, 169 (1981).
66. H. Lavayssière, G. Dousse, J. Satgé, J. Barrau, and M. Traore, *J. Organomet. Chem.* (in press) (1982); *Angew. Chem.* **94**, 455 (1982).
67. R. Huisgen, *Angew. Chem., Int. Ed. Engl.* **2**, 565 (1963).
68. P. Rivière, A. Cazès, A. Castel, M. Rivière-Baudet, and J. Satgé, *J. Organomet. Chem.* **155**, C58 (1978).
69. M. Rivière-Baudet, P. Rivière, and J. Satgé, *J. Organomet. Chem.* **154**, C23 (1978).

70. M. Rivière-Baudet, P. Rivière, J. Satgé, and G. Lacrampe, *Recl. Trav. Chim. Pays-Bas* **98,** 42 (1979).
71. M. Elsheikh, N. R. Pearson, and L. H. Sommer, *J. Am. Chem. Soc.* **101,** 2491 (1979).
72. A. Baceiredo, G. Bertrand, and P. Mazerolles, *Tetrahedron Lett.* p. 2553 (1981).
73. G. Lacrampe, H. Lavayssière, M. Rivière-Baudet, and J. Satgé, *Recl. Trav. Chim. Pays-Bas* (in press).
74. M. Rivière-Baudet, G. Lacrampe, P. Rivière, A. Castel, and J. Satgé, unpublished results.
75. P. Rivière, J. Satgé, and A. Castel, *C. R. Hebd. Seances Acad. Sci., Ser. C* **282,** 971 (1976).
76. M. Rivière-Baudet, P. Rivière, A. Castel, G. Lacrampe, and J. Satgé, *Recl. Trav. Chim. Pays-Bas* (submitted for publication).
77. J. D. Andriamizaka, Thesis, University Paul Sabatier, Toulouse (1981).
78. J. Escudie, C. Couret, J. Satgé, and J. D. Andriamizaka, *Int. Symp. Organosilicon Chem., 6th, 1981* Abstracts, p. 223 (1981).
79. C. Couret, J. Escudié, J. Satgé, and J. D. Andriamizaka, *Bull. Soc. Chim. Fr.* No. 11–12, 34 (1980).
80. C. Couret, J. Escudié, J. Satgé, J. D. Andriamizaka, and B. Saint-Roch, *J. Organomet. Chem.* **182,** 9 (1979).
81. J. Escudié, C. Couret, and J. Satgé, *Recl. Trav. Chim. Pays-Bas* **98,** 461 (1979).
82. C. Couret, J. Satgé, J. D. Andriamizaka, and J. Escudié, *J. Organomet. Chem.* **157,** C35 (1978).
83. C. Couret, J. D. Andriamizaka, J. Escudié, and J. Satgé, *J. Organomet. Chem.* **208,** C3 (1981).
84. J. Escudie, C. Couret, J. Satgé, and J. D. Andriamizaka, *J. Organomet. Chem.* **228,** C76 (1982).
85. C. A. Kraus, *J. Chem. Educ.* **29,** 417 (1952).
86. K. Triplett and M. D. Curtis, *J. Organomet. Chem.* **107,** 23 (1976).
87. P. Rivière, A. Castel, and J. Satgé, *J. Organomet. Chem.* **212,** 351 (1981).
88. P. Rivière, J. Satgé, and D. Soula, *C. R. Hebd. Seances Acad. Sci., Ser. C* **277,** 895 (1973).
89. P. Mazerolles, M. Joanny, and G. Tourrou, *J. Organomet. Chem.* **60,** C3 (1973).
90. G. A. Razuvaev and M. N. Bochkarev, *J. Organomet. Chem. Libr.* **12,** 241 (1981).
91. M. N. Bochkarev, N. I. Gur'ev, and G. A. Razuvaev, *J. Organomet. Chem.* **162,** 289 (1978).
92. W. Gäde and E. Weiss, *J. Organomet. Chem.* **213,** 451 (1981).
93. B. G. Gowenlok and J. A. Hunter, *J. Organomet. Chem.* **111,** 171 (1976).
94. B. G. Gowenlok and J. A. Hunter, *J. Organomet. Chem.* **140,** 265 (1977).
95. Ph. Durand and J. C. Barthelat, *Theor. Chim. Acta* **38,** 283 (1975).
96. G. Trinquier, Thesis, University Paul Sabatier, Toulouse (1981).
97. G. Trinquier, J. C. Barthelat, and J. Satgé, *J. Am. Chem. Soc.* (in press) (1982).
98. V. W. Laurie, *J. Chem. Phys.* **30,** 1210 (1959).
99. A. L. Allred and E. G. Rochow, *J. Inorg. Nucl. Chem.* **5,** 269 (1958).
100. A. L. Allred, *J. Inorg. Nucl. Chem.* **17,** 215 (1961).
101. T. Kudo and S. Nagase, *Chem. Phys. Lett.* **84,** 375 (1981).
102. C. Glidewell, D. W. H. Rankin, and A. G. Robiette, *J. Chem. Soc. A* p. 2935 (1970).
103. M. Ziegler and J. Z. Weiss, *Z. Naturforsch. B: Anorg. Chem., Org. Chem., Biochem., Biophys., Biol.* **26B,** 735 (1971).
104. J. C. Barthelat, private communication.
105. D. W. H. Rankin, A. G. Robiette, G. M. Sheldrick, B. Beagley, and T. G. Hewitt, *J. Inorg. Nucl. Chem.* **31,** 2351 (1969).

106. G. Trinquier, J. P. Malrieu, and P. Rivière, *J. Am. Chem. Soc.* (submitted for publication).
107. J. S. Ogden and M. J. Ricks, *J. Chem. Phys.* **52,** 352 (1970).
108. P. Weinberger and E. Wimmer, *Ber. Bunsenges. Phys. Chem.* **83,** 890 (1979).
109. M. Rivière-Baudet, A. Marchand, P. Rivière, A. Castel, G. Lacrampe, and J. Satgé, unpublished results.
110. J. Dubac, Thesis, University Paul Sabatier, Toulouse (1969).
111. V. Yu. Orlov, L. E. Gusel'nikov, E. Sh. Finkel'shtein, and V. M. Vdovin, *Izv. Akad. Nauk SSSR, Ser. Khim.* p. 1984 (1973).
112. F. Johnson, R. S. Gohlke, and W. A. Nasutavicus, *J. Organomet. Chem.* **3,** 233 (1965).
113. L. N. Gorokhov, *Zh. Strukt. Khim.* **6,** 766 (1965).
114. F. Glockling and R. C. Light, *J. Chem. Soc. A* p. 717 (1968).

Index

A

Aluminum alkyls, reaction with 1,4-diaza-1,3-butadienes, 222–225
Aluminum, borane complexes, 63–66

B

Berry pseudorotation, in metal carbonyl substitution mechanism, 134
Beryllium, borane complexes, 63–65, 82
Bismuth, germanium compound, 274
Boranes, *see also* individual compounds
  as cluster fragments, 92–103
  hydridic character, 57
  as ligands, 71–92
    hydrocarbon analogs, 71–80
  and metal clusters, 58–59, 92

C

Cadmium complexes
  of borane, 65
  of 1,4-diaza-1,3-butadienes, 167, 193
Chromium carbonyls, substitution mechanisms
  effect of hydroxide ion, 138–139
  hexacarbonyl derivatives, 115–118, 121–122, 127–130
    of amines, 127–130
    CO dissociation rate parameters, 117
    ligand dissocation rate parameters, 117
    relative ligand lability, 117–118
    with trans bis-ligands, 121
  phenanthroline derivatives, 143
  stereoselectivity, 140–141
  tetracarbonyl norbornadiene complex, 132–134
  ^{13}CO substitution, 133
Chromium complexes
  of boranes, 63–64
  of 1,4-diaza-1,3-butadienes
    as catalyst, 233
    with chelated ligand, 170, 172–174
    with monodentate ligand, 165
    radical anion, 217
  of substituted cyclopentadienyls, 9, 36

Cluster complexes, and boranes, 58–59
Cobalt complexes
  bis(dicarbonyl)fulvalene, 28
  of boranes, 63–66, 68–70, 82, 102–103
    isomerism, 94–95
    multinuclear species, 98–102, 104
  of 1,4-diaza-1,3-butadienes
    with chelated ligand, 168–169, 181–182
    mixed binuclear carbonyls with manganese, 162–163, 177, 195–198, 205, 208–209
  dicarbonyl cyclopentadienyls, 9, 10
    associative substitution mechanism, 124
    with halogen substituents, 26–28
    polymer-bound catalyst, 48–51
    reactions at ring, 10
    vinyls, 41–43
  tetracarbonyl radical, 143
Cobaltocenes
  bis(dimethylaminomethyl), 31
  bis(diphenylphosphide), 34, 36
  1,1′-diacetyl, 8
  1,1′-dicarbomethoxy, 8
Copper complexes
  of boranes, 63–69, 75, 82, 91, 94, 102
  percyanovinylcyclopentadienyl, 45
Cyclobutadiene,
  metal complexation, 77
  tetraborane analogy, 77–79
Cyclopentadienyl ligand, with functional substituents,
  see also individual metals, 1–55
  acid chloride, 5
  air-stability, 17
  aldehydes, 4–5, 14, 19
  amide, 5, 6
  azo derivatives, 32
  cyanide, 5–6, 19
  dicyanides, 14–16, 19
  diesters, 13, 19
  diformyl, 14, 19
  N,N′-dimethylamine, 30
  dimethylaminomethyl, 30–31
  fused lactone, 6

Cumulative List of Contributors

Cumulative List of Titles

Acetylene and Allene Complexes: Their Implication in Homogeneous Catalysis, **14**, 245

Activation of Alkanes by Transition Metal Compounds, **15**, 147

Alkali Metal Derivatives of Metal Carbonyls, **2**, 157

Alkali Metal–Transition Metal π-Complexes, **19**, 97

Alkyl and Aryl Derivatives of Transition Metals, **7**, 157

Alkylcobalt and Acylcobalt Tetracarbonyls, **4**, 243

Allyl Metal Complexes, **2**, 325

π-Allylnickel Intermediates in Organic Synthesis, **8**, 29

1,2-Anionic Rearrangement of Organosilicon and Germanium Compounds, **16**, 1

Application of ^{13}C-NMR Spectroscopy to Organo-Transition Metal Complexes, **19**, 257

Applications of ^{119m}Sn Mössbauer Spectroscopy to the Study of Organotin Compounds, **9**, 21

Arene Transition Metal Chemistry, **13**, 47

Arsonium Ylides, **20**, 115

Aryl Migrations in Organometallic Compounds of the Alkali Metals, **16**, 167

Biological Methylation of Metals and Metalloids, **20**, 313

Boranes in Organic Chemistry, **11**, 1

Boron Heterocycles as Ligands in Transition-Metal Chemistry, **18**, 301

Carbene and Carbyne Complexes, On the Way to, **14**, 1

Carboranes and Organoboranes, **3**, 263

Catalysis by Cobalt Carbonyls, **6**, 119

Catalytic Codimerization of Ethylene and Butadiene, **17**, 269

Catenated Organic Compounds of the Group IV Elements, **4**, 1.

Chemistry of Carbon-Functional Alkylidynetricobalt Nonacarbonyl Cluster Complexes, **14**, 97

Chemistry of Titanocene and Zirconocene, **19**, 1

Chiral Metal Atoms in Optically Active Organo-Transition-Metal Compounds, **18**, 151

^{13}C NMR Chemical Shifts and Coupling Constants of Organometallic Compounds, **12**, 135

Compounds Derived from Alkynes and Carbonyl Complexes of Cobalt, **12**, 323

Conjugate Addition of Grignard Reagents to Aromatic Systems, **1**, 221

Coordination of Unsaturated Molecules to Transition Metals, **14**, 33

Cyclobutadiene Metal Complexes, **4**, 95

Cyclopentadienyl Metal Compounds, **2**, 365

1,4-Diaza-1,3-butadiene (α-Diimine) Ligands: Their Coordination Modes and the Reactivity of Their Metal Complexes, **21**, 151

Diene-Iron Carbonyl Complexes, **1**, 1

Dyotropic Rearrangements and Related σ-σ Exchange Processes, **16**, 33

Electronic Effects in Metallocenes and Certain Related Systems, **10**, 79

Electronic Structure of Alkali Metal Adducts of Aromatic Hydrocarbons, **2**, 115

Fast Exchange Reactions of Group I, II, and III Organometallic Compounds, **8**, 167

Fischer–Tropsch Reaction, **17**, 61

Flurocarbon Derivatives of Metals, **1**, 143

Fluxional and Nonrigid Behavior of Transition Metal Organometallic π-Complexes, **16**, 211

Free Radicals in Organometallic Chemistry, **14**, 345

Functionally Substituted Cyclopentadienyl Metal Compounds **21**, 1

Heterocyclic Organoboranes, **2**, 257